A. S. Wightman
Palmer Physical Laboratory
3 July 1962

# FUNDAMENTAL PROBLEMS IN STATISTICAL MECHANICS

# FUNDAMENTAL PROBLEMS IN STATISTICAL MECHANICS

*Proceedings*
*of the NUFFIC International Summer Course in Science*
*at Nijenrode Castle, The Netherlands*
*August, 1961*

COMPILED BY

E. G. D. COHEN

*Institute for Theoretical Physics, University of Amsterdam*

1962

NORTH-HOLLAND PUBLISHING COMPANY – AMSTERDAM

INTERSCIENCE PUBLISHERS, INC. – NEW YORK

PUBLISHERS: NORTH-HOLLAND PUBLISHING CO.

SOLE DISTRIBUTORS FOR U.S.A.:

INTERSCIENCE PUBLISHERS INC., NEW YORK

PRINTED IN THE NETHERLANDS

# PREFACE

After the successful first course on "Present Problems Concerning the Structure and Evolution of the Galactic System" which was held at Breukelen from July 28 - August 16, 1960, the NUFFIC International Summer Course in Science 1961 dealt with "Fundamental Problems in Statistical Mechanics" and was again held at Nijenrode Castle, Breukelen, from August 1 - 16. It was the first summer school on theoretical physics in this country and was attended by 62 young research workers and students coming from 15 different countries. The organization of the school was in the hands of NUFFIC (Netherlands Universities Foundation for International Co-operation) while financial support was given by NATO and by the Netherlands Organization for Pure Scientific Research (Z.W.O.).

The scientific organizing committee was composed of:

| | |
|---|---|
| J. de Boer, Amsterdam | N.M. Hugenholtz, Groningen |
| E.G.D. Cohen, Amsterdam | N.G. van Kampen, Utrecht |
| H.J. Groenewold, Groningen | P. Mazur, Leiden |
| S.R. de Groot, Leiden | B.R.A. Nijboer, Utrecht |

"Statistical mechanics" was chosen as the subject of this year's school because of the tradition existing in this field in the Netherlands, dating back to J.D. van der Waals and H.A. Lorentz, and continued by P. Ehrenfest and others. The program of the school comprised 42 hours of regular lectures given during the morning by 9 lecturers and, in addition, a number of seminars on related but more incidental topics, given during some of the afternoons.

In this volume the regular lectures are published with the exception of those given by G.E. Uhlenbeck (dealing mainly with the theory of condensation). On the other hand, one of the seminar talks given by E.W. Montroll is added. In some cases the text of the lectures as published was not provided by the lecturers themselves, but composed from the notes of some of the participants. The rapid publication which was aimed at naturally somewhat influenced the presentation of the material. Moreover for none of the lecturers (except Montroll) English is his native tongue. For these reasons the readers' indulgence is asked.

CONTENTS

# GENERAL INTRODUCTION

B.R.A. NIJBOER
Institute for theoretical physics,
Utrecht, Netherlands

## 1. *Literature*

It is my task to give you in 4 hours an introduction to the following lectures. As a consequence I am going to talk about a number of questions which many of you will know perfectly well already. Nevertheless it was thought that such an introduction could serve a useful purpose:

1. it may provide a kind of common background for people coming from various countries and various universities,
2. it will discuss a number of things, which, though you will have heard about them before, some of you may have forgotten in the mean time,
3. it may serve as a kind of warming up for what is going to follow. Many people like to wet their chest before jumping into the water.

Let me begin by writing down some references where the topics to be treated in this introduction are discussed in more detail:

1. P. and T. Ehrenfest, Begriffliche Grundlagen der statistischen Auffassung in der Mechanik, Enc. d. Math. Wiss. IV, 32, 1912, reprinted in P. Ehrenfest, Collected Scientific Papers, Amsterdam 1959. An English translation, The conceptual foundations of the statistical approach in mechanics, has been published by Cornell University Press 1959.
2. R.C. Tolman, The principles of statistical mechanics, Oxford 1938.
3. D. ter Haar, Elements of statistical mechanics, New York 1954. Foundations of statistical mechanics, Rev. Mod. Phys. 27, 289 (1955).
4. G.E. Uhlenbeck, Over statistische methoden in de theorie der quanta, Leiden, thesis 1927. Higgins Lectures, Princeton University 1954.
5. R. Becker, Theorie der Wärme, Springer 1955.
6. J. de Boer, Reports on Progress in Physics 12, 305 (1949).
7. S.R. de Groot, Lecture notes, Utrecht 1952.
8. L. Van Hove, Lecture notes, Utrecht 1955.

## 2. *Aim of statistical mechanics*

Let me first remind you that statistical mechanics aims at establishing relations between the observed properties of large systems (i.e. systems consisting of a very large number of particles) and the properties of the particles and their interactions as given by quantum mechanics. It therefore tries to provide an atomistic foundation of the phenomenological laws established in thermodynamics and other fields.

The macroscopic system considered may be a gas, a liquid, a solid, a plasma etc. It may be in equilibrium and then we wish to explain quantitatively e.g. the pressure as a function of density and temperature, the specific heat and other caloric properties, dielectric, magnetic or optical properties. Or we may want to consider an irreversible process and e.g. try to calculate transport coefficients such as viscosity, heatconductivity, electrical conductivity.

Generally speaking, the statistical mechanics of equilibrium states is in satisfactory shape. The basic equations are known, though due to great mathematical difficulties really quantitative calculations are in general still impossible at least for dense systems. To this category belongs also the theoretical treatment of phase transitions. This is a difficult subject to which we will pay some attention these days. The statistical treatment of non-equilibrium processes on the contrary has proceeded much less. A very fundamental problem in this field is how to understand the irreversible behaviour at all, because as is well-known the equations of motions of the molecules (considered either classically or quantummechanically) are completely reversible. With regard to this problem progress has been made during the last few years and some of the following lectures will be devoted to it.

In addition to and independent of the distinction between equilibrium and non-equilibrium statistical mechanics one can distinguish classical statistics and quantum statistics. In quantum statistics one has to take the quantisation of energy into account, so that the integration over phase space is replaced by a summation over the energy levels. In some problems e.g. when dealing with the behaviour of electrons in a metal or with liquid helium, the indistinguishability of the particles also plays an essential role and the so-called Boltzmann statistics is replaced by Bose-Einstein or Fermi-Dirac statistics. For ideal systems (with non-interacting particles) the implications of these statistics have been familiar since a long time, but only recently progress has been made in the quantum statistics of interacting particles.

A macroscopic quantity of matter contains of the order of $10^{20}$ molecules. Suppose that our system may be considered classically, then, if we should know coordinates and velocities of all molecules at a certain instant of time $t_0$, we could in principle calculate the exact state of the system at any other time $t$ ($t > t_0$ or $t < t_0$). Obviously we do not have such a detailed information and moreover the mathematical calculations would be hopelessly complicated. But we are not interested in position and velocity of every molecule at some later time at all. We are interested only in a relatively small number of so-called macroscopic quantities, which can be measured. Consider e.g. the pressure of a gas. The force exerted by the molecules on the walls of the container varies extremely rapidly. It does not interest us, it can neither be measured nor calculated. What we do measure is some average of this force over small space and time intervals. And this we would like to calculate. It is evident that for this purpose probability or statistical arguments have to be invoked.

### 3. *Kinetic derivation of the H-theorem*

I should now like to remind you briefly by way of introduction of the considerations connected with *Boltzmann's H-theorem*, which occupies a central position in the historical development of statistical mechanics.

Let us consider a gas at low density, consisting of molecules all of one kind. We introduce a function $f(\vec{r}, \vec{v}, t)$, the so-called kinetic distribution function, such that $f(\vec{r}, \vec{v}, t)\, d\vec{r}\, d\vec{v}$ represents the number of molecules which at time $t$ occupies the element $d\vec{r}\, d\vec{v}$ of the 6-dimensional position-velocity space. Let us for simplicity first suppose that the molecules are uniformly distributed over configuration space and that no external forces are present. In this simplified case the distribution function will depend only on the velocity components and on time, $f(\vec{v},t)$. If we now divide velocity space in equal finite cells $\omega_i$, then we may call $f_i(t)$ the number of molecules whose velocity vector ends at time $t$ in cell $\omega_i$.

For the rate of change of $f_i$ we may put

$$\frac{df_i}{dt} = \sum_{j,kl} (a^{ij}_{kl} f_k f_l - a^{kl}_{ij} f_i f_j) \tag{1}$$

an equation which dates back to *Clausius* about 100 years ago. The R.H.S. of (1) is often called the *Stosszahlansatz* (collision number hypothesis). The number of collisions per unit time of molecules from the cell $\omega_i$ with those from cell $\omega_j$, which yield molecules in $\omega_k$ and $\omega_l$ is put equal to $a^{kl}_{ij} f_i f_j$. The number of inverse (or restituting) collisions that yields a molecule in $\omega_i$ is determined by

$$\sum_{j,kl} a^{ij}_{kl} f_k f_l \, .$$

One usually assumes that

$$a^{ij}_{kl} = a^{kl}_{ij} \qquad \text{(microscopic reversibility)} \, . \tag{2}$$

For the simple case of a spherical symmetric force law between the molecules the latter property can be proved.

In the more general case that $f$ depends also on $\vec{r}$ and that external forces (giving an acceleration $\vec{a}(\vec{r})$) are present, *Boltzmann*

(1872) put forward his famous integro-differential equation, which we write down in somewhat more detail than (1):

$$\frac{\partial f(\vec{r},\vec{v},t)}{\partial t} = -\vec{v}\cdot\text{grad}_{\vec{r}} f - \vec{a}\cdot\text{grad}_{\vec{v}} f + \int d\vec{v}_1 \int d\Omega\, g\, I(g,\Theta)$$

$$\{f(\vec{r},\vec{v}',t)f(\vec{r},\vec{v}_1',t) - f(\vec{r},\vec{v},t)f(\vec{r},\vec{v}_1,t)\} \qquad (3)$$

Equation (3), the Boltzmann equation, is used to calculate transport coefficients for gases of not too high density where the interaction between molecules is known (Chapman-Enskog theory). The first two terms on the R.H.S. represent the rate of change of $f$ due to streaming, i.e. the representative points in position-velocity space move as a consequence of the velocities and accelerations of the molecules. The remaining term represents the change of $f$ due to collisions, it is the mathematical formulation of the Stosszahlansatz. Only binary collisions are taken into account; hence the density of the gas is supposed to be small i.e. the average distance between particles is large compared to the range of the interaction. One distinguishes again the direct collisions ($\vec{v},\vec{v}_1 \rightarrow \vec{v}',\vec{v}_1'$) which lead to a decrease of $f(\vec{r},\vec{v},t)$ and the restituting collisions ($\vec{v}',\vec{v}_1' \rightarrow \vec{v},\vec{v}_1$) which make $f(\vec{r},\vec{v},t)$ increase. $g = |\vec{v} - \vec{v}_1| = |\vec{v}' - \vec{v}_1'|$ is the magnitude of the relative velocity of the colliding molecules and $I(g,\Theta)$ is the differential cross-section. Notice that $\vec{v}'$ and $\vec{v}_1'$ should be considered in the integration as functions of $\vec{v}$, $\vec{v}_1$ and the scattering angle $\Theta$. On writing down this equation it has been assumed that no correlations exist between the colliding molecules and also that the latter are so close that the functions $f$ in the product may be taken at the same position.

When will the distribution function $f$ be stationary? Let us consider first again the simple equation (1). A sufficient condition for $df_i/dt = 0$ is that

$$f_i f_j = f_k f_l \qquad \text{whenever} \qquad \begin{cases} \vec{v}_i + \vec{v}_j = \vec{v}_k + \vec{v}_l \\ v_i^2 + v_j^2 = v_k^2 + v_l^2 \end{cases}$$

(momentum and kinetic energy are conserved in a collision). This means that

$$\log f(\vec{v}_i) + \log f(\vec{v}_k) = \text{const. for} \qquad \begin{cases} \vec{v}_i + \vec{v}_k = \text{const.} \\ v_i^2 + v_k^2 = \text{const.} \end{cases}$$

i.e. $\log f$ must depend linearly on the quantities which are conserved in a collision. This leads to

$$f(\vec{v}) = \text{const.}\, e^{-\frac{1}{2}m\beta\left|\vec{v}-\vec{v}_o\right|^2} \qquad (4)$$

where $\beta$ and $\vec{v}_o$ are constants. $\vec{v}_o$ corresponds to a motion of the system as a whole; further, as is well-known, one can show by calculating the pressure of an ideal gas that $\beta = 1/kT$. We have found here the so-called *Maxwell-Boltzmann distribution* of the velocities (Maxwell 1859).

Boltzmann then showed that the M.B.-distribution is the only stationary distribution and moreover that any initial distribution will approach the M.B.-distribution monotonically. We introduce the so-called $H$-function of Boltzmann

$$H = \sum_i f_i \log f_i .$$

From (1) it follows that

$$\frac{dH}{dt} = \sum_i \frac{df_i}{dt} \log f_i + \sum_i \frac{df_i}{dt} = \sum_i \frac{df_i}{dt} \log f_i$$

$$= \sum_{ij,kl} a_{ij}^{kl} \log f_i (f_k f_l - f_i f_j) \qquad \text{(because } a_{kl}^{ij} = a_{ij}^{kl} \text{ (2))}$$

$$= \tfrac{1}{2} \sum_{ijkl} a_{ij}^{kl} (\log f_i + \log f_j)(f_k f_l - f_i f_j) \qquad \text{(because } a_{ij}^{kl} = a_{ji}^{kl}\text{)}$$

$$= \tfrac{1}{4} \sum_{ijkl} a_{ij}^{kl} (\log f_i f_j - \log f_k f_l)(f_k f_l - f_i f_j) \qquad \text{(because } a_{kl}^{ij} = a_{ij}^{kl} \text{ (2))}$$

$$\leqslant 0$$

We find that $dH/dt < 0$ except if $f_i f_j = f_k f_l$ for all $i,j,k,l$, which correspond to each other, i.e. which can be connected in a collision. That is $H$ will decrease monotonically until the stationary M.B.-distribution is reached.

In the case of the more general equation (3) one can proceed in an analogous way. Here $H = \int f \log f\omega \, d\vec{r} \, d\vec{v}$ where $\omega$ is an arbitrary constant of dimension length$^3$ . velocity$^3$. Then we have from (3)

$$\frac{dH}{dt} = \left(\int \frac{\partial f}{\partial t} \log f\omega \, d\vec{r} \, d\vec{v}\right)_{\text{streaming}} + \left(\int \frac{\partial f}{\partial t} \log f\omega \, d\vec{r} \, d\vec{v}\right)_{\text{collisions}} .$$

The first term on the R.H.S. can (e.g. by partial integrations) be shown to vanish. The second term is again always negative, except that it is zero if

$$ff_1 = f' f_1' .$$

This again leads to

$$f = A \, e^{-\frac{1}{2} m\beta |\vec{v} - \vec{v}_o|^2}$$

where $A$, $\beta$ and $\vec{v}_o$ are still functions of coordinates and of time. In every volume element of velocity space $f$ approaches therefore monotonically the local M.B.-distribution. In configuration space the approach to a stationary distribution is much slower and also more complicated. The stationary distribution must obey the equation

$$\vec{v}\,.\,\mathrm{grad}_{\vec{r}}\, f + \vec{a}\,.\,\mathrm{grad}_{\vec{v}}\, f = 0\ .$$

If the external forces are derived from a potential $V(x,y,z)$ one finds for a gas in rest:

$$\vec{v}_0 = 0 \qquad \beta = \text{const.} \qquad A = \text{const.}\ e^{-\beta V(x,y,z)}\ .$$

Remarks: 1. In equilibrium we have $a_{ij}^{kl} f_i f_j = a_{kl}^{ij} f_k f_l$ for all corresponding $i,j,k,l$, i.e. the number of direct collisions equals the number of restituting collisions. This is called the *principle of detailed balancing.*

2. Boltzmann's $H$-function is closely related to the entropy, in fact we have for an ideal gas $S = -kH$. $-H$ measures the degree of disorder of the system. That $H$ approaches towards a minimum corresponds to the thermodynamic observation that $S$ approaches a maximum.

## 4. *Objections against the H-theorem*

Serious criticism has been raised against the $H$-theorem as discussed up till now. As is well-known the equations of classical mechanics $m\ddot{\vec{r}}_i = -\,\mathrm{grad}_i\, V(\vec{r}_i \ldots \vec{r}_n)$, which describe the motion of the molecules, are perfectly reversible in time. Nevertheless $H$ would always decrease, a typical irreversible phenomenon. One has the impression that the irreversibility has somehow been smuggled in. If in a process such as considered above and where $H$ decreases, one were to reverse at an arbitrary instant all velocities, the system would pass all states of motion in reverse order. On changing the sign of all velocities $H$ itself would not change. We should then have obtained a succession of states where $H$ would increase continuously. This is the so-called *Umkehreinwand of Loschmidt*. In addition *Zermelo* raised the so-called *Wiederkehreinwand*, based on a theorem due to *Poincaré*. It states that any mechanical system enclosed in a finite volume will return after a certain time (the Poincaré-cycle) practically to its original state. The recurrence time may be extremely long. Boltzmann estimated that a system consisting of $10^{18}$ atoms in 1 $\mathrm{cm}^3$ with average velocity of $5.10^4$ cm/sec would reproduce all its coordinates within $10^{-7}$ cm and all its velocities within $10^2$ cm/sec in a time of the order of $10^{10^{19}}$ year. If one does not distinguish the molecules amongst each other this time would naturally be much shorter. Anyhow the function $H$ has to increase again eventually.

These objections have made it clear that the $H$-theorem in the above form cannot be generally valid in the sense that for any system starting from any initial state $H$ would always decrease. By means of the Stosszahlansatz we have introduced a statistical element in the derivation, thus it is against this Ansatz that the criticism should be directed.

Some examples may be illuminating: Consider a cubic volume in which the particles all move with the same velocity parallel to an edge, at such a mutual distance that collisions do not occur. Then we have obviously a stationary state different from the M.B.-distribution. Evidently the Boltzmann equation is not applicable to such a singular situation.

In the case of the well-known wind-tree model (Ehrenfest) one can show explicitly that the Stosszahlansatz cannot be correct outside equilibrium for both the direct and the inverse collisions.

Clearly the solution of the difficulties is that the Stosszahlansatz only holds for the average number of collisions of a certain kind. But there will be fluctuations. The Boltzmann equation and the $H$-theorem therefore describe the average behaviour of a large number of identical systems. The average behaviour of an ensemble of systems will be irreversible. The first to reach a clear understanding of this point was Boltzmann himself. It has been stated that at this moment kinetic theory changed into statistical mechanics.

## 5. *The H-theorem considered statistically*

Besides $\mu$-space (with coordinates $x, y, z, p_x, p_y, p_z$ of a single molecule) we introduce a $6N$-dimensional $\Gamma$-space, i.e. the phase space for the whole system consisting of $N$ molecules. A point in $\Gamma$-space specifies the microscopic state of the whole system and represents at a given time the coordinates and momenta of all particles. This phase point describes an orbit in phase space, which in principle is determined as soon as one of its points is given. Through every point only one orbit passes, two orbits never cross. At every collision of a pair of molecules the phase point jumps suddenly to quite a different region of $\Gamma$-space. In particular because of the large number of dimensions the orbit in phase space is complicated beyond all comprehension. If the system is isolated thermally and mechanically, the total energy will be a constant and the phasepoint moves on the so-called energy surface.

Let $\mu$-space be divided in small, finite equal cells $o_i$. A given point in $\Gamma$-space corresponds to a definite distribution ($n_i$ molecules in cell $o_i$), but on the other hand with a given distribution $Z$ corresponds an extended volume in $\Gamma$-space, the so-called $Z$-star of Ehrenfest, whose size is given by

$$W(Z) = \frac{N!}{n_1! n_2! \ldots} o_1^{n_1} o_2^{n_2} \ldots . \tag{6}$$

The volume $o_1^{n_1}\, o_2^{n_2} \ldots$ corresponding to a given distribution of $N$ numbered molecules among the cells has to be multiplied by the factor $N!/\prod_i n_i!$ which indicates the number of ways in which $N$ molecules can be divided into groups $n_1, n_2 \ldots$ .

If we now start from the assumption of equal a priori probability of equal volume elements in $\Gamma$-space, i.e. from the assumption that the probability to find the phase point in a certain region of $\Gamma$-space is proportional to the volume of that region, then it follows that the probability of a certain distribution $Z$ of the molecules over the cells in $\mu$-space is proportional to the volume $W(Z)$.

Boltzmann now defines the equilibrium distribution $Z^{(o)}$ as the distribution for which the probability is a maximum. We restrict ourselves to those distributions for which

$$\sum_i n_i = N \tag{7}$$

and

$$\sum_i n_i \, \varepsilon_i = E \, , \tag{8}$$

where $\varepsilon_i$ is the energy corresponding to the cell $o_i$ in $\mu$-space. Notice that we consider again the case of a dilute gas only, for interacting molecules the energy $\varepsilon_i$ has no meaning.

We will now show that $W(Z)$ is a maximum for the Maxwell-Boltzmann distribution and further that the maximum is a very sharp one, so that a slightly deviating distribution is already much less probable.

We have:

$$\log W(Z) = \log N! \; - \sum_i \log n_i! \; + N \log o \; .$$

We now apply Stirlings asymptotic formula for $\log n!$, (which is accurate within 1% already for $n = 10$) $\log n! \sim n \log n - n$.

$$\begin{aligned} \log W(Z) &= N \log N - N - \sum_i n_i \log n_i + \sum n_i + N \log o \\ &= - \sum n_i \log n_i + \text{const.} \\ &= - H + \text{const.} \end{aligned}$$

Further:

$$\begin{aligned} \delta \log W = \log \frac{W + \Delta W}{W} &= \sum_i n_i \log n_i - \sum_i (n_i + \Delta n_i) \log (n_i + \Delta n_i) \\ &= \sum_i n_i \log n_i - \sum_i (n_i + \Delta n_i) \left\{ \log n_i + \log (1 + \frac{\Delta n_i}{n_i}) \right\} \\ &= - \sum_i \left\{ n_i \log (1 + \frac{\Delta n_i}{n_i}) + \Delta n_i \log n_i + \Delta n_i \log (1 + \frac{\Delta n_i}{n_i}) \right\} \end{aligned}$$

$$= - \sum_i \left\{ n_i \left( \frac{\Delta n_i}{n_i} - \frac{(\Delta n_i)^2}{2\, n_i^2} \right) + \Delta n_i \log n_i + \frac{(\Delta n_i)^2}{n_i} \right\}$$

$$= - \sum_i \Delta n_i \log n_i - \tfrac{1}{2} \sum_i \frac{(\Delta n_i)^2}{n_i} .$$

Use has been made of the restriction $\sum_i \Delta n_i = 0$ which follows from (7). Further from (8) we have $\sum_i \varepsilon_i \Delta n_i = 0$. According to the method of the Lagrange multipliers we have:

$$\delta \log W = \sum_i \Delta n_i \left(- \log n_i + \log A - \beta\varepsilon_i\right) - \tfrac{1}{2} \sum_1 \frac{(\Delta n_i)^2}{n}$$

For suitably chosen parameters $A$ and $\beta$ we find for the most probable distribution

$$\log n_i^{(o)} = \log A - \beta\varepsilon_i$$

$$n_i^{(o)} = A\, e^{-\beta\varepsilon_i} \qquad (9)$$

This again is the M.B.-distribution, which thus leads to maximum probability. We have indeed a maximum because the second variation is seen to be negative.

In the maximum we have:

$$\log \frac{W + \Delta W}{W} = - \tfrac{1}{2} \sum_i \frac{(\Delta n_i)^2}{n_i^{(o)2}}\, n_i^{(o)} .$$

For $\Delta n_i / n_i^{(o)} \sim 10^{-5}$ this is $- \tfrac{1}{2}\, 10^{-10}\, N$. For 1 $cm^3$ gas at room temperature and atmospheric pressure $N \sim 10^{19}$, so $W{+}\Delta W / W \sim e^{-10^9}$. Hence a slightly deviating distribution is already overwhelmingly less probable.

Remark: For cells with a few molecules only Stirling's formula is strictly inapplicable; these cells, however, contribute very little to the properties of the gas.

Above we have characterized the equilibrium distribution as the one with maximum probability. Alternatively we may characterize the equilibrium distribution as the average distribution, defined as follows:

$$\bar{n}_i = \frac{\sum'' n_i\, W(Z)}{\sum'' W(Z)}$$

Here the summations are taken over all possible distributions (all possible series of $n_i$-values) satisfying the conditions

$$\sum_i n_i = N$$
$$\sum_i n_i \, \varepsilon_i = E \, .$$

*Darwin* and *Fowler* have developed methods to evaluate this type of summation. It is found that the average distribution is again the M.B.-distribution, i.e.

$$\bar{n}_i = n_i^{(o)} = A \, e^{-\beta \varepsilon_i} \, .$$

By similar methods one can calculate the average deviations $\sqrt{\overline{(\Delta n_i)^2}}$, which turn out to be of order $\sqrt{\bar{n}_i}$.

For an ideal gas at equilibrium one can now calculate the thermodynamic quantities. E.g. for the entropy we have:

$$S = - k \sum_i n_i^{(o)} \log n_i^{(o)}$$

With $n_i^{(o)} = A \, e^{-\varepsilon_i/kT}$, $A = N/\sum_l e^{-\varepsilon_i/kT}$ we find

$$S = - k \sum_i A \, e^{-\varepsilon_i/kT} \left(\log A - \frac{\varepsilon_i}{kT}\right)$$

$$E = A \sum_i \varepsilon_i \, e^{-\varepsilon_i/kT}$$

$$F = E - TS = - kTN \log \sum_i e^{-\varepsilon_i/kT} \qquad (10)$$

($\sum_i e^{-\varepsilon_i/kT}$ is called the partition function for 1 molecule).

Boltzmann has also discussed the $H$-theorem from this statistical point of view. One can no longer maintain that $H$ will always decrease monotonically. Fluctuations will always occur. Rigorous and general proofs about the behaviour of $H$ do not seem to have been given, but also from an accurate discussion of simplified models the following may be asserted with great plausibility: If we start from a situation where $H$ is larger than its minimum value $H$ will almost always decrease. This holds for both directions of time. Furthermore $H$ will almost always remain close to its minimum value. Or in other words: If the gas has not the M.B.-distribution, it will reach it almost always very rapidly, and once the distribution is the M.B.-distribution, the deviations will almost always remain very small. Formulated in this way the objections of Loschmidt and Zermelo no longer apply. One should remember, as Ehrenfest has pointed out, that $H(Z)$ is a discontinuous function of the phase. This is because the cells in $\mu$-space are taken finite. As long as the phase point remains in the same $Z$-star, the $n_i$'s and therefore also $H$ do not change. $H$ will therefore be a step function of time. Actually one should consider the values of $H$ at successive times $\tau$, $2\tau$, $3\tau$ ... where $\tau$ is large enough for several collisions to occur. The values of $H$ so obtained form a discrete manifold.

A process will be considered to be reversible or irreversible depending on whether the Poincaré-cycle of the initial state is short or long compared to the time during which the system is under observation. This recurrence time increases extremely rapidly when the deviations from equilibrium become larger. Irreversibility is a human illusion (*Smoluchowski*).

*Becker* in his book considers as a simple example the case of density fluctuations in a small subvolume (1 $cm^3$) of a gas of normal conditions. He estimates the average recurrence time connected to a given relative deviation from the equilibrium density.

| Relative deviation | Recurrence time |
|---|---|
| $2.10^{-10}$ | $4.10^{-3}$ sec |
| $3.10^{-10}$ | 1 sec |
| $4.10^{-10}$ | 21 minutes |
| $5.10^{-10}$ | 5 months |
| $6.10^{-10}$ | $3.10^{4}$ years |
| $7.10^{-10}$ | $2.10^{10}$ years |

This table illustrates that observable fluctuations in general require times much longer than the age of the universe.

A macroscopically observable deviation will practically never occur spontaneously, it will always have been conditioned by external means. From then $H$ will practically always decrease again.

## 6. *The Ehrenfest-model*

A very illuminating model originates from the Ehrenfests. It is the so-called dog-flee model. We have 2 dogs and say $2R$ flees, each of which carry a number. Let all flees be first on dog A. Draw an arbitrary number from 1 - $2R$. The flee with the number drawn jumps to the other dog B. This procedure can be continued. A similar experiment has been performed by Schrödinger and Kohlrausch (1926). If $N_A + N_B = 2R$ and $N_A - N_B = 2k$, one can plot $k$ as a function of time in a certain experiment. One obtains a result as follows:

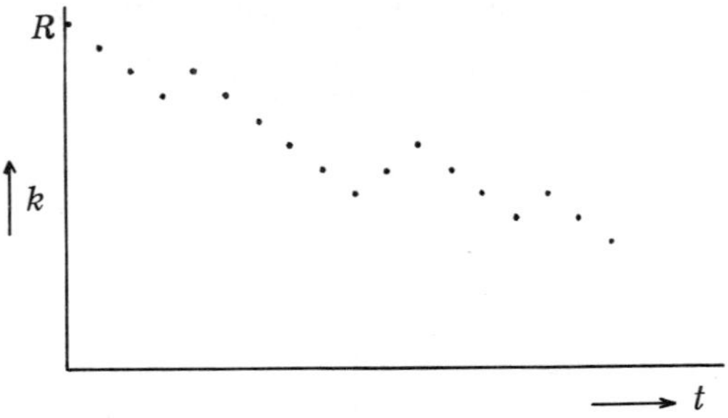


For given $k$ one can prove that

| no. of maxima | : | no. of decreases | : | no. of increases | : | no. of minima |
|---|---|---|---|---|---|---|
| $= \dfrac{R+k}{R-k}$ | : | 1 | : | 1 | : | $\dfrac{R-k}{R+k}$ |

For large $k$ the no. of maxima is much larger than the no. of minima.

In particular *M. Kac* has considered this model in great mathematical detail. In analogy to the Boltzmann formula (6) we have for the a priori probability of a value $k$:

$$W(k) = \frac{(2R)!}{N_A!\ N_B!} \left(\tfrac{1}{2}\right)^{2R} . \tag{11}$$

This expression has a very sharp maximum for large $R$ if $k = 0$, i.e. $N_A = N_B = R$. This corresponds to the equilibrium M.B.-distribution. Small equilibrium fluctuations occur with a Gaussian probability distribution. In fact one has for large $R$ (a proof is to be found e.g. in Becker on p. 64)

$$W(k) \approx \frac{1}{\sqrt{\pi R}}\, e^{-k^2/R} . \tag{11a}$$

The Ehrenfest-model provides an example of a so-called *Markoff*-process, where the probability to find after $s$ steps a value of $k$ depends only on the value of $k$ found after $s - 1$ steps. Let us introduce the probability $P(n,m;s)$, i.e. the probability of finding $m$ after $s$ steps starting from $n$. We then have for a Markoff-process:

$$P(n, m; s) = \sum_k P(n, k; s-1)\ P(k, m; 1) .$$

Call $P(k,m;1) = Q(k,m)$, the transition probability. We have $\sum_m Q(k,m) = 1$. Therefore:

$$\begin{aligned} P(n, m; s) &= \sum_k P(n,k;s-1)\ Q(k,m) \\ &= \sum_{k\neq m} P(n,k;s-1)\ Q(k,m) + P(n,m;s-1)\ \{1 - \sum_{l\neq m} Q(m,l)\} \end{aligned}$$

or: $P(n,m;s) - P(n,m;s-1) =$

$$= \sum_{k \neq m} P(n,k;s-1)\, Q(k,m) - P(n,m;s-1) \sum_{l \neq m} Q(m,l) \,. \quad (12)$$

This equation has a certain analogy with the Boltzmann-equation. However, in the case of a gas the variable $m$ is replaced by the whole series $(n_1 n_2 \ldots)$, the equation describes how the *probability* of a *distribution* $(n_1 n_2 \ldots)$ changes with time. It is called the *master equation*. Note that it is a linear equation. The transition probability $Q$ is supposed to be known. For the Ehrenfest-model we have

$$Q(k_o, k) = \frac{R + k_o}{2R}\, \delta(k_o - 1, k) + \frac{R - k_o}{2R}\, \delta(k_o + 1, k) \qquad (13)$$

so that (12) becomes

$$P(n, m; s) = \frac{R + m + 1}{2R} P(n, m+1; s-1) + \frac{R - m + 1}{2R} P(n, m-1; s-1) \,. \quad (14)$$

If one starts with the value $n$ ($P(n, m; 0) = \delta(n, m)$) one proves easily:

$$\overline{m(s)} = \sum_m m\, P(n, m; s) = \left(1 - \frac{1}{R}\right) \overline{m(s-1)} = n \left(1 - \frac{1}{R}\right)^s .$$

For large values of $R$ and $s$ this is $n\, e^{-s/R}$ so that the average value of $m$ decreases to its equilibrium value 0 exponentially. In fact we can go further and show that any distribution $P(n,m;s)$ approaches for large $s$ to $W(m)$ given above by (11). The proof, given by Kac, starts from the observation (see above) that $P(n, m; s)$ is obtained by successive matrix multiplication by $Q(k,m)$. From a consideration of the eigenvalue problem of the matrix $Q(k,m)$ the statement given above follows. One can further calculate the mean recurrence time and it is found that it is proportional to $1/W(m)$, which again is very large for large $m$ as can be seen from (11) or (11a).

We have talked here about probabilities. This means that we should repeat an experiment as considered in the Ehrenfest-model a large number of times, in other words that we should consider an ensemble of experiments. If every time we start out from a given value $n$, we obtain for every member of the ensemble a series of successive $m$-values. Their distribution after $s$ steps or at time $t$ is described by $P(n,m;s)$ or $P(n,m;t)$.

Every member of the ensemble behaves quasi-periodically, but the probability distribution over the ensemble approaches the equilibrium distribution. We are thus led to the idea that also in the case of a gas we should consider an ensemble of equal systems differing only in phase. The concept of an ensemble is used here in the sense of Gibbs. If in the case of a gas we start out with a given distribution $(n_1, n_2 \ldots)$, then there corresponds to it a region on the energy surface in $\Gamma$-space. The various points of this region move quite differently. At some later time we obtain a whole series of different distributions. By analogy with the Ehrenfest-model we then expect that for arbitrary initial distribution the probability of a distribution $P(n_1, n_2 \ldots)$ will for $t \to \infty$ approach to (6), i.e. to $N!/\prod_i n_i!\,(0/\Omega)^N$ where $\Omega$ is the total volume of $\mu$-space, so that the overwhelming majority of the members of the ensemble will ultimately have a M.B. distribution.

## 7. *Equilibrium represented by a microcanonical ensemble*

To derive the above expression for the probability of a distribution over the cells of $\mu$-space, Boltzmann started from the assumption of equal a priori probability of equal volume elements in $\Gamma$-space, in other words that the probability for the phase point to lie in a certain volume element in $\Gamma$-space is equal for equal volume elements. We considered only the case that the total energy of the system was given, so that we only considered elements on the energy surface or rather between two neighbouring energy surfaces $E$ and $E + \delta E$, where $\delta E$ represents the accuracy with which the total energy can be measured. For an ensemble Boltzmann's assumption means that the phase points of the systems in the ensemble are uniformly distributed between the neighbouring energy surfaces, i.e. the ensemble corresponding to Boltzmann's probability assumption (6) for a distribution is *Gibbs' microcanonical ensemble.*

In terms of an ensemble we can state our expectation about the behaviour of a gas as mentioned above in the following words: If we consider an ensemble in $\Gamma$-space we expect a largely arbitrary initial density on the energy surface (or rather between two neighbouring energy surfaces) to approach for $t \to \infty$ to a uniform density between these energy surfaces, i.e. every ensemble will approach the microcanonical ensemble. We will come back to this point later. It is well-known that a uniform density on the hypersphere $\sum_i \frac{p_i^2}{2m} = E$ in $\Gamma$-space leads for large $N$ to the M.B.-distribution for the momenta of one molecule. For we have by projection on the $p_1$-axis:

$$f(p_1)\,\mathrm{d}p_1 = \frac{\sigma_{3N-2}(r)\,\mathrm{d}p_1 \frac{R}{r}}{\sigma_{3N-1}(R)},$$

where $r^2 = R^2 - p_1^2$, $R^2 = 2\,m\,E = 3\,N\,\overline{p_1^2}$ and $\sigma_n(R) \propto R^n$ is the volume of a hypersphere with radius $R$. Therefore:

$$\begin{aligned} f(p_1)\,\mathrm{d}p_1 &= \text{const.}\,\frac{r^{3N-3}}{R^{3N-2}}\,\mathrm{d}p_1 \\ &= \text{const.}\,\Big(1 - \frac{p_1^2}{3\,N\,\overline{p_1^2}}\Big)^{\frac{3N-3}{2}} \to \text{const.}\; e^{-\frac{p_1^2}{2\overline{p_1^2}}}. \end{aligned} \tag{15}$$

It should be noted

1. that in the above discussion we did not make statements about the behaviour of one system as in the original discussion of the $H$-theorem, but only about an ensemble of systems,
2. and even so our statements about the behaviour of an ensemble were not founded on rigorous proofs as those for the Ehrenfest- and similar models, but were only based on plausibility and analogy arguments.

Of course in our experiments we make observations on a single system. For a justification of statistical mechanics it would be necessary to show that the value measured on a single system of some macroscopic quantity could be calculated by taking the average over a suitably chosen ensemble. In particular the equilibrium properties of an isolated system should be shown to follow from averages over a microcanonical ensemble.

You know that Boltzmann has tried to give a justification of equilibrium statistical mechanics by proving that the time average (taken over an infinite time) of any phase function is equal to its (microcanonical) ensemble average. He makes use of the so-called *ergodic hypothesis*, i.e. the hypothesis that the phase point of a system passes through every point of the energy surface, in other words that the energy surface is completely covered by one phase orbit. Indeed we then have:

$$\text{time average of } f(q_1 \ldots p_N) = \lim_{\tau\to\infty} \frac{1}{\tau} \int_0^\tau f(q_1 \ldots p_N)\, dt =$$

microcanonical ensemble average of time average (because of the ergodic hypothesis) = time average of microcanonical ensemble average = microcanonical ensemble average (because the microcanonical ensemble is a stationary ensemble). However, the ergodic hypothesis has in 1912 been shown by *Rosenthal* and *Plancherel* to be untenable. The Ehrenfests have then introduced the so-called quasi-ergodic hypothesis, this is the assumption that the phase orbit through $P$ passes through an arbitrary small region around any other point $Q$. This hypothesis could be proved for a large number of mechanical systems. But the equivalence of time average and ensemble average does not follow from it. We should also remark here that the time average was calculated for an infinitely large time interval. The measurement of a macroscopic quantity, however, lasts only a relatively short time. It is not clear that even on the assumption of ergodicity the average over the time of observation would equal the ensemble average.

I will not finish this introductory discussion of the justification of statistical mechanics and the explanation of the approach to equilibrium without mentioning the $H$-theorem in the form as discussed by Gibbs. But before I am able to do that I must remind you of *Liouville's theorem*, which occupies a central position in ensemble theory.

## 8. *Liouville's theorem*

We consider a large number $\mathcal{N}$ of identical mechanical systems. In classical mechanics the exact state of every system is represented by a point in $\Gamma$-space. An ensemble can therefore be represented by a cloud of phase points moving in $\Gamma$-space. It is characterized by a density $\rho(q_1 \dots p_n, t)$ where $\int \rho \, d\tau = \mathcal{N}$, the total number of systems in the ensemble.

The systems of the ensemble move independently, they all have the same hamiltonian $\mathcal{H}$, but they differ in phase. The ensemble average of a phase function $f(q_1 \dots p_n)$ is determined by

$$\overline{f(q_1 \dots p_n)} = \frac{1}{\mathcal{N}} \int f\rho \, dq_1 \dots dp_n \tag{16}$$

The fundamental assumption of statistical mechanics is that the macroscopic quantities as measured on a system correspond to the ensemble averages defined above, taken over a suitably chosen ensemble. For a system in equilibrium $\rho$ may not explicitly depend on time, i.e. equilibrium ensembles should be stationary. But even for a nonquilibrium system $\rho$ cannot be chosen arbitrary. First of all we have of course $\rho \geqslant 0$. And further $\rho$ should obey the so-called Liouville equation.

We have namely:

$$\frac{\partial \rho}{\partial t} + \sum_k \frac{\partial}{\partial q_k} (\rho \dot{q}_k) + \sum_k \frac{\partial}{\partial p_k} (\rho \dot{p}_k) = 0 . \tag{17}$$

This is the equation of continuity, expressing the fact that the increase of the number of phase points in an arbitrary volume element of phase space equals the number of phase points entering through its boundaries as a result of the motion of the phase points.

We may write then:

$$\frac{\partial \rho}{\partial t} + \sum_k \dot{q}_k \frac{\partial \rho}{\partial q_k} + \sum_k \dot{p}_k \frac{\partial \rho}{\partial p_k} + \sum_k \rho \frac{\partial \dot{q}}{\partial q_k} + \sum_k \rho \frac{\partial \dot{p}}{\partial p_k} = 0 . \tag{18}$$

Now according to the equations of motion of Hamilton:

$$\frac{\partial \dot{q}_k}{\partial q_k} = \frac{\partial}{\partial q_k} \left(\frac{\partial \mathcal{H}}{\partial p_k}\right) \qquad \text{and} \qquad \frac{\partial \dot{p}_k}{\partial p_k} = \frac{\partial}{\partial p_k} \left(- \frac{\partial \mathcal{H}}{\partial q_k}\right)$$

so that the last two terms in (18) cancel. We therefore find:

$$\frac{d\rho}{dt} \equiv \frac{\partial \rho}{\partial t} + \sum_k \dot{q}_k \frac{\partial \rho}{\partial q_k} + \sum_k \dot{p}_k \frac{\partial \rho}{\partial p_k} = \frac{\partial \rho}{\partial t} + (\rho, \mathcal{H}) = 0 . \tag{19}$$

This is the *equation of Liouville* . $d\rho/dt$ is the rate of change of the density as seen when moving with the phase points (i.e. the convected rate of change), $(\rho, \mathcal{H})$ is the Poisson bracket

$$(\rho, \mathcal{H}) = \sum_k \left(\frac{\partial \rho}{\partial q_k} \frac{\partial \mathcal{H}}{\partial p_k} - \frac{\partial \rho}{\partial p_k} \frac{\partial \mathcal{H}}{\partial q_k}\right) .$$

The equation of Liouville states therefore that $\rho$ is conserved as we move with the phase points or also that the phase points move as an incompressible fluid. It can also be expressed as the conservation of extension in phase space. If we consider a given volume in $\Gamma$-space and follow its phase points in their motion, then the volume which they occupy at some later time is equal in size, though generally very different in shape.

For a stationary ensemble, representing a system in equilibrium, we require as mentioned above that $\partial\rho/\partial t = 0$. A necessary and sufficient condition for $\partial\rho/\partial t = 0$ is that $\rho$ only depends on integrals of motion. In practice one only considers the case that $\rho$ depends exclusively on the value of $E$, the total energy. (Other integrals of motion are usually not known.) For a system in equilibrium we have therefore $\rho = \rho(E)$. The best-known stationary ensembles are:

1. *Microcanonical ensemble:*
   $\rho(E)$ = const., for $E_0 < E < E_0 + \delta E$, i.e. uniform density between 2 neighbouring energy surfaces,
   $\rho(E) = 0$, outside the energy shell.
   Symbolically one often writes $\rho(E) = \text{const.}\ \delta(E - E_0)$.
2. *Canonical ensemble:* $\rho(E) = e^{\frac{\psi - E(p,q)}{\Theta}}$
   $\Theta$ is called the modulus, it is closely connected to the temperature $T$. The constant $\psi$ should be chosen so as to satisfy the normalisation condition.
3. *Grand-canonical ensemble* , to be briefly discussed later.

## 9. *Gibbs' H-theorem*

We have stated above, that it is plausible for a density which at time $t_0$ deviates from the microcanonical one to approach in the course of time towards the microcanonical density, i.e. to a uniform density within the energy shell.

It seems reasonable to generalize Boltzmann's $H$-function in ensemble theory as follows:

$$H = \int \rho \log(\rho\omega)\, d\tau \tag{20}$$

where $d\tau$ is the volume element in $\Gamma$-space and $\omega$ is an arbitrary constant of dimension such that $\rho\omega$ is dimensionless. For an ideal gas without interaction between the molecules this quantity indeed reduces to the $H$-function of Boltzmann, except for an additional constant. The expression (20) has the property to take its minimum value for $\rho =$ $= \rho_0 =$ const. For

$$H - H_0 = \int \rho \log (\rho\omega)\, d\tau - \int \rho_0 \log (\rho_0\omega)\, d\tau \ .$$

Because we have

$$\int \rho\, d\tau = \int \rho_0\, d\tau = \rho_0 \int d\tau = \mathcal{N} ,$$

$$H_0 = \rho \int \log (\rho_0\omega)\, d\tau$$

hence

$$H - H_0 = \int \left(\rho \log \frac{\rho}{\rho_0} + \rho_0 - \rho\right) d\tau$$

or putting $\rho/\rho_0 = x$

$$H - H_0 = \rho_0 \int (x \log x + 1 - x)\, d\tau \geqslant 0$$

because $x \log x - (x - 1) \geqslant 0$ and $= 0$ only for $\rho = \rho_0$.

If now we could prove that $dH/dt \leqslant 0$ we would have proved the approach to the microcanonical ensemble. However, because we know that $d\rho/dt = 0$ the quantity $H$ defined above remains constant in the course of time. An arbitrary function of $\rho$, when integrated over the whole of $\Gamma$-space, leads to a constant value. The quantity $-kH$ cannot therefore be used to define entropy outside equilibrium. Still we can well imagine that ultimately a kind of uniform distribution will arise. If we put a drop of ink into water and stir the liquid thoroughly we shall obtain a uniformly coloured liquid. Only when looking at a microscopic scale we can still distinguish ink and water particles, but our general impression is that of a homogeneous mixture.

Similarly in $\Gamma$-space $\rho$ remains constant during the motion. If we start with a density everywhere 0 except in a small region where $\rho = \rho_1$, then at some later time $\rho$ is still either 0 or $\rho_1$, but the region where $\rho = \rho_1$ has changed in shape considerably. It will practically extend over the whole of the energy surface.

Ehrenfest has pointed out already that the distribution can only become uniform in a coarse-grained sense, i.e. we would expect that the average density of phase points averaged over small but finite volume elements will approach uniformity.

If we introduce $\tilde{\rho}_J = \int_J \rho d\tau / \int_J d\tau$, averaged over a cell $J$ in $\Gamma$-space, we can define a coarse-grained $H$-function by

$$\tilde{H} = \sum_J \tilde{\rho}_J \log (\tilde{\rho}_J\omega)\, \Delta\tau_J = \int \rho \log (\tilde{\rho}_J\omega)\, d\tau \ . \qquad (21)$$

$\tilde{H}$ does change with time, however, it has not been proved that $d\tilde{H}/dt \leqslant 0$. But one can prove that starting out at $t = 0$ with a particular non-uniform density, namely one where the density is constant, but different, in the various cells, $\tilde{H}$ is less at any later time $t$. One expects that the time needed for the density to become uniform in the coarse-grained sense will be the shorter the larger are the cells to be averaged over. It is, however, not so obvious which coarse-grained density corresponds with our knowledge with respect to an initial non-equilibrium state.

About this point we shall hear more in later lectures. In recent years advances have been made in the problem how to obtain a master equation or a Boltzmann-type equation (both of which lead to an irreversible behaviour) starting from Liouville. In any case some statistical assumptions have to be introduced in one way or another.

## 10. *Classical equilibrium ensembles*

I now would like to stop this introductory discussion about how equilibrium is approached and how the fundamental assumptions of equilibrium statistical mechanics can or cannot be justified. In the following I wish to restrict myself to equilibrium statistics and first give a more detailed discussion of the usual classical equilibrium ensembles.

Boltzmann in his statistical discussion of the $H$-theorem considered the case of ideal gases only. Then it makes sense to introduce $\mu$-space and to talk about the energy of individual molecules. The interaction energy between molecules was not taken into account explicitly. Collisions were considered only in so far as they are responsible for establishing equilibrium. In systems of higher density the interaction between molecules plays a significant role. Then the whole gas or liquid or crystal, has to be considered as one single system. We then must adopt the viewpoint of Gibbs and introduce ensembles in $\Gamma$-space. In equilibrium statistics the best known ensembles are the canonical, the microcanonical and the grand ensemble. The results obtained are largely independent of the precise choice of the ensemble used.

a. A *canonical ensemble* is assumed to give a correct representation of a system in equilibrium at a given temperature, i.e. in contact with a large thermostat. Consider for a moment an ideal gas of $N$ molecules. The probability to find a subsystem of $n$ molecules ($n \ll N$) in the volume element $d\vec{r}_1 \ldots d\vec{r}_n \, d\vec{p}_1 \ldots d\vec{p}_n$ will be the product of the individual probabilities and hence proportional to

$$e^{-\sum_i \frac{p_i^2}{2mkT}} d\vec{r}_1 \ldots d\vec{p}_n = e^{-\mathcal{H}/kT} \, d\vec{r}_1 \ldots d\vec{p}_n \, .$$

As a generalization one expects for a general system in a thermostat the probability to find the system in volume element $d\tau$ of $\Gamma$-space to be proportional to $e^{-\mathcal{H}/kT} d\tau$. And this corresponds to a canonical ensemble. The occurrence of an exponential seems reasonable if one considers 2 systems without interaction, so that $\mathcal{H} = \mathcal{H}_1 + \mathcal{H}_2$. The combined probability is then indeed the product of the two partial probabilities. For the canonical ensemble the probability density is given by

$$P = \frac{\rho}{\mathcal{N}} = \frac{1}{N!\, h^{3N}}\, e^{\frac{\psi - \mathcal{H}}{\Theta}} . \tag{22}$$

2 factors $1/h^{3N}$ and $1/N!$ have been added here: The first for dimensional reasons (it could be $1/h_0^{3N}$ where $h_0$ is an arbitrary constant having the dimension of an action, but to get correspondence with quantum statistics we put here already $h$, though the results are independent of it), the second in order that the free energy will be an extensive quantity (otherwise one meets with the so-called Gibbs paradox). Division by $N!$ is natural, because every state is counted $N!$ times. The particles should be considered indistinguishable.

The Hamiltonian $\mathcal{H}(q_i\, p_i\, \xi_k)$ depends not only on coordinates and momenta of the particles, but also on external parameters $\xi_k$ (e.g. volume of the vessel, magnetic field etc.). We have

$$\int \rho \, d\tau = \mathcal{N} \qquad \text{or} \qquad \int P \, d\tau = 1 ,$$

so that

$$e^{-\frac{\psi}{\Theta}} = Z_N = \frac{1}{N!\, h^{3N}} \int e^{-\frac{\mathcal{H}}{\Theta}}\, d\vec{r}_1 \ldots d\vec{p}_N . \tag{23}$$

This is the *canonical partition integral* or canonical integral over states (Zustandsintegral). As a consequence of a variation of the external parameters $\xi_k$ work is performed by the external forces. If $p_i\, q_i$ remain constant the work is given by

$$\sum_k \frac{\partial \mathcal{H}}{\partial \xi_k} d\xi_k .$$

Averaged over all states this is

$$\sum_k \overline{\frac{\partial \mathcal{H}}{\partial \xi_k}} d\xi_k .$$

Now from (22) it is seen that

$$\overline{\frac{\partial \mathcal{H}}{\partial \xi_k}} = -\Theta \frac{\partial}{\partial \xi_k} \log Z_N = \frac{\partial \psi}{\partial \xi_k} .$$

This shows already, that $\psi$ must represent the free energy $F$, because the work done by the external forces is given by the change in free energy. Or in some more detail, on varying the external parameters we have:

$$\int e^{\frac{\psi-\mathcal{H}}{\Theta}} \left\{ \frac{\Theta\, d(\psi - \mathcal{H}) - (\psi - \mathcal{H})\, d\Theta}{\Theta^2} \right\} d\tau = 0$$

$$\int e^{\frac{\psi-\mathcal{H}}{\Theta}} \left\{ \frac{d\psi}{\Theta} - (\psi - \mathcal{H}) \frac{d\Theta}{\Theta^2} - \frac{d\mathcal{H}}{\Theta} \right\} d\tau = 0$$

or
$$\frac{d\psi}{\Theta} - \frac{\psi - \overline{\mathcal{H}}}{\Theta^2}\, d\Theta - \frac{\overline{d\mathcal{H}}}{\Theta} = 0 .$$

Now
$$\overline{d\mathcal{H}} = - \sum_k \overline{K}_k\, d\xi_k \qquad \text{with} \qquad \overline{K}_k = - \frac{\partial \mathcal{H}}{\partial \xi_k}$$

so that
$$d\psi = \frac{\psi - \overline{\mathcal{H}}}{\Theta}\, d\Theta - \sum_k \overline{K}_k\, d\xi_k . \tag{24}$$

Let us compare this with the thermodynamic relations

$$dF = dU - T\, dS - S\, dT \qquad \text{and} \qquad T\, dS = dU + p\, dV$$

from which we have

$$dF = - \frac{S}{k} d(kT) - p\, dV .$$

$p\, dV$ may alternatively be $H\, dM$, in general it is the work term $\sum_k K_k\, d\xi_k$. On identifying terms we see that:

$F = \psi$, the free energy

$U = \overline{\mathcal{H}}$, the internal energy

$kT = \Theta$, $\Theta$ and $T$ give the temperature in energy units resp. in degrees

$K_k = \overline{K}_k$

$-\frac{S}{k} = \frac{\psi - \overline{\mathcal{H}}}{\Theta}$

Further one has

$$p = - \left(\frac{\partial F}{\partial V}\right)_T \qquad \text{and} \qquad S = - \left(\frac{\partial F}{\partial T}\right)_V$$

so that with

$$Z_N = \frac{1}{N!\, h^{3N}} \int e^{-\mathcal{H}/kT}\, d\vec{r}_1 \ldots d\vec{p}_N =$$
$$= \frac{1}{N!} \left(\frac{2\pi m k T}{h^2}\right)^{\frac{3N}{2}} \int e^{-\frac{V(\vec{r}_1 \ldots \vec{r}_N)}{kT}}\, d\vec{r}_1 \ldots d\vec{r}_N \tag{26}$$

where the integration over momenta has been performed, we find:

$$F = -kT \log Z_N \tag{27}$$

$$U = kT^2 \frac{\mathrm{d} \log Z_N}{\mathrm{d}T} \tag{28}$$

$$p = kT^2 \frac{\mathrm{d} \log Z_N}{\mathrm{d}V} \tag{29}$$

$$S = k \frac{\mathrm{d}}{\mathrm{d}T}(T \log Z_N) \, . \tag{30}$$

For the case where the interaction between molecules is repulsive at short distances and decreases rapidly at large distances one can prove (Van Hove) that $\lim_{N\to\infty} 1/N \log Z_N$ becomes independent of $N$ if we keep $N/V$ (the density) constant, so that it indeed represents an *intensive* quantity, namely $-kTf$ where $f$ is the free energy per particle. The formula for $f$

$$e^{-\frac{Nf(v,T)}{kT}} = \frac{1}{N!\, h^{3N}} \int e^{-\mathcal{H}/kT} \mathrm{d}\vec{r}_1 \ldots \mathrm{d}\vec{p}_N \tag{31}$$

is basic in the statistical mechanics of interacting systems. Its apparent simplicity, however, is misleading. Evaluation of the integral is for dense systems extremely complicated.

b. *The microcanonical ensemble*

Here we have $P = \rho/\mathcal{N} = P_o$, for $E$ between $E$ and $E + \mathrm{d}E$.
$= 0$, outside this shell.

We have

$$\int \rho \,\mathrm{d}\tau = \mathcal{N} \, , \qquad \int P \,\mathrm{d}\tau = 1 \, , \qquad \text{so} \qquad P_o^{-1} = \int_E^{E+\mathrm{d}E} \!\!\ldots \int \mathrm{d}\tau \, .$$

Let us introduce

$$\Omega(E) = \int_0^E \!\!\ldots \int \mathrm{d}\tau \, , \qquad \text{then} \qquad P_o^{-1} = \Omega'(E)\, \mathrm{d}E \, .$$

In a similar way as above for the canonical ensemble one can investigate the thermodynamical analogies. It is then found that the entropy is given by

$$S = k \log \frac{\Omega(E)}{N!\, h^{3N}} \tag{32}$$

and of course $U = E$.

Instead of $\log \Omega(E)$ one may also take $\log \Omega'(E)$ for a very large system. For $\Omega(E) \propto E^{3N/2}$, $\Omega'(E) \propto E^{3N/2-1}$ and $\log \Omega \sim \log \Omega'$.

In the canonical ensemble we find $F(T,V)$ from $Z(T,V)$, similarly in the microcanonical ensemble one finds $S(U,V)$ from $\Omega(E,V)$. And further one has

$$\frac{1}{T} = \left(\frac{\partial S}{\partial U}\right)_T , \qquad \frac{p}{T} = \left(\frac{\partial S}{\partial V}\right)_U . \tag{33}$$

In this way one finds $p$ and $T$ as functions of $U$ and $V$. Elimination of $U$ yields the equation of state. For an ideal gas one finds with both ensembles:

$$F = - NkT \log \frac{V}{N} - 3/2\ NkT \log 2\pi mkT - NkT \tag{34}$$

$$S = Nk \log \frac{V}{N} + 3/2\ Nk \log T + 5/2\ Nk + 3/2\ Nk \log 2\pi mk .$$

The canonical and the microcanonical ensemble are of course closely connected. In the canonical ensemble we can write:

$$\bar{E} = \frac{\int E\, \Omega'(E)\ ^{-E/kT}\, \mathrm{d}E}{\int \Omega'(E)\, e^{-E/kT}\, \mathrm{d}E} . \tag{35}$$

From this formula one finds immediately:

$$\text{specific heat } \gamma = \frac{\mathrm{d}\bar{E}}{\mathrm{d}T} = \frac{\overline{E^2} - \bar{E}^2}{kT^2}$$

In the simple case that $\gamma$ does not depend on $T$ one has $\bar{E} = \gamma T$, hence

$$\frac{\overline{E^2} - \bar{E}^2}{\bar{E}^2} = \frac{\gamma kT^2}{\gamma^2 T^2} = \frac{k}{\gamma} ,$$

which is of order $1/N$. This comes about because $\Omega'(E)\, e^{-E/kT}$ has a very steep maximum for $E = \bar{E}$. Already slightly deviating values of $E$ practically do not occur. Only in this sense the energy becomes a uniquely defined function of temperature.

For an ideal gas $\Omega'(E) = \text{const. } E^{\frac{3}{2}N-1}$. The function $f(E) = E^{\nu} e^{-E/kT}$ has a very steep maximum for $E = \nu kT$. For $E = x\nu kT$, $f(E) = (\frac{\nu kT}{e})^{\nu}\ (xe^{1-x})^{\nu}$. Now $xe^{1-x}$ is 1 for $x = 1$ and $< 1$ for other $x > 0$. The large power makes the maximum extremely sharp.

If $x = 1 + \xi$,

$$xe^{1-x} = (1 + \xi)(1 - \xi + \tfrac{1}{2}\xi^2 \ldots) = 1 - \tfrac{1}{2}\xi^2 + \ldots ; \quad (xe^{1-x})^{\nu} \sim e^{-\frac{\nu}{2}\xi^2} .$$

One could also write (35) as a sum over successive energy shells. The sum may then be replaced by its maximum term. Both ensembles are equivalent as regards the calculation of thermodynamic quantities. The microcanonical ensemble is usually said to give a correct representation for an isolated system, the canonical ensemble for a system in a heat bath. Darwin and Fowler consider the latter as a mathematical trick to calculate averages in an easy way. They have, as was mentioned before, developed methods to calculate averages for $E$ = const.

c. *The grand canonical ensemble*

This can be considered as a collection of canonical ensembles with varying numbers of particles, in a similar way as the canonical ensemble could be considered as a collection of microcanonical ensembles with varying energies.

Consider a macroscopic subvolume $V$ in a canonical system enclosed in a large volume $\mathcal{V}$ ($V \ll \mathcal{V}$). One may then ask for the probability to find in the subvolume $V$, $N$ particles in the phase $\vec{r}_1 \ldots \vec{r}_N$, $\vec{p}_1 \ldots \vec{p}_N$. One finds for this probability:

$$P(N, \vec{r}_1 \ldots \vec{p}_N) = \frac{1}{N!\, h^{3N}}\, e^{\frac{(\chi + N\mu - \mathcal{H}(\vec{r}_1 \ldots \vec{p}_N))}{kT}} \tag{36}$$

Here $e^{\chi}$ is a normalizing factor chosen so that

$$\sum_N \frac{1}{N!\, h^{3N}} \int e^{\frac{(\chi + N\mu - \mathcal{H})}{kT}}\, d\vec{r}_1 \ldots d\vec{p}_N = 1\,. \tag{37}$$

Further it is found that

$$\chi = -pV \qquad \text{and} \qquad \mu = f(v,T) - v\,\frac{\partial f(v,T)}{\partial v}$$

where $v$ and $f$ are volume and free energy per particle. $\mu$ is the thermodynamic potential per particle or specific Gibbs function.

As a generalization we call a grand canonical ensemble an ensemble such that the probability to pick a system of $N$ particles in the phase $\vec{r}_1 \ldots \vec{p}_N$ is given by the expression (36). The ensemble average of a physical quantity $M$ is given by

$$\overline{M} = \sum_N \int M\, P\, d\tau_N\,.$$

The advantage of this ensemble is that the combined summation + integration often can be carried out more simply than the integration alone. In the grand ensemble the overwhelming majority of systems again have a number of particles very close to $\overline{N}$, so that it leads to the same results as a canonical ensemble. Again the summation in the calculation of averages may be replaced by the maximum term.

The grand canonical partition function is given by

$$Z_{gr} = e^{pV/kT} = \sum_N \frac{1}{N!\ h^{3N}} \int e^{\frac{(N\mu - \mathcal{H})}{kT}} d\vec{r}_1 \ldots d\vec{p}_N \tag{38}$$

so that

$$pV = kT \log Z_{gr} \tag{39}$$

and further one has the thermodynamic relations

$$S = \left(\frac{\partial pV}{\partial V}\right)_{V,\mu} , \qquad p = \left(\frac{\partial pV}{\partial V}\right)_{T,\mu} , \qquad n = \frac{N}{V} = \left(\frac{\partial p}{\partial \mu}\right)_{T,V} . \tag{40}$$

From (39) one obtains the pressure $p$ as a function of $T$ and $\mu$, or of $T$ and the so-called activity

$$z = \left(\frac{2\pi mkT}{h^2}\right)^{3/2} e^{\mu/kT} .$$

To find the equation of state $\mu$ (or $z$) has to be eliminated with the help of the 3rd relation (40).

With the grand ensemble one can compute the fluctuation in the number of particles present in a subvolume. One finds:

$$\sqrt{\overline{(\Delta N)^2}} = \overline{N}^{\frac{1}{2}} \sqrt{\frac{kT}{v^2 \dfrac{\partial^2 f}{\partial v^2}}} . \tag{41}$$

The relative fluctuations are very small except in the region of condensation where

$$\frac{\partial^2 f}{\partial v^2} \longrightarrow 0 .$$

## 11. *Molecular distribution functions*

For an arbitrary ensemble ($\mathcal{N}$ systems each consisting of $N$ molecules) we have introduced the number density $\rho_N$ and the probability density $P_N = \rho/\mathcal{N}$ in $\Gamma$-space, where

$$\int \rho_N(\vec{r}^N, \vec{p}^N)\ d\vec{r}^N\ d\vec{p}^N = \mathcal{N}$$

$$\int P_N(\vec{r}^N, \vec{p}^N)\ d\vec{r}^N\ d\vec{p}^N = 1 .$$

Here $\vec{r}^N$ stands for all coordinates $\vec{r}_1, \vec{r}_2 \ldots \vec{r}_N$ and similarly $\vec{p}^N$.

Let us now introduce the specific distribution functions $P_h(\vec{r}^\eta \vec{p}^\eta)$, i.e. the probability to find $h$ numbered molecules (collectively denoted by $\eta$) in the phase point $\vec{r}_{\eta_1} \ldots \vec{r}_{\eta_h}$, $\vec{p}_{\eta_1} \ldots \vec{p}_{\eta_h}$ (denoted by $\vec{r}^\eta \vec{p}^\eta$), then

$$P_h(\vec{r}^{\eta}\vec{p}^{\eta}) = \int P_N \, d\vec{r}^{N-\eta} \, d\vec{p}^{N-\eta} \; . \tag{42}$$

In particular $\quad P_2(\vec{r}_1,\vec{r}_2,\vec{p}_1,\vec{p}_2) = \int P_N \, d\vec{r}^{N-2} \, d\vec{p}^{N-2}$

$$P_1(\vec{r}_1,\vec{p}_1) = \int P_N \, d\vec{r}^{N-1} \, d\vec{p}^{N-1} = \int P_2 \, d\vec{r}_2 \, d\vec{p}_2 \; .$$

We further introduce the *generic distribution functions* giving the probability to find $h$ molecules, whatever their numbers, in the phasepoint $\vec{r}^{\eta} \, \vec{p}^{\eta}$,

$$f_h = \frac{N!}{(N-h)!} P_h \, .$$

The factor $N!/(N-h)!$ gives the number of ways to choose $h$ molecules out of $N$. In particular $f_1(\vec{r}_1,\vec{p}_1) = NP_1$, in equilibrium it can be identified with the kinetic distribution function $f(\vec{r}_1,\vec{p}_1)$ introduced before.

On integrating over momenta one obtains the specific or generic distribution functions in configuration space.

$$n_h(\vec{r}^{\eta}) = \frac{N!}{(N-h)!} P_h(\vec{r}^{\eta}) = \int f_h \, d\vec{p}_1 \ldots d\vec{p}_h \; .$$

In particular $n_2(\vec{r}_1,\vec{r}_2) = N(N-1)\,P_2(\vec{r}_1,\vec{r}_2)$ is the pair distribution function and $n_1(r_1)$ = number density.

The above definitions are perfectly general. Let us now specialize to the canonical equilibrium ensemble. As we have seen we have here:

$$P_N(\vec{r}^N,\vec{p}^N) = \frac{1}{N!\,h^{3N}\,Z_N} e^{-\mathcal{H}/kT} \quad \text{and} \quad P_N(\vec{r}^N) = \frac{1}{N!\,Q_N} e^{-\frac{\Phi(\vec{r}^N)}{kT}}$$

with

$$Z_N = \frac{1}{N!\,h^{3N}} \int e^{-\mathcal{H}/kT} \, d\vec{r}^N \, d\vec{p}^N \quad \text{and} \quad Q_N = \frac{1}{N!} \int e^{-\frac{\Phi(\vec{r}^N)}{kT}} \, d\vec{r}^N \tag{43}$$

Here $Q_N = \lambda^{3N} Z_N$, $\lambda^2 = h^2/2\pi mkT$, $\Phi(\vec{r}^N)$ is the potential energy of the system.

The caloric and thermal equations of state can be expressed in terms of the pair function $n_2$, if the potential energy is a sum of pair interactions.

Let

$$\mathcal{H} = \sum_i \frac{p_i^2}{2m} + \tfrac{1}{2} \sum\sum_{i \neq k} \varphi(r_{ik})$$

then we find for the internal energy

$$U = \frac{1}{N!\, h^{3N}\, Z_N} \int e^{-\mathcal{H}/kT} \left(\sum_i \frac{p_i^2}{2m} + \tfrac{1}{2} \sum_{i \neq k}\sum \varphi(r_{ik})\right) d\vec{r}^N\, d\vec{p}^N .$$

Performing the integrations as far as possible we obtain the so-called *caloric equation of state*

$$U = 3/2\, RT + \tfrac{1}{2} \iint n_2(\vec{r}_1, \vec{r}_2)\, \varphi(r_{12})\, d\vec{r}_1\, d\vec{r}_2 . \tag{44}$$

On the other hand the virial theorem yields:

$$2 \sum_i \frac{\overline{p_i^2}}{2m} = 3\, p\, V + \tfrac{1}{2} \overline{\sum_{i \neq k}\sum r_{ik} \frac{d\varphi}{dr_{ik}}}$$

this leads to:

$$p\, V = \frac{1}{N!\, h^{3N}\, Z_N} \iint e^{-\mathcal{H}/kT} \left\{ \frac{2}{3} \sum_i \frac{p_i^2}{2m} - \frac{1}{6} \sum_{i \neq k}\sum r_{ik} \frac{d\varphi}{dr_{ik}} \right\} d\vec{r}^N\, d\vec{p}^N$$

or
$$p\, V = RT - \tfrac{1}{6} \iint n_2(\vec{r}_1, \vec{r}_2)\, r_{12} \frac{d\varphi}{dr_{12}}\, d\vec{r}_1\, d\vec{r}_2 , \tag{45}$$

the *thermal equation of state.*

The pair distribution occurring in these equations of state is in the canonical ensemble given by

$$n_2(\vec{r}_1, \vec{r}_2) = \frac{N(N-1)}{N!\, Q_N} \int e^{-\frac{\Phi(\vec{r}^N)}{kT}}\, d\vec{r}_3 \ldots d\vec{r}_N . \tag{46}$$

It can be shown to be the functional derivative of the free energy with respect to the interaction potential.

The results (44) and (46) can alternatively be obtained from the formula (27) for the free energy by application of

$$p = -\left(\frac{\partial F}{\partial V}\right)_T \qquad U = \left(\frac{\partial\left(\frac{F}{T}\right)}{\partial\left(\frac{1}{T}\right)}\right)_V$$

$Q_N$ (cf. 43) may be expanded in a so-called clusterexpansion.

If we put $e^{-\frac{\varphi(r_{ik})}{kT}} = 1 + f_{ik}$, then $Q_N = \frac{1}{N!} \int \prod_{i<k} (1 + f_{ik})\, d\vec{r}^N$.

Expanding the product one finds finally

$$p\, V = RT \left(1 - \sum_k \frac{k}{k+1} \beta_k\, n^k\right) \tag{47}$$

$$U = \tfrac{3}{2} RT + RT^2 \sum_k \frac{1}{k+1} \frac{d\beta_k}{dT}\, n^k \tag{48}$$

where $n$ is the number density and $\beta_k$ are the so-called irreducible clusterintegrals:

$$V\beta_1 = \iint f_{12}\, d\vec{r}_1\, d\vec{r}_2 \qquad \text{or schematically}$$ 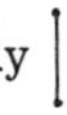

$$V\beta_2 = \iiint f_{12}\, f_{13}\, f_{23}\, d\vec{r}_1\, d\vec{r}_2\, d\vec{r}_3\,, \text{ or}$$ 

$V\beta_3 =$ 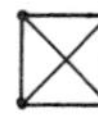 $+ 6$  $+ 3$  , etc.

The virial coefficients are given by $B_k = -\dfrac{k-1}{k}\,\beta_{k-1}$.

For the pair distribution function an expansion in powers of the density can be derived in an analogous way.

From the definition of the distribution function $n_h$ in a canonical ensemble one finds by simple differentiation

$$kT\frac{\partial n_h(\vec{r}^{\,h})}{\partial \vec{r}_l} = -\sum_{k=1}^{h}\frac{\partial\varphi(\vec{r}_{lk})}{\partial \vec{r}_l}\, n_h(\vec{r}^{\,h}) - \int \frac{\partial\varphi(r_{l,h+1})}{\partial \vec{r}_l}\, n_{h+1}(\vec{r}^{\,h+1})\, d\vec{r}_{h+1}\,. \tag{49}$$

This equation connects 2 successive distribution functions. Kirkwood, by making for $n_3$ a simple assumption,

$$n_3 = \frac{n_2(r_{12})\, n_2(r_{13})\, n_2(r_{23})}{n^3}\,,$$

which expresses $n_3$ in terms of $n_2$, derived for $n_2$ an approximate integral equation, which has given rise to some discussion.

The above relation is a particular case of very general relations which, also for non-stationary ensembles, connect $P_h$ and $P_{h+1}$. They can be derived from Liouville by integration and are often called the Bogoliubov-Born-Green-Yvon-Kirkwood hierarchy.

## 12. *Quantumstatistics*

Finally I should like to remind you very briefly of some important formulae in quantumstatistics. Let us assume that the energy states of the whole system are $E_\alpha$ with eigenfunctions $\Phi_\alpha$. Then the probability to find a system in a canonical ensemble in the stationary state $E_\alpha$ is given by

$$P_\alpha = \frac{e^{-\frac{E_\alpha}{kT}}}{\sum_\alpha e^{-\frac{E_\alpha}{kT}}} \tag{50}$$

and the free energy per particle $f$ is given by

$$e^{-\frac{Nf}{kT}} = Z_N = \sum_{\alpha} e^{-\frac{E_\alpha}{kT}} . \tag{51}$$

This sum over states may alternatively be written as

$$Z_N = \sum_{\alpha} e^{-\frac{E_\alpha}{kT}} = \sum_{r} (\chi_r , e^{-\frac{\mathcal{H}}{kT}} \chi_r) = Sp(e^{-\frac{\mathcal{H}}{kT}}) \tag{52}$$

where $\chi_r$ denotes an arbitrary, complete orthonormal system. In the classical limiting case (51) leads back to (31).

In the particular case of particles without interaction, we can introduce 1-particle energies $\varepsilon_i$ and 1-particle eigenfunctions $\varphi_i$ (e.g. plane wave states if there is no external field). For bosons the eigenfunctions of the whole system are represented by

$$\Phi_s = \sum_{P} \varphi_{i_1}(\vec{r}_{k_1}) \ \varphi_{i_2}(\vec{r}_{k_2}) \dots \varphi_{i_N}(\vec{r}_{k_N}) \tag{53}$$

and for fermions by

$$\Phi_a = \sum_{P} (-1)^P \varphi_{i_1}(\vec{r}_{k_1}) \ \varphi_{i_2}(\vec{r}_{k_2}) \dots \varphi_{i_N}(\vec{r}_{k_N}) \tag{54}$$

where $P$ denotes the permutations of the particles.

We then have:

$$N = \sum n_i$$

$$E = \sum n_i \ \varepsilon_i$$

where $n_i$ is the number of particles in the 1-particle state $\varepsilon_i$ (for fermions $n_i = 0$ or 1).

The canonical sum over states

$$e^{-\frac{Nf}{kT}} = \sum_{n_i} \exp\left(-\frac{\sum_i n_i \ \varepsilon_i}{kT}\right) \tag{55}$$

cannot be evaluated very simply because of the restriction $\sum_i n_i = N$. The grand canonical ensemble turns out to be much more suitable in this case. Here the probability to find a system of $N$ particles, with $n_i$ particles in the 1-particle state $\varepsilon_i$ is given by

$$P(N, n_1 \dots n_i \dots) = \exp\left(\frac{-pV + \sum_i n_i (\mu - \varepsilon_i)}{kT}\right) \tag{56}$$

so that
$$e^{\frac{pV}{kT}} = \sum_N e^{\frac{N\mu}{kT}} \sum_{n_i} \exp\left(-\sum_i \frac{n_i \varepsilon_i}{kT}\right) \tag{57}$$

which becomes simply

$$e^{\frac{pV}{kT}} = \sum_{n_i} \exp\left(\sum_i \frac{(\mu - \varepsilon_i)\, n_i}{kT}\right) \tag{58}$$

without restriction in the summation over the $n_i$'s. One finds:

$$e^{\frac{pV}{kT}} = \prod_i \sum_{n_i} \exp\left(\frac{(\mu - \varepsilon_i)\, n_i}{kT}\right) = \prod_i \frac{1}{1 - \exp\frac{\mu - \varepsilon_i}{kT}} \qquad \text{B.E.} \tag{59}$$

$$= \prod_i \left(1 + \exp\left(\frac{\mu - \varepsilon_i}{kT}\right)\right) \qquad \text{F.D.} \tag{60}$$

In this way one obtains:

$$\frac{pV}{kT} = \pm \sum_i \log\left(1 \pm \exp\frac{\mu - \varepsilon_i}{kT}\right) \tag{61}$$

with a + sign for F.D. statistics, a - sign for B.E. statistics. On use of $\partial p/\partial\mu = 1/v$ it is possible to eliminate $\mu$ and one finds the equation of state. For small quantum corrections one is led to

$$pV = NkT\left(1 \pm \frac{N}{V}\frac{\lambda^3}{2^{5/2}} + \ldots\right) \tag{62}$$

where $\lambda^2 = h^2/2\pi mkT$.

The mean occupation number of a 1-particle level is

$$\bar{n}_i = \sum_{N, n_1 \ldots} P(N, n_1 \ldots n_i \ldots)\, n_i\, .$$

This yields

$$\bar{n}_i = \frac{\sum_{n_i} n_i \exp\left(\sum_i (\mu - \varepsilon_i)\, n_i/kT\right)}{\sum_{n_i} \exp\left(\frac{\sum_i n_i(\mu - \varepsilon_i)}{kT}\right)} = \frac{\partial}{\partial\left(\frac{\mu - \varepsilon_i}{kT}\right)} \log e^{\frac{pV}{kT}}\, .$$

Hence:

$$\bar{n}_i = \frac{1}{\exp\left(\frac{-\mu + \varepsilon_i}{kT} - 1\right)} \qquad \text{B.E.}$$

$$\bar{n}_i = \frac{1}{\exp\left(\frac{-\mu + \varepsilon_i}{kT} + 1\right)} \qquad \text{F.D.} \tag{63}$$

Starting from these formulae one may discuss in the well-known way e.g. Bose-Einstein condensation or the properties of a free electron gas.

More generally, in quantum statistics the *density matrix* takes the role of the ensemble density in Γ-space in classical statistics. It is defined in the following way: Let a system be described by a state function $\psi(\vec{r},t)$, which can be expanded into an arbitrary orthonormal system $u_k(\vec{r})$

$$\psi(\vec{r},t) = \sum_k a_k(t)\, u_k(\vec{r}) \; .$$

The density matrix in $u$-language is defined by

$$\rho_{nm} = \frac{1}{\mathcal{N}} \sum_\alpha a_m^{*(\alpha)}(t)\, a_n^{(\alpha)}(t) = \overline{a_m^* a_n} \; . \tag{64}$$

averaged over an ensemble consisting of $\mathcal{N}$ systems, having state functions $\psi^{(\alpha)}$ with expansioncoefficients $a_k^{(\alpha)}$. If state $\alpha$ has statistical weight $p^{(\alpha)}$, then

$$\rho_{nm} = \sum_\alpha p^{(\alpha)}\, a_m^{*(\alpha)}\, a_n^{(\alpha)} \; .$$

We see that $\rho_{nm} = \rho_{mn}^*$ i.e. $\rho$ is a Hermitian matrix. Further we have

$$Sp\,\rho = 1 \qquad 0 \leqslant \rho_{nn} \leqslant 1 \; .$$

The probability to find an individual system in a state $u_n$ is

$$P_n = \rho_{nn} = \overline{a_n^* a_n} \, .$$

For an arbitrary quantity $F$ the expectation value in the state $\psi$ is

$$\langle F \rangle = \int \psi^* F \psi \, d\vec{r} = \sum_{mn} F_{mn}\, a_m^* a_n$$

averaged over the ensemble:

$$\overline{\langle F \rangle} = \sum_{m,n} F_{mn}\, \rho_{nm} = Sp(F\rho) \; . \tag{65}$$

This expression is naturally independent of the representation used. As a special representation we can take for $u_k$ the eigenfunctions ($\delta$-functions) of the coordinates:

$$\rho(\vec{r},\vec{r}\,') = \overline{\psi^*(\vec{r}\,',t)\, \psi(\vec{r},t)} = \sum_{n,m} \rho_{nm}\, u_n(\vec{r})\, u_m^*(\vec{r}\,') \; . \tag{66}$$

$\rho(\vec{r},\vec{r})$ is the probability to find the system in the configuration $\vec{r}$.

The quantum analogue of the equation of Liouville is, as follows from the Schrödinger equation,

$$\frac{\partial \rho}{\partial t} = \frac{-i}{\hbar} [\mathcal{H}\,\rho]_- \; . \tag{67}$$

In the canonical ensemble the density matrix takes the following form:

$$\rho = \frac{e^{-\mathcal{H}/kT}}{Z_N} \tag{68}$$

where the sum over states $Z_N$ has been defined already in (52).

In energy representation $\rho$ is given by

$$\rho_{nm} = \frac{e^{-E_n/kT}\,\delta_{nm}}{Z_N} \tag{69}$$

and in coordinate representation by

$$\rho(\vec{r}, \vec{r}\,') = \frac{\sum_{\alpha} e^{-E_\alpha/kT}\,\varphi_\alpha(\vec{r})\,\varphi_\alpha^*(\vec{r}\,')}{Z_N} \tag{70}$$

where $\varphi_\alpha(\vec{r})$ are the eigenfunctions of the hamiltonian.

In a similar way as in classical statistics one can of course introduce density matrices for a reduced number of particles, but we will not go further into this matter now.

# FLUCTUATIONS, STOCHASTIC PROCESSES, BROWNIAN MOTION

H. WERGELAND

## I. FLUCTUATIONS

### a. *Gibbs' theory*

Gibbs' method of calculating fluctuations is contained in the formula[1])

$$\overline{(A - \overline{A})^2} = \Theta \left( \overline{\frac{\partial^2 E}{\partial a^2}} - \frac{\partial^2 \psi}{\partial a^2} \right) . \tag{1}$$

Here $a$ means an external coordinate and $A$ the conjugate force so that the work performed by the system in an infinitesimal change $a \to a + \delta a$ is $-\delta E = A\, \delta a$. The other letters have their usual meaning: $\psi(\Theta, a \ldots)$ is the Free Energy *under the conditions given* so that

$$\begin{aligned} -\frac{\partial \psi}{\partial a} &= \text{average force} \\ \frac{\partial^2 \psi}{\partial a^2} &= \text{isothermal tension coefficient with respect to } a. \end{aligned} \tag{2}$$

Therefore the last term of formula (1) is immediately given in terms of thermodynamical quantities. But the interpretation of the average $\overline{\partial^2 E / \partial a^2}$, on the other hand, has been the subject of some discussion[2]).

In Gibbs' words (l.c. p. 83) $\partial E^2 / \partial a^2$ "represents an elasticity measured under the condition that while $a$ is varied the internal coordinates $q_1\, q_2 \ldots q_f$ remain fixed".

Now a force $\mathscr{A} = -(\partial E / \partial a)_{p,q}$ would correspond to the reaction against a change that occurs so fast that the molecular variables stay essentially constant. It is clear that this force may acquire enormous magnitude, and although it turns out that its average will have the correct thermodynamical value:

$$\overline{\mathscr{A}} = -\partial \psi / \partial a = \overline{A} \tag{3}$$

its mean square will not be normal.

In order to understand this better we must recall the origin of Gibbs' formula (1). It belongs to a series of identities which *formally* are obtained by successive differentiations of the normalization integral. Assuming two states, $a$ and $a + \delta a$, both to be equilibria at the same temperature $\Theta$ we must have

$$\int e^{\frac{\psi+\delta\psi-E-\delta E}{\Theta}} (\mathrm{d}p\ \mathrm{d}q) = \int e^{\frac{\psi-E}{\Theta}} (\mathrm{d}p\ \mathrm{d}q) \tag{4}$$

Expanding in powers of $\delta a$ we obtain then

$$\int e^{\frac{\psi-E}{\Theta}} (\mathrm{d}p\ \mathrm{d}q)\ \{ c_1\ \delta a + c_2\ \delta a^2 + \ldots \} = 0$$

or

$$\bar{c}_1 = \bar{c}_2 = \ldots = 0\ ,$$

where

$$\bar{c}_1 = \frac{1}{\Theta}\left(\frac{\partial\psi}{\partial a} - \overline{\frac{\delta E}{\delta a}}\right)$$

$$\bar{c}_2 = \frac{1}{2\ \Theta^2}\overline{\left(\frac{\delta E}{\delta a} - \frac{\partial\psi}{\partial a}\right)^2} + \frac{1}{2\ \Theta}\left[\frac{\partial^2\psi}{\partial a^2} - \overline{\frac{\delta}{\delta a}\left(\frac{\delta E}{\delta a}\right)}\right] \tag{5}$$

and so on.

$\bar{c}_2 = 0$ gives us the fluctuation formula (1) when the difference quotients are replaced by fluxions. But I have for the time being written them in this way to indicate that they may not be unique. The typical case in point is when $a$ is the volume and $A$ the pressure. One finds then that although the average $\overline{\delta E/\delta a}$ has an unambigous meaning, $\overline{(\delta/\delta a)(\delta E/\delta a)}$ has not.

However, we must remember the physical idea underlying the procedure (4) - (5). And that is to exhibit how a small reversible change $a \to a + \delta a$ upon the system is represented by the ensemble. In each stage of this process then, the distribution is to retain its canonical form and in particular the normalization (4) shall stay constant. For this to be true it is in general not sufficient that the process be infinitesimal, it must also be quasi-static. Certainly all the averages implied in Gibbs' "Thermodynamical Analogies" (l.c. p. 168) must be understood in this sense. For example:

$$\Theta\ \mathrm{d}\bar{\eta} = \delta\bar{E} - \overline{\delta E}_{\text{quasi-static}} \tag{6}$$

in general different from

$$\delta Q = \delta\bar{E} - \overline{\delta E}_{\text{arbitrary}} \tag{7}$$

The change $\delta a$ is therefore to be considered slow compared to molecular motions, or adiabatic in the sense of Ehrenfest.

In the simple case when the energy is known in terms of action integrals or quantum numbers we have only to keep these constant in the differentiations of $E$ with respect to $a$. But we need not have this detailed knowledge of the Hamiltonian. For an adiabatic transformation performed upon the members of the ensemble, corresponds to a change at constant entropy of the system:

$$\overline{\delta E}_{I...} = \delta \overline{E}_{\overline{\eta}} = \delta U_S \ . \tag{8}$$

We can therefore identify the averages occurring in the fluctuation formula with known thermodynamical quantities:

$$\overline{\frac{\partial E}{\partial a}} = \left(\frac{\partial U}{\partial a}\right)_S \qquad \overline{\frac{\partial^2 E}{\partial a^2}} = \left(\frac{\partial^2 U}{\partial a^2}\right)_S = -\left(\frac{\partial \overline{A}}{\partial a}\right)_S$$

that is

$$\overline{(A - \overline{A})^2} = \Theta \left\{ \left(\frac{\partial \overline{A}}{\partial a}\right)_\Theta - \left(\frac{\partial A}{\partial a}\right)_{\overline{\eta}} \right\} . \tag{9}$$

In this way we can for example calculate the pressure fluctuation as has been discussed in the recent paper of Martin J. Klein [2)].

b. *Einstein's method*

Consider again a system at given temperature, and let $\lambda$ be an observable parameter which is not fixed by the equilibrium, but fluctuates. The probability to find it in a specified interval $(\lambda, \lambda + d\lambda)$ can then be written

$$f(\lambda)\, d\lambda = \int_\lambda^{\lambda+d\lambda} (dp\, dq)\, e^{\frac{\psi - E}{\Theta}} = e^{\frac{\psi}{\Theta}}\, d\lambda \int \omega(E,\lambda)\, dE \, . \, e^{-\frac{E}{\Theta}} . \tag{10}$$

Here the phase integral

$$\int \omega(E,\lambda)\, dE \, . \, e^{-\frac{E}{\Theta}} = e^{-\frac{\psi(\lambda)}{\Theta}}$$

defines a subensemble which is again canonically distributed in phase; but has a fixed value of the parameter $\lambda$. It does not represent an equilibrium under the external conditions actually given, but it would have done so if by some kind of control on the system we had clamped $\lambda$. We can thus express the fluctuation probability by

$$f(\lambda) = e^{-\frac{\Delta\psi}{\Theta}} \tag{11}$$

where $\Delta\psi = \psi(\lambda) - \psi_{\text{equilibrium}}$ is the *minimum work* which acoording to thermodynamics would be required to produce the fluctuation $\Delta\lambda = \lambda - \overline{\lambda}$.

As the free energy is a minimum at equilibrium, the Taylor expansion

$$\Delta\psi = \tfrac{1}{2} \frac{d^2\psi}{d\lambda^2} (\lambda - \overline{\lambda})^2 + \dots$$

normally begins with the second power. Usually it is enough to consider only this first term, readjusting the normalization factor. For the mean square one obtains then

$$\overline{(\lambda - \bar{\lambda})^2} = \Theta \Big/ \frac{d^2\psi}{d\lambda^2} . \qquad (12)$$

This looks somewhat simpler than Gibbs' formula 1) or its pendant

$$\overline{(a - \bar{a})^2} = \Theta \left(\overline{\frac{\partial^2 E}{\partial A^2}} - \frac{\partial^2 \psi}{\partial A^2}\right) \qquad (\text{when } \delta E = a\ \delta A) . \qquad (13)$$

Actually it is more difficult to apply because it requires circumspection of which variables are to be kept constant in the differentiation.

As a simple example we consider volume fluctuations at constant pressure by both methods. In the Einstein formula (12) we then take

$\psi = G =$ free energy at constant pressure

$\lambda = V$

$$\Delta\psi = (\Delta G)_{p,\Theta} = (\Delta F)_{\Theta} + p\ \Delta V$$

$$= 0\ .\ \Delta V + \tfrac{1}{2}\frac{\partial^2 F}{\partial V^2}(V - \bar{V})^2 + \dots$$

Accordingly by (12)

$$\overline{(V - \bar{V})^2} = -\ kT \Big/ \left(\frac{\partial p}{\partial V}\right)_T$$

$$= \frac{\bar{V}^2}{N} \qquad \text{for an ideal gas.}$$

Using Gibbs' formula we must choose the constant pressure ensemble, that is an energy function $E = E(p, q) + pV$ then

$$\overline{\frac{\partial E}{\partial p}} = \bar{V} \qquad \frac{\partial^2 E}{\partial p^2} = 0 \qquad \psi = G .$$

$$\therefore\ \overline{(V - \bar{v})^2} = \Theta \left[0 - \frac{\partial^2 G}{\partial p^2}\right] = -\ kT \left(\frac{\partial v}{\partial p}\right)_T .$$

c. *Decay of fluctuations*

The equilibrium theory gives us a full account of the distribution of fluctuations; but not of their temporal behaviour. This question is connected with the theory of irreversible processes, but a preliminary sketch of the decay of fluctuations we can make as follows:

Let $\lambda$ be a fluctuating variable with $\bar{\lambda} = 0$. If it is away from zero it will on the average tend to return to equilibrium with some velocity:

$$\frac{d\lambda}{dt} = f(\lambda \ldots) .$$

If $\lambda$ is small we can develop: $f(\lambda\ldots) = 0 - \beta\lambda + \ldots$ where $\beta = -f'(\bar{\lambda}\ldots)$. In this way we will expect an exponential law:

$$\lambda(t) = \lambda(0)\, e^{-\beta t} . \tag{15}$$

Now, this piece of mathematics is not absolutely convincing, but there is a little more to substantiate the form of (15). Let us make a statistics of the following kind: We observe a fluctuation $\lambda$, watch its decay in a time $\tau$, and form the product $\lambda(t_i) \cdot \lambda(t_i + \tau)$ a great number of times $t_i$.

If (15) is true we must for the average find

$$<\lambda(t)\ \lambda(t+\tau)> = \overline{\lambda^2}\ e^{-\beta\tau} .$$

The average on the left is called the correlation function of the random proces $\lambda_t$. And we shall see in the next lecture that the exponential time dependence has a very important meaning.

*References*

1. J.W. Gibbs, "Elementary Principles of Statistical Mechanics, Chapter VII, p. 81.
2. See for example:
Martin J. Klein, "Pressure Fluctuations", Physica 26 (1961) 1073.
A. Münster, Physica 26 (1961) 1117.

## II. STOCHASTIC PROCESSES

The classic example is the displacement of a Brownian particle with time $x_t$. It is traditional to conceive a stochastic process as defined by the totality of joint distribution functions for the coordinate values $x_{t_1}\, x_{t_2} \ldots x_{t_n}$ at any $n$ instants. We can then state the probability

$$f_n \begin{pmatrix} x_1 & x_2 & \ldots & x_n \\ t_1 & t_2 & & t_n \end{pmatrix} dx_1\, dx_2 \ldots dx_n$$

for the path to pass any number of specified "gates" 1)

$$\begin{aligned} x_1 &< x_{t_1} < x_1 + dx_1 \\ x_2 &< x_{t_2} < x_2 + dx_2 \\ &\vdots \\ x_n &< x_{t_n} < x_n + dx_n . \end{aligned}$$

One particular path - $x_\alpha(t)$ say, exhibited for example by an observation - is called a *realization* of the process. (It is an ordinary function of $t$.) Considering each realization as an elementary event we can label them by a parameter $\alpha$. When we let $\alpha$ assume various values we get a collection of paths. The question thus arises to form an ensemble, that is to represent $\alpha$ as point of a suitable set and give it a probability measure which engenders all the distributions $f_n$. If we have a discrete evolution parameter: $t_j = j \cdot \tau$ ; $j = 0, 1, 2, \ldots n$ there is no difficulty at all, $\alpha$ could be taken a vector $\{\alpha_1 \alpha_2 \ldots \alpha_n\}$ in an $n$-dimensional space. By this observation we can jump the abyss to the continuum by carrying out an actual subdivision of our time interval in $n$ parts and let $n \to \infty$ at an appropriate place in the calculations. And this procedure will be all right for a continuous process. We need not here, however, try to formulate any actual parametrization $\alpha$. But it is helpful to keep in mind that we have to consider the process $x_t$ - not as a function of $t$ - but as a function, $x(t, \alpha)$ of $t$ *and* the random variable $\alpha$.

So far the distribution functions $f_n$ should only satisfy the general conditions of

compatibility: $\int f_n \, dx_j = f_{n-1}(x_1 x_2 \ldots x_{j-1} x_{j+1} \ldots x_n)$

symmetry: with respect to permutations $\binom{x_i}{t_i} \rightleftarrows \binom{x_k}{t_k}$

and normalization: $\int f_n \, dx_1 \, dx_2 \ldots dx_n = 1$

But physics has until now used only a very small part of this vastly general field of possibilities.

## a. *Stationary processes*

By *stationarity* is meant that all the probability distributions are unaffected whether we translate the time scale by an arbitrary step $\tau$:

$$f\begin{pmatrix} x_1 & x_2 & \ldots & x_n \\ t_1+\tau, & t_2+\tau & & t_n+\tau \end{pmatrix} = f\begin{pmatrix} x_1 & x_2 & \ldots & x_n \\ t_1 & t_2 & & t_n \end{pmatrix} . \tag{1}$$

Clearly this implies some kind of permanence in the conditions of the physical system from which we draw our statistics. But is not that an awkward limitation. After all, what we want to study is evolution. Now, it is not quite as bad as it looks at first sight. Sometimes the process can be decomposed into two parts:

$$x_t = \underset{\substack{\uparrow \\ \text{systematic}}}{m(t)} + \underset{\substack{\uparrow \\ \text{stochastic stationary}}}{\xi_t}$$

Or it is of the type

$$x_t = x_{t-\tau} + \xi_\tau$$

where the *increments* $\xi$ are stationary.

So we see that stationarity still leaves us a fairly wide scope.

## b. *Gaussian processes*

Due to the central limit theorem of probability - which states that the distribution of the sum of a great number of independent quantities will tend to the Gaussian form under rather wide conditions - stochastic processes in physics will often be governed by this law. A process is called Gaussian if all its distributions are of the form

$$f_n = \text{const. exp } \{ \sum_{i,k}^{n} a_{ik}(t_1 \ldots t_n)\, x_i\; x_k \} \, . \tag{2}$$

A Gaussian process need not be stationary, but when that is the case then the functions $a_{ik}$ can only depend upon the differences $t_i - t_k$, $t_l - t_m$, ... .

If, however, a process is *both* Gaussian and stationary we need not bother ourselves with the array of many dimensional distribution functions. It will namely then be completely determined by two quantities only:
the *mean* $\langle x_{t_1} \rangle = \langle x_{t_2} \rangle = \ldots = m$ independent of $t$, and
the *correlation function* $\langle x_t \cdot x_s \rangle = R(t - s)$.
The latter can then only be a function of $|t - s|$.

We shall soon have more to say about the correlation function; but first we must introduce another specialization which will also bring a strong condensation of the scheme.

## c. *Markov processes*

Instead of defining the process by the joint probabilities $f_1 \ldots f_n \ldots$ we could also do it by the conditional probabilities to find $x_{t_n}$ in the interval $(x_n, x_n + dx_n)$ given that $x_{t_1}\, x_{t_2} \ldots x_{t_{n-1}}$ have preassigned values $x_1\, x_2 \ldots x_{n-1}$ :

$$P\left( \begin{matrix} x_1 & x_2 & \ldots & x_{n-1} \\ t_1 & t_2 & & t_{n-1} \end{matrix} \middle| \begin{matrix} x_n \\ t_n \end{matrix} \right) dx_n \, .$$

If this probability is independent of all the knowledge $x_1\, x_2 \ldots x_{n-2}$, prior to the next preceding instant $t_{n-1}$, we have a *Markov process*. The probability of a step $x_i \to x_f$ depends then only upon the two consecutive states $i$ and $f$: the prehistory $t < t_i$ has no *after effect*.

Such processes are clearly exemplified by various kinds of "random walk" and hazard games etc. On the other hand, looking at a process like the displacement in Brownian motion, it can certainly not be quite correct to disregard the past positions since the particle has some inertia. In this case the situation is remedied by including the velocity along with the position into the description. One gets then again a Markov process. In some cases therefore processes with after effect can be established as Markoffian in an extended number of variables.

Apart from the normalization condition:

$$\int_{-\infty}^{\infty} P\left(\begin{array}{c|c} x_1 & x_2 \\ t_1 & t_2 \end{array}\right) dx_2 = 1 \tag{3}$$

a Markov process must also fulfill the evident condition

$$P(x_1 \mid x_2) \to \delta(x_1 - x_2) \qquad \text{when} \qquad t_2 \to t_1 \,.$$

Further it must satisfy the so-called Chapman Kolmogorov equation[2].

$$P\left(\begin{array}{c|c} x_1 & x_2 \\ t_1 & t_2 \end{array}\right) = \int P\left(\begin{array}{c|c} x_1 & x' \\ t_1 & t' \end{array}\right) dx' \, P\left(\begin{array}{c|c} x' & x_2 \\ t' & t_2 \end{array}\right) ; \qquad t_1 < t' < t_2 \,. \tag{4}$$

When a Markov process is also stationary, $P$ can obtain the time only in the difference $t_2 - t_1$ and $P(x_1 \mid x_2) \to f_1(x_2)$ when $t_2 \to \infty$ regardless of initial state. The equation (4) indicates that $P$ has some sort of matrix property and this comes out clear if we discretize both $x$ and $t$. Already Markov himself utilized this property extensively [3], and I should also mention in this connection some beautiful solutions of stochastic problems by Kac and Montroll [4].

### d. *The correlation function*

We have already mentioned the correlation function $R(s,t) = \langle x_s \,.\, x_t \rangle$. For a stationary process it can only be a function of $|t - s|$. To simplify we consider the "reduced" process:

$$\xi_t = (x_t - m)/\sigma \,, \qquad m = \langle x_t \rangle \,, \qquad \sigma^2 = \langle x_t^2 \rangle - m^2 \tag{5}$$

writing

$$\langle \xi_s \,.\, \xi_t \rangle = \rho(s - t) \,. \tag{6}$$

The function $\rho$ has by its definition the following properties:

$$\begin{aligned} \rho(0) &= 1 \\ \rho(-\tau) &= \rho(\tau) \\ |\rho(\tau)| &\leq \rho(0) \end{aligned} \tag{7}$$

Proof: Consider the random variable,

$$y = \xi(0, \alpha) - \rho(\tau)\, \xi(\tau, \alpha)$$

$$<y^2> = <\xi_0^2> - 2 <\xi(0)\ \xi(\tau)>\ \rho(\tau) + \rho(\tau)^2 <\xi(\tau)^2>$$
$$= 1 - \rho(\tau)^2 \geq 0 \qquad \text{q.e.d.}$$

In fact $\rho$ has the property of being positive in the sense that the quadratic forms:

$$\sum_{i,k}^{n} a_i\, a_k\ \rho(t_i - t_k) \geq 0 \tag{8}$$

(the $a_i$ and $n$ arbitrary) because they are average squares $<s^2>$ of $s = \Sigma\, a_i\, x_{t_i}$. This can be used [5] to find an orthogonal expansion of the process itself. If we consider $\rho$ as the kernel of a Fredholm integral equation,

$$\varphi(t) = \lambda \int \rho(t - s)\ \varphi(s)\ ds \tag{9}$$

then it can be expanded in the "bilinear series"

$$\rho(s - t) = \sum_j \frac{\varphi_j(s)\ \varphi_j(t)}{\lambda_j} \tag{10}$$

of the eigenfuncties $\varphi_j$; $\lambda_j$ real and positive.

If we now imagine the process itself in these orthonormal functions

$$\xi(t, \alpha) = \sum_j \alpha_j\, \varphi_j(t) \tag{11}$$

and take the coefficients $\alpha_j$ to be random variables with the properties:

$$<\alpha_i\ \alpha_j> = \frac{1}{\lambda_j}\ \delta_{ij}\ , \qquad <\alpha_i> = 0 \tag{12}$$

we have already satisfied the requirement:

$$<\xi_s\ \xi_t> = \rho(t - s)\ , \qquad <\xi_t> = 0\ .$$

But it must be remembered of course that this need not yet entirely specify the process. In the case of a stationary Markov process (in one dimension) even the functional form of the correlation function is fixed [6] viz. $\rho(\tau) = \exp(-\beta|\tau|)$.

e. *Spectral resolution of a stationary process*

It is often convenient to represent the process by a Fourier expansion:

$$x(t, \alpha) = \int_{-\infty}^{\infty} g(\omega, \alpha)\ e^{i\omega t}\ d\omega\ , \qquad g(-\omega) = g(\omega)^*\ . \tag{13}$$

The correlation function is then expressed as

$$<x_t\ x_s> = \iint_{-\infty}^{\infty} d\omega\ d\omega' <g(\omega, \alpha)\ g(\omega', \alpha)> . e^{i\omega t + i\omega' s} , \tag{14}$$

and if this is to be a function of $|s - t|$ only, the random functions $g$ must have the property:

$$<g(\omega, \alpha)\ g(\omega', \alpha)> = h(\omega)\ \delta(\omega + \omega') . \tag{15}$$

Then

$$R(t - s) = \int_{-\infty}^{+\infty} h(\omega)\ e^{i(s-t)\omega}\ d\omega .$$

(To include sharp spectral lines we should have used Stieltjes integral.) Since $h(\omega)$ is by (13) (14) not only real but also positive, we can write the correlation function in the form[7]:

$$R(\tau) = 2 \int_0^{\infty} h(\omega) \cos \omega\tau\ d\omega . \tag{16}$$

Conversely it can be shown [7] that any reasonable positive function $h(\omega)$ can be considered as the *spectral function* of the correlation of a stationary process. For a Markov process we must evidently have

$$h(\omega) = \sigma^2 \frac{\beta}{\pi} \frac{1}{\beta^2 + \omega^2} . \tag{17}$$

### f. *Ergodicity*

At this point it may be right to say something about time averages and ensemble averages.

Consider the time average,

$$\bar{x}^T = \frac{1}{2T} \int_{-T}^{T} x(t, \alpha)\ dt = \int d\omega\ g(\omega, \alpha) \frac{\sin \omega T}{\omega T} .$$

This is still a random function with some fluctuations around the ensemble average $<\bar{x}^T> = 0$. Its mean square is:

$$<(\bar{x}^T)^2> = \int d\omega\ h(\omega) \frac{\sin^2 \omega T}{\omega^2 T^2}$$

$$\xrightarrow[T\to\infty]{} \frac{\pi}{T} h(0) = \frac{1}{2T} \int_{-T}^{T} R(\tau)\ d\tau .$$

Therefore if

$$\lim \frac{1}{T} \int_0^T R(\tau)\ d\tau = 0 , \tag{18}$$

then the time average of $x_t$ converges to the ensemble average (in mean square) [8].

The condition (18) is amply satisfied for stationary Markov processes. So far, we have only indicated a proof for $x_t$ itself and linear functionals of $x_t$. However, we shall in the sequel assume ergodicity throughout and thus when convenient, replace the ensemble average by the time average which can in principle be obtained from one single realization of the process.

*References*

1. See L. Onsager and S. Machlup, Phys. Rev. 91 (1953) 1509.
2. M. Smoluchowski (1913), See Ostwalds Klassiker no. 207, p. 30
3. S. Chapman, Roy. Soc. A 119 (1928) 34.
   A. Kolmogorov, Math. Ann. 104 (1931) 415.
3. See M. B. Hostinsky, "Méthodes Générales du Calcul des Probabilitées", Paris 1931.
4. M. Kac, Am. Math. Soc. LIV (1947) 369.
   E. Montroll, Comm. P. and Appl. Math. V (1952) 415.
5. M. Kac and A. Siegert, Annals of Math. Stat. XVIII (1947) 439.
6. See Min Chen Wang and G. E. Uhlenbeck, Rev. of Mod. Phys. 17 (1945) 330. and J. L. Doob, "Stochastic Processes", p. 233, New York (1952).
7. N. Wiener, Acta Math. 55 (1930) 117.
   A. Kintchine, Math. Ann. 109 (1934) 604.
8. E. E. Slutsky, "Sur les Fonctions Aleatoires", p. 33, Paris (1938).

## III. IONIZATION TRACKS

As application of the preceding concepts I shall take multiple scattering of charged particles. This is somewhat aside from our dear statistical theory of heat; but nevertheless ionization tracks make pretty good examples of stochastic processes.

When multiple scattering is fully developed the mean square deflection $\langle\vartheta^2(l)\rangle$ after traversal of a path $l$ is given by an expression of the form

$$\langle\vartheta^2(l)\rangle = l \cdot \vartheta^2 \; . \tag{1}$$

Here $\vartheta^2 = \langle\vartheta^2(l=1)\rangle$ = m. sq. deflection/unit path.

In a given scattering medium $\vartheta^2$ is a function of the particle's energy and charge (roughly $\sim (z/pv)^2$) and is therefore eagerly studied by high energy physicists. Now there are several experimental methods of extracting this information from a given ionization track. And it is sometimes important to know the relation between them.

### a. *The spectrum*

A convenient way of collecting the various techniques under one common point of view is to consider the track as part of a homogeneous stochastic process $y(x, \alpha)$ with the spectral resolution:

$$g(\omega) = \frac{1}{2\pi} \int_{-L}^{+L} dx\; y(x)\; e^{-i\omega x} . \qquad (2)$$

What is measured is always a path ("time") average over some quadratic functional

$$\overline{\Phi^2} = \frac{1}{2L} \int_{-L}^{+L} dx\; \Phi^2[y(x),\; y'(x), y(x+l), \ldots] , \qquad (L \to \infty) . \qquad (3)$$

In the "Tangent Method" of Goldschmidt-Clermont et al. one measures the deflection directly

$$\Phi_{\tan}(l) = y'(x + l) - y'(x) . \qquad (4)$$

Inserting Fourier expansion one obtains in this case

$$\overline{\Phi^2_{\tan}} = \int d\omega\, d\omega'\; g\, g'\; \omega\, \omega'\; [1 - e^{i\omega l}] . [e^{i\omega' l} - 1] \frac{\sin(\omega + \omega') L}{(\omega + \omega') L} \qquad (5)$$

$$\xrightarrow{U \to \infty} 2\pi \int d\omega\, |g(\omega)|^2 \omega^2 (1 - \cos \omega l) .$$

Now we do not directly know the spectrum of $y$ but we may assume that the curvature $d\vartheta/ds$ and hence, for small angles, also $y''$ will be a purely random [1)] Gaussian process, that is one whose correlation is of $\delta$-function type. Therefore - if the spectral function of $y$ be $|g|^2$ and that of $y''$ accordingly $\omega^4 . |g|^2$ - the latter should be constant (white spectrum).

Evaluating (5) we have then

$$\overline{\Phi^2_{\tan}} = 2\pi \int |g|^2 . \omega^4\, d\omega \frac{1 - \cos \omega l}{\omega} = 2\pi^2 . l . |g\omega^2|^2 .$$

In the direct method we have by definition $\Phi^2 = l . \vartheta^2$ accordingly we may in our calculations use the spectral function

$$|g|^2 = \frac{\vartheta^2}{2\pi^2} . \frac{1}{\omega^4} . \qquad (6)$$

Now consider one of the other methods, for example the "Center of Gravity Method" (Rochat). According to this method the track is divided into cells of length $l$ and the ancle $\Phi_{c.g.}$ is defined by the second difference of the "center of gravity ordinates"

$$Y_x = \frac{1}{l} \int_x^{x+l} y(x)\, dx \qquad (7)$$

as:

$$\Phi_{c.g.} = \frac{1}{l} [2\, Y_x - Y_{x-l} - Y_{x+l}]$$

$$= \frac{1}{l^2} \int g(\omega) \frac{d\omega}{i\omega} e^{i\omega x} [e^{-i\omega l} - 3 + 3\, e^{i\omega l} - e^{2i\omega l}] .$$

For the mean square one obtains

$$\overline{\Phi^2_{\tan}} = \frac{\pi}{l^4}\int |g|^2\, d\omega \,\frac{1}{\omega^2}\,|[\quad]|^2$$

$$= \frac{\vartheta^2}{2\pi l^4}\int_{-\infty}^{\infty}\frac{d\omega}{\omega^6}\{10 - 15\cos\omega l + 6\cos 2\omega l - \cos 3\omega l\}$$

$$= \frac{11}{20}\cdot\vartheta^2 . l . \tag{8}$$

So we see that this method gives only about 50% of the ideal value. Similarly the other methods can be compared.

b. *Distribution of the ordinates*

Another question in this connection is the following: Draw a chord of length $L$ to the track. What is the probability to find an ordinate $y_x$, given that $y(0) = y(L) = 0$?

We can again use the Markoffian property of the deflection. Divide the chord into $N$ parts

$$x_n = n\,\Delta\ , \qquad \Delta = L/N\ . \tag{9}$$

Let the initial direction $y'(0)$ be $\eta_o$. The probability density for the track to pass:

| | | |
|---|---|---|
| $x_1$ | with inclination | $y'(x_1) = \eta_1$ |
| $x_2$ | with inclination | $y'(x_2) = \eta_2$ |
| ⋮ | | ⋮ |
| $x_j$ | with inclination | $y'(x_j) = \eta_j$ |

is then given by the chain:

$$f_n\left(\begin{array}{c|cccc} \eta_o & \eta_1 & \cdots & \eta_j & \cdots \\ 0 & \Delta & & j\Delta & \end{array}\right) = P(\eta_o\,|\,\eta_1)\,P(\eta_1\,|\,\eta_2)\ldots P(\eta_{N-1}\,|\,\eta_N)$$

where

$$P(\eta_j\,|\,\eta_{j+1})\,d\eta_{j+1} = \frac{d\eta_{j+1}}{\sqrt{2\pi\vartheta^2\Delta}}\cdot e^{-(\eta_{j+1}-\eta_j)^2/2\vartheta^2\Delta} \tag{10}$$

is the probability to find $y'(x_j + \Delta)$ in the interval $d\eta_{j+1}$ around $\eta_{j+1}$ when $y'(x_j) = \eta_j$.

The underlying assumption for this to be valid is of course that the particle suffers many independent small deflections in traversing a layer $\Delta$. However, as the single scattering resp. the fine structure of the path does not interest us now this will not prevent us from making $\Delta$ infinitesimal in the mathematical model. (As you may know even the assumption of many single collisions does not always grant a Gaussian multiple scattering due to the conditions on the central limit theorem; but we shall ignore that here.) Now we want the distribution - not of the inclinations $\eta_j$ but of the ordinates $y_j$: To the first order of $\Delta$ we have,

$$y_n = \Delta \sum_{j=0}^{n-1} \eta_j \ ; \qquad 0 = \sum_{0}^{N-1} \eta_j \ . \tag{11}$$

The conditional probability to find the $n$'th ordinate equal to $y_n$ will therefore be

$$I(\eta_0, y_n, x_n, L) = \\ = \underset{\text{(with restrictions (11))}}{\int \ldots \int} P(\eta_0 | \eta_1)\, d\eta_1 \ P(\eta_1 | \eta_2)\, d\eta_2 \ldots P(\eta_{N-1} | \eta_N)\, d\eta_N \ . \tag{12}$$

This is easily carried out and the transition to the continuum $N, n \to \infty$, $n.\Delta = X$ is straight forward. Averaging over the initial directions one finds

$$f(\tfrac{Y}{X}) = \left(\frac{3\,L}{2\,\pi\,\vartheta^2}\right)^{\frac{1}{2}} \frac{\exp\left\{\dfrac{-3\,L\,Y^2}{2\,\vartheta^2\,X^2(L-X)^2}\right\}}{X\,(L-X)} \tag{13}$$

For the "Sagitta", that is the ordinate to the middle of the chord, one has in particular:

$$\overline{Y^2(\tfrac{L}{2})} = \frac{1}{48} L^3 \vartheta^2 \ , \tag{14}$$

which is also much used for the determination of $\vartheta^2$.

### c. *Improvement of small angle approximation*

Even if we overlook the error of our Gaussian hypothesis we have still made a much debayed step viz. "the small angle approximations".

Onsager and Machlup [2)] have used a method for the evaluation of Markov chains which - although it is exact only in the Gaussian case - may be of use in this old problem.

We assume as before a Gaussian law for the deflection in a path element $\Delta s$

$$P\left(\begin{matrix}\vartheta \\ s\end{matrix} \middle| \begin{matrix}\vartheta + \Delta\vartheta \\ s + \Delta s\end{matrix}\right) = \frac{1}{\sqrt{2\,\pi\,\vartheta^2\,\Delta s}}\, e^{-(\Delta\vartheta)^2/2\vartheta^2\Delta s} \ . \tag{15}$$

In order to explain the method we consider first the small angle approximation, replacing for the moment $\vartheta$ by $y'$ and $s$ by $x$. We had to evaluate the integral (12) with the restrictions (11). Onsager and Machlup wrote such integrals in the form

$$I = \int dy_0'\, dy_1' \ldots dy_N'\, e^{-J[y']} \ . \text{ Normalisation factor} \tag{16}$$

where

$$J[y'] = \frac{1}{2\,\vartheta^2} \sum_{i=0}^{N-1} \left(\frac{y'_{i+1} - y'_i}{\Delta}\right)^2 . \Delta . \tag{17}$$

By the properties of quadratic forms and Gauss integrals one has then exactly

$$I = e^{-\text{Min.}\,J[y']} . \tag{18}$$

Making now the transition to the continuum, replacing Riemann sums by integrals, the evaluation of the Markov chain (12) is seen to be equivalent to the solution of the variational problem:

$$J[y'] = \frac{1}{2\,\vartheta^2} \int_0^L (y'')^2 \, dx = \text{Minimum} \tag{19}$$

$$y(X) = Y^\circ , \qquad y(0) = y(L) = 0 \; ^{*}) .$$

Equating the first variation of $J$ to zero and demanding the "natural boundary conditions" $y''= 0$ at the end points,

$$y,\ y' \text{ and } y'' \text{ continuous at } x = X , \tag{20}$$

one obtains the Euler equation

$$y'''' = 0 \qquad \text{for} \qquad x \neq X . \tag{21}$$

In the point $x = X$ the right hand side of (21) will be infinite. The extremal - or "most probable path" through $(0,0)$ $(X,Y)$ and $(L,0)$ - has therefore the character of a Greens function. Actually its form is just that of an elastic rod supported at the end points and bent by a concentrated load at $x = X$:

$$\begin{aligned} y(x < X) &= a_1\,x^3 + b_1\,x^2 + c_1\,x \\ y(x > X) &= a_2(x - L)^3 + b_2(x - L)^2 + c_2(x - L) \end{aligned} \tag{22}$$

where the constants are determined by the boundary conditions (19). But we need only

$$a_1 = \frac{Y}{2\,X^2(L - X)} \qquad \text{and} \qquad a_2 = \frac{-\,Y}{2\,X\,(L - X)^2}$$

to find the minimum value

$$J_{\min} = \frac{3\,L}{2\,\vartheta^2}\,\frac{Y^2}{X^2(L - X)^2} \tag{23}$$

in agreement with (13) as it must be.

We shall now try to remove the limitation to small angles, retaining the Gaussian law (15) for the elementary stochastic process.

* In the small angle approximation the restrictions have the form of linear equations in the $y_i'$ . Each such condition will reduce the number of variables by one and bring in a constant factor.

It may be noted, however, that the latter assumption could perhaps also be relaxed since the Onsager Machlup saddlepoint method may be approximately correct also for non-quadratic functionals.

The Euler equation takes the simplest form in natural coordinates. We write therefore

$$J = \frac{1}{2\,\vartheta^2} \int_0^T \dot{\vartheta}^2 \, ds \,. \tag{24}$$

After some calculation one obtains then

$$J_{\min} = \frac{4\,\alpha}{\vartheta^2}\left[E - (1 - k^2)\, K\right]. \tag{25}$$

Here: $K(k)$, $E(k)$ = the complete elliptic integrals of first and second kind
$k$ = modulus
$\alpha$ = a further integration constant.

The elimination of $k$, $\alpha$ is not possible in closed form, but for the Sagitta, $X = L/2$, one has:

$$\alpha = \frac{2\,k}{Y} \qquad \text{exactly, and} \qquad k = \frac{\pi}{2}\frac{Y}{L} + O\left(\frac{Y^3}{L^3}\right).$$

In this approximation one obtains

$$\overline{Y^2}\left(\frac{L}{2}\right) = \frac{2}{\pi^4}\, L^3\, \vartheta^2 \tag{26}$$

that is $\frac{1}{2}\,\pi^4 = 45.7$ instead of the 48 in the formula (14) of the small angle approximation [3].

*References*

1. For definition of this limiting case, see Wang and Uhlenbecks article l.c.[6] preceding lecture.
2. L. Onsager and S. Machlup l. c. [1] preceding lecture.
3. The detauls of the calculation can be found in an unpublished note by H. Wergeland and H. Øveraas: "Proceedings of the Physics Seminar in Trondheim" no. 5, 1955/56.

## IV. BROWNIAN MOTION

The theory of how a large system overall approaches equilibrium is perhaps still in "statu nascendi" but we possess for a long time the beautiful theory of Brownian Motion - founded by Einstein and Smoluchowski and deepened by Langevin, Ornstein, Uhlenbeck and others[1] which describes the stochastic motion of a single particle.

Although the fine structure - or dynamical substratum - of

Brownian Motion changes from system to system, it is of interest to see how it all comes about in a special case. What I propose to you is:

a. To consider the 2 $N$ integrals $p(t)$, $q(t)$ ... of the motion of a large system as functions of the time and the initial values $p^o q^o \ldots$ .

b. Give the initial values a distribution $f(\ldots p^o q^o \ldots)$ and look at stochastic processes based on this ensemble.

We shall thus for a while return to the old aspiration of statistical mechanics, originating in Gibbs' ideas of equilibrium theory; which is to introduce the statistics through inderterminateness of the initial values and nothing more - except eventually the Ehrenfest coarse graining in phase space.

## a. *Dynamics of the linear chain*

The only known many body system which is amenable to such a preposterous program is a set of linearly coupled harmonic oscillators. Such systems are patently non ergodic, which is probably the reason why we can manage the dynamics at all. But the lack of ergodicity is veiled as long as we do not look at the motion of the normal modes and especially if we look only at a single particle. For our present purpose we choose the one dimensional chain of mass points tied neighbour to neighbour by Hookes law force. That is an evergreen which has been cultivated from the times of Daniel Bernoulli and Euler. Among the many treatments I will therefore only refer to Hemmers thesis and to the recent article by Montroll and Mazur [3]) from which the present example is drawn. The mechanics of the system is contained in the following:

*Lagrangean*

$$\mathcal{L} = \sum_{o}^{N-1} \{\frac{m}{2} \dot{x}_k^2 - \frac{\alpha}{2} (x_{k+1} - x_k)^2\}$$

$x_k$ = displacement of $k$ atom from eq. position.

*Normal coordinates*

$$Q_j = \sum_{o}^{N-1} C_{jk} \, x_k$$

with $$C_{jk} = \sqrt{\frac{\varepsilon_j}{N}} \cos\left[(k + \tfrac{1}{2}) \frac{\pi_j}{N}\right], \qquad \varepsilon_j = \begin{cases} 2 & j \neq 0 \\ 1 & j = 0 \end{cases}$$

$$\longrightarrow \mathcal{L} = \frac{m}{2} \sum_{j=1}^{N-1} \{\dot{Q}_j^2 - \omega_j^2 Q_j^2\} + \text{uniform translation.} \qquad (1)$$

*Proper frequencies* (free ends)

$$\omega_j = \omega_o \sin \frac{\pi_j}{2N} \qquad \omega_o = 2\sqrt{\frac{\alpha}{m}} .$$

From the motion in normal coordinates

$$Q = Q^0 \cos \omega t + \dot{Q}^0 \frac{\sin \omega t}{\omega}$$

we can transform back again to get

$$x_k = \sum_{l=0}^{N-1} \{A_{kl}\, x_l^0 + B_{kl}\, \dot{x}_l^0\} \tag{2}$$

$$A_{kl} = \sum_{j=1}^{N-1} C'_{jk}\, C_{jl} \cos \omega_j t \; ; \qquad B = \int_0^t A \, dt \; .$$

In the classical case this scheme can be further simplified by an elegant transformation of Schrödinger:

$$\xi_{2n} = \sqrt{m}\, \dot{x}_n \qquad \xi_{2n+1} = \sqrt{\alpha}\, (x_n - x_{n-1}) \; .$$

In these variables the equations of motion

$$m\, \ddot{x}_n = \alpha\, (x_{n+1} + x_{n-1} - 2\, x_n) \tag{3}$$

take the form

$$d\xi_n / d(\omega_0 t) = \tfrac{1}{2}\, (\xi_{n-1} - \xi_{n+1})$$

which is just the differential-difference equation for a cylinder function $\mathcal{C}_n(\omega_o t)$. The general solution in Schrödingers coordinates can be written

$$\xi_n(t) = \sum_{-(N-1)}^{N-1} a_{nk}(t)\, \xi_k^0 \tag{3'}$$

where

$$a_{nk}(t) = \sum_{r=-\infty}^{+\infty} \{J_{n-k+4rN}(\omega_o t) + (-1)^k J_{n+k+2N-4rN}\} \; .$$

For the limiting case of an infinite chain this simplifies to:

$$\lim_{N\to\infty} a_{nk} = J_{n-k}(\omega_o t) \qquad (J = \text{Bessel function}) \; . \tag{4}$$

The evolution in time $\xi(0) \to \xi(t) = a \cdot \xi^0$ is an orthogonal transformation as can also be seen from the conservation of energy:

$$E = \tfrac{1}{2} \sum (\xi_n^0)^2 = \tfrac{1}{2} \sum \xi_n(t)^2 \; .$$

b. *Velocity distribution of a single particle*

When the motion is completely determined the phase density entering in Liouvilles theorem will be a $\delta$-function.

$$\rho(\xi_1 \xi_2 \ldots, t) = \prod_n \delta(\xi_n - \sum a_{nk}\, \xi_k^o)\ . \tag{5}$$

We want now the probability to find a selected particle - in the middle of the chain say - with a velocity $v = \xi_o/\sqrt{m}$ in the interval $(v,\ v + dv)$ at time $t$ when its velocity was $v^o$ at time $t = 0$ and all the other coordinates and velocities were undetermines in a way corresponding to thermal equilibrium. The natural way of doing it is to consider an ensemble of chains whose initial coordinates have a distribution

$$f(\ldots \xi^o \ldots) = \delta(\xi_o - \xi_o^o) \prod_{n\neq 0} (2\pi \cdot \Delta\xi^2)^{-\frac{1}{2}}\ . \ e^{-(\xi_n^o)^2/2\Delta\xi^2}\ . \tag{6}$$

But it may be noted that rather different ensembles such as uniform distribution on the energy shell give essentially the same results when $N$ is large. The required probability is then the relative number of systems for which $\xi_o(t)$ lies in the interval $(\xi,\ \xi + d\xi)$ irrespective of the other coordinates:

$$P\left(\begin{matrix}\xi^o \\ 0\end{matrix}\middle|\begin{matrix}\xi \\ t\end{matrix}\right) = \int \rho(\ldots\xi\ldots t)\, f(\ldots \xi^o \ldots) \prod_{n\neq 0} (d\xi_n) \prod_n (d\xi_n^o)\ {*})\ . \tag{7}$$

Using the orthogonality (3) one finds:

$$P = \frac{e^{-(\xi - a\xi^o)^2/2\Delta\xi^2(1-a^2)}}{\sqrt{2\pi\ \Delta\xi^2\ (1 - a^2)}}$$

where

$$a(t) = \sum_{r=-\infty}^{+\infty} J_{2rN}\ (\omega_o t) \qquad (= a_{oo}(t))\ ; \tag{8}$$

that is a Gaussian distribution with a time dependent mean

$$<\xi(t)>_{\xi(0)=\xi^o} = a(t)\ \xi^o$$

and fluctuation

$$<(\xi - <\xi>)^2> = \Delta\xi^2(1 - a^2)\ . \tag{9}$$

Consider first the infinite chain. We have then by (3):

$$<\xi> = \xi^o\ J_o(\omega_o t) \qquad <(\xi - <\xi>)^2> = \Delta\xi^2(\ 1 - J_o^2)$$

as sketched in Fig. 1.

It lies at hand to compare this with the well-kn[illegible]

* In this case of a [illegible]
need[illegible]

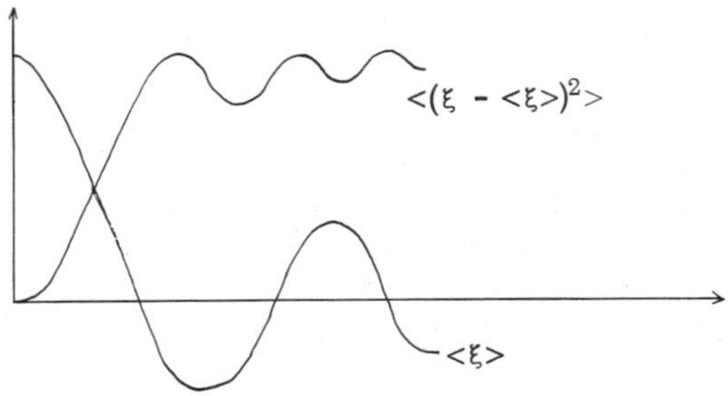


Fig. 1. Mean motion and fluctuation of $\xi(t)$ conditional on $\xi(0) = \xi^{o}$.

$$\langle v \rangle_{v(o)=v^{o}} = v^{o}\, e^{-\beta t}\,, \qquad \langle (v - \langle v \rangle)^2 \rangle = \Delta v^2 \cdot (1 - e^{-2\beta t}) \qquad (9')$$

for the Brownian Motion of a free particle [1].

The probability distribution $P(v^{o}\,|\,v)$ is then not only Gaussian but furthermore it satisfies a Fokker-Planck equation in velocity space or more generally the Smoluchowski-Chapman-Kolmogorov equation (II 4). This latter is certainly not the case when $\exp(-\beta t)$ is replaced by $J(\omega_o t)$ as in our probability distribution (8).

Therefore $\xi_t$ is not quite a Markov process, at least not as long as the selected particle is equal to all the others. On the other hand if we had embedded a heavier particle $(M)$ in a chain of lighter ones $(m)$ then we would expect its motion to be more "genuinely" Brownian. This is indeed the case. Hemmer l.c. and R. Rubin [5] have shown that the stochastic motion of the heavy particle will tend to a Markov process as the mass ratio $M/m$ increases.

c. *Correlations and spectrum*

Let us now look at the correlation functions of the $\xi_n(t, \ldots \xi^{o})$. Our random parameter is just the initial values and the realizations are given by (3) explicitly. If we take the ensemble to be such that

$$\langle \xi_k^{o}\, \xi_l^{o} \rangle = \Delta\xi^2 \cdot \delta_{kl}\,, \qquad \langle \xi_k^{o} \rangle = 0$$

we have

$$\rho_{nn}(\tau) = \sum_{k=-\infty}^{+\infty} a_{nk}(t)\, a_{nk}(t+\tau)\,. \qquad (10)$$

The result is again simple for the infinite chain eq. (4). We can then use a well-known addition theorem of the Bessel functions to obtain

$$\rho_{nn}(\tau) = J_o(\omega_o \tau)$$

and also

$$\rho_{nm}(\tau) = J_{|n-m|}(\omega_o\, |\tau|)\,. \qquad (10')$$

It may also be illuminating to write down the corresponding spectral function. Using (II 16) one finds

$$h_{nn}(\omega) = \frac{1}{2\pi} \cdot \frac{1}{\sqrt{\omega_o^2 - \omega^2}} \cdot \tag{10''}$$

This is just the frequency spectrum of the infinite chain. For a finite chain we obtain of course a discrete spectrum. It coincides again with the mechanical eigen frequencies but every second line $\omega_{2j+1}$ is missing (see eq. (14)).

By means of eq. (4) and the definition of the Bessel functions we can also write down explicitly the random Fourier amplitudes of the $\xi_n$:

$$g_n(\omega, \ldots \xi^o \ldots) = \frac{1}{2\pi \sin\varphi_o} \sum_{k=-\infty}^{+\infty} (-i)^{n-k} \cdot \xi_k^o \cdot e^{i(n-k)\varphi_o} \tag{11}$$

$$\varphi_o = \arc\cos\frac{\omega}{\omega_o}$$

and check how the stationarity condition (II, 14) is assured for this particular process.

d. *Finite chain*

Next we must discuss the case of a finite chain. The mean motion $\langle\xi\rangle = a\,\xi$ will then change in a characteristic way. We see from (6) that in times $t \ll N/\omega_o$, $a(t)$ can still be replaced by its limiting form $J_o(\omega_o t)$, decreasing by damped oscillations $\mathrm{Max}|a| \sim t^{-\frac{1}{2}}$. However, in the course of long times, $a(t)$ will return to any value in a certain interval. In order to see this one can rewrite (6) by means of Poissons summation formula which gives

$$a(t) = \sum_{r=-\infty}^{+\infty} J_{2rN}(\omega_o t) = \frac{1}{N} \sum_{-(N-1)/2}^{(N-1)/2} \cos\Omega_j t$$

where

$$\Omega_j = \omega_o \sin\frac{\pi j}{N} \,. \tag{12}$$

The maximum of the right hand side is + 1 and the minimum is - 1 + $2/N$, and the sum will assume all intermediate values over and over again. In fact $a(t)$ belongs to a special type of "Almost Periodic Functions" [6].

We get therefore the following picture of the mean motion: A large initial value $\xi^o$ will first decay in a seemingly irreversible way. During sufficiently long times, nevertheless, the probable motion must occasionally stray away from, instead of approaching equilibrium. But when $N$ is large such behaviour will be rare and the more so the larger the bout $\xi$.

This can best be seen from the value distribution of $a(t)$. The probability density to find $a \leq a(t) \leq a + da$ can evidently be written:

$$f(a) = \lim_{T\to\infty} \frac{1}{2T} \int_{-T}^{T} dt \int_{-\infty}^{+\infty} \frac{ds}{2\pi} e^{is[a-a(t)]}$$

$$= \frac{1}{2\pi} \int ds\, e^{is(a-1/N)} \lim \frac{1}{2T} \int dt \prod_{j=1}^{\frac{N-1}{2}} e^{\frac{2is}{N}\cos\Omega_j t}$$

$$= \frac{1}{2\pi} \int ds\, e^{is(a-1/N)} \lim \int \frac{dt}{2T} \prod_j \sum_{m_j} i^{m_j} J_{m_j}(\frac{2s}{N})\, e^{im_j \Omega_j t}$$

$$= \sum_{(\dots m_1 m_2 \dots)} \prod_j (J_{m_j} i^{m_j}) \lim_{T\to 0} \frac{\sin \sum m_j \Omega_j T}{\sum m_j \Omega_j T}$$

In the limes only one single term will remain, namely the one from all $m_j = 0$ (incommensurability of the $\Omega_j$ presumed! ). And so we have

$$f(a) = \frac{1}{2\pi} \int ds\, e^{is(a-1/N)} [J_0(\frac{2s}{N})]^{\frac{N-1}{2}}$$

$$\approx \sqrt{\frac{N}{2\pi}}\, e^{-\frac{N}{2}a^2} . \qquad (15)$$

This shows that for large $N$ the distribution of $a$ is sharply peaked around the long time average, zero.

It is clear from Poincarés theorem that the dynamical motion of a finite chain must be recurrent. We see now that this festure is not even erased in the stochastic motions we construct by ignoring a great number of coordinates and distributing their initial values.

## e. *Quantum mechanics*

A corresponding quantum mechanical treatment of the linear chain is equally simple [7]. One can then represent the initial state by a many dimensional wave packet and ask for the probability to find a selected particle with momentum $p$ at time $t$ when it is known that at $t = 0$ the quantum mechanical expectation value of $p$ was $p^o$. The result is again a Gauss distribution with mean, $\langle\langle p \rangle\rangle = p^o a(t)$ but the fluctuation is now a little more elaborate:

$$\langle\langle (p - \langle\langle p \rangle\rangle)^2 \rangle\rangle = \Delta p^2 (1 - a^2) + \delta p^2\, b(t) + m^2 \omega_o^2\, \delta q^2\, c(t) . \qquad (16)$$

The first term on the right is just the classical one.
$\delta p^2$ = initial uncertainty of the position.
$\delta q^2$ = initial uncertainty of the momentum.

$b(t)$ and $c(t)$ are again two Almost Periodic Functions. They have the property that, when $a(t)$ is close to 1 so that $\langle\langle p \rangle\rangle \approx p^0$ then $b(t) \approx 1$ and $c'(t) \approx 1/N$.

$$\therefore \langle\langle (p - \langle\langle p \rangle\rangle^2 \rangle\rangle \approx \delta p^2$$

whereby the initial wave packet is nearly resurrected.

*References*


1. Ming Chen Wang and G. E. Uhlenbeck l. c. [6] lecture II.
   S. Chandrasekhar, Rev. Mod. Phys. 15 (1945) 1.
2. P. Chr. Hemmer, Nordita Publications no. 2 (1959).
   E. W. Montroll and P. Mazur, J. Math. Phys. 1 (1960) 70.
5. Hemmer l.c. [3].
   R. Rubin, J. Math. Phys. 2 (1961) 373.
6. H. Bohr, "Fastperiodische Funktionen", Erg. d. Math. I (1932) 421.
7. P. Chr. Hemmer and H. Wergeland, Festskrift til E. A. Hylleraas p. 137, Kgl. Norske Vidensk. Selsk. Trondheim (1958).


## V. RECURRENCE TIME

We have already touched the question of the recurrence of states many times. It is of course inherent as well in the concept of stationary stochastic processes as in the dynamics of finite systems [1]. In Pointcarés great memoir: *Sur les problemes des trois corps, et les equations de la dynamique* [2] we find the first clear recognition of the fact that all finite mechanical motions have a repetitive character. So, the phenomenon of recurrence in the evolution of dynamical systems is a general one which must certainly be taken quite seriously in statistical physics.

For example we have seen yesterday that if probability is introduced by indeterminateness of the dynamical initial conditions - and nowhere else - the phenomenon of recurrence will survive also in the ensuing statistical equations except of course for the limiting case of infinitely large systems. So, it is in any stage a check upon the theory whether it preserves to some extent this quasiperiodic feature of its dynamical origin. If this has been completely erased, it may mean that some tacit step has been committed in general equivalent to the introduction of a further probalistic element into the edifice. Cases in point are the equations of Brownian Motion and - especially interesting - the theory of heat conduction in crystals; both of which lack the required feature ab initio.

I shall now consider recurrence of a strictly determined mechanical motion in our favourite model the linear chain. But the results we shall obtain are for that matter applicable to any separable mechanical system. It will be convenient to compound normal coordinates and momenta pairwise into complex vectors.

$$Z_j = P_j + i\, m\, \omega_j\, Q_j = a_j\, e^{i\omega_j t} . \tag{1}$$

A mechanical state will then be specified by the situation of **these** vectors $Z_j$, which rotate with uniform velocities $\omega_j$ in their respective planes $j = 0, 1, 2, \ldots N\text{-}1$.

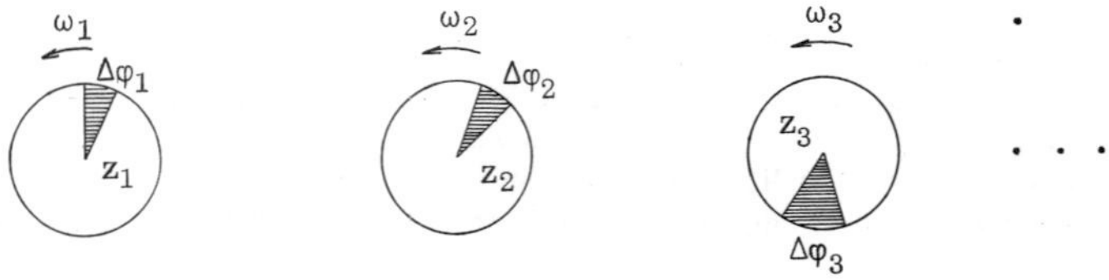


Now according to a famous theorem of Kronecker it is always possible to find times $t$ for which all the vectors return as closely as desired to any previous configuration

$$\varphi_j \leq \arg Z_j \leq \varphi_j + \Delta\varphi_j$$

provided the frequencies are rationally independent; that is the equation

$$n_1\, \omega_1 + n_2\, \omega_2 + \ldots\, n_{N-1}\, \omega_{N-1} = 0 \tag{2}$$

insoluble in integers $n_j$ not all zero.

Clearly reentrance into some specified situation of the angles $\varphi_j = \omega_j t + \delta_j$ means recurrence of mechanical state within some latitude whatever the coordinates used for its description. Kroneckers theorem tells us that there will be recurrences. The question is how long on the average one must wait for such an event to occur.

Hemmers [3] definition of this time is a kind of "law of large numbers" average:

$$T_{\text{Rec}} = \lim_{r=\infty} (t_r / r) . \tag{3}$$

He considered first the simplified case of only two "clocks", $Z_j$ and $Z_k$ say, with periods

$$T_j = \frac{2\,\pi}{\omega_j} < T_k = \frac{2\,\pi}{\omega_k} \tag{4}$$

Consider now the times $n_j\, T_j$ and $n_k T_k$; $n_j n_k = 1, 2, \ldots$ where the hands pass the initial position, and especially the *delay*

$$\Delta = n_j\; T_j - m\; T_k \tag{5}$$

between the $n$'th passage of $Z_j$ and the immediately preceding one of $Z_k$ which is denoted by $m$. This delay is then smaller than $T_k$:

$$\Delta = \varepsilon\, T_k \qquad \text{with} \qquad 0 \leq \varepsilon \leq 1$$

or

$$n \frac{\omega_k}{\omega_j} - m = \varepsilon\ . \tag{6}$$

Thus $m$ is the greatest integer not exceeding $n \frac{\omega_k}{\omega_j}$. In customary notation this is written:

$$n \frac{\omega_k}{\omega_j} - [n \frac{\omega_k}{\omega_j}] = \varepsilon\ , \tag{7}$$

and now a theorem of Weyl comes in handy:

It states that $\varepsilon(n)$ will be uniformly distributed in the interval $[0,1]$ is $\omega_j$ and $\omega_k$ are incommensurable.

Hence all relative angles $\varphi_j - \varphi_k$ must be equally probably over long times ($n = 1, 2, \ldots \infty$), provided

$$\frac{\omega_k}{\omega_j} = \frac{\sin (\pi k/2N)}{\sin (\pi j/2N)} \tag{8}$$

is irrational which we shall here only assume *).

Due to this independence of the angles we may argue as follows:

$$\text{Prob}\ \{Z_j \in \Delta\varphi_j\} = \prod_{j=1}^{N-1} (\frac{\Delta\varphi_j}{2\pi})\ . \tag{9}$$

But this probability is again equal to

$$\frac{\text{time in coincidence}}{\text{total time}} \qquad (\lim_{t\to 0})$$

Dividing both numerator and denominator of this ratio by the number of recurrences $r$ we obtain the relation

$$\frac{T_c}{T_R} = \prod (\frac{\Delta\varphi_j}{2\pi}) \tag{10}$$

where $T_c$ = mean duration of the recurrent state.

The point is that this latter is more easy to obtain, whereby we have also the recurrence time $T_R$.

It turns out that **

* The condition for rational independence was also worked out by Hemmer l.c. 4. lecture; it is: $\varphi(2N) > N - 1$ ($\varphi(n)$ = Eulers function $= n \prod_i (1 - 1/p_i)$, $i$ = different prime factors of n).

** This can be shown in various ways for examplw by considering the $\varphi_j$ as the polar angles and azimuth on an N-dimensional sphere. A particularly simple proof is given by Hemmer [3].

$$\frac{1}{T_c} = \frac{\omega_1}{\Delta\varphi_1} + \frac{\omega_2}{\Delta\varphi_2} + \ldots = \frac{1}{t_1} + \frac{1}{t_2} + \ldots \frac{1}{t_{N-1}} \tag{11}$$

$t_j = \Delta\varphi_j / \omega_j$ = passage time of $Z_j$ through its interval and accordingly we have for the recurrence time:

$$T_R = \prod_{j=1}^{N-1} \left(\frac{2\pi}{\Delta\varphi_j}\right) / \sum_{j=1}^{N-1} \frac{\omega_j}{\Delta\varphi_j} . \tag{12}$$

In order to get an idea of the times involved, take a chain consisting of $N = 10$ atoms, which is not a very large system. With $\Delta\varphi = \pi/100$ and $\omega_o = 10/\text{sec.}$ one obtains then already $T_R \sim 10^{10}$ years. So we see that these recurrence times are truly enormous even for a moderately sharp definition of the state.

## *References*

1. M. Kac, various papers; the most recent one Phys. Rev. 120 (1961).
2. H. Poincaré, Acta Math. 13 (1890) 67.
3. P. Chr. Hemmer, Diploma essay, Trondheim 1956.
   Kgl. Norske Vidensk. Selsk. Forh. 33 (1960) 101.

# LIQUID HELIUM

K. HUANG

Notes by P. P. J. M. Schram and J. M. J. van Leeuwen *

## 1. *Experimental facts*

A remarkable property of $He^3$ and $He^4$ is that they remain liquids, probably down to absolute zero. Furthermore $He^4$ has two liquid phases. The transition from one to the other is called the λ-transition.

The phase diagrams of $He^4$ and (for comparison) of water are shown in Fig. 1 and 2.

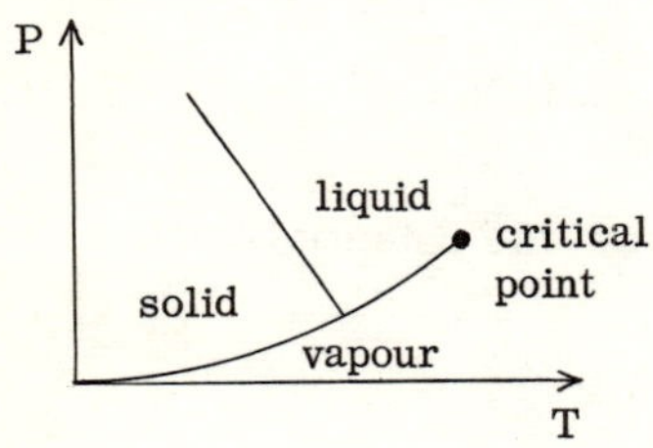


Fig. 1. P-T diagram of water.

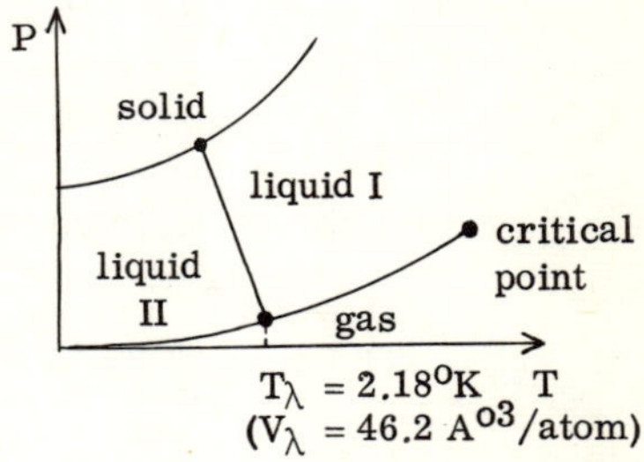


Fig. 2. P-T diagram of $He^4$.

$He^3$ is also a liquid but has no λ-transition. The difference is due to the fact that $He^4$ has spin zero and should be trated by means of Bose-Einstein statistics, whereas $He^3$ has spin $\frac{1}{2}$ and thus obeys Fermi-Dirac statistics. Probably there exists a connection between the λ-transition and the Bose-Einstein condensation.

The specific heat $C_v$ has a logarithmic singularity on the gas-liquid line at $T = T_\lambda$, cf. Fig. 3.

Near $T = 0$ $C_v \propto T^3$ as in the Debije theory of crystals.

Why does helium not solidify? The answer consists of two points:

1. weak interaction (He is a noble gas),
2. small mass.

* In preparing these notes the compilers took advantage of the notes of a course given by Prof. Huang on statistical mechanics at the Massachusetts Institute for Technology (to be published). Since Prof. Huang has not seen these notes prior to publication, the compilers are sole responsible for any errors.

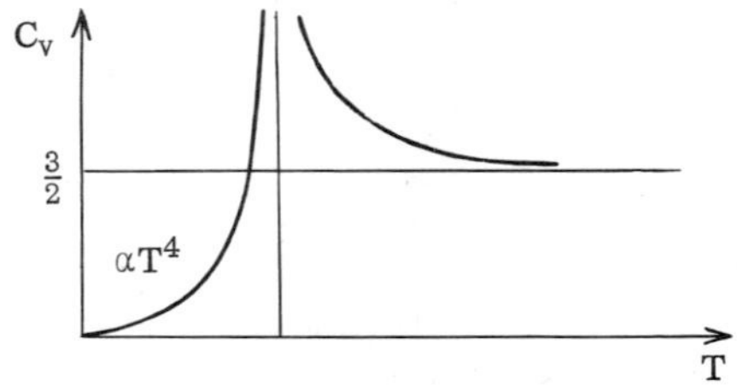


Fig. 3. Specific heat of $He^4$.

The thermal wavelength $\lambda$ and the average distance between the particles $r_0$ are of the same order of magnitude; so one cannot specify the position of the particles with an uncertainty less than $r_0$. Therefore He is a liquid and not a crystal. This argument can be made more quantitative by plotting the potential energy of two particles as a function of their distance (Fig. 4).

Suppose that He were a crystal. This would mean that one particle is in the potential well of the other.

Estimating the uncertainty in position as $\Delta x \approx 0.5$ Å one finds:

$$\Delta E \approx \frac{(\Delta p)^2}{2\ m} \approx \frac{\hbar^2}{(\Delta x)^2\ 2\ m} \approx 10^{\circ}\mathrm{K} \tag{1}$$

and therefore the particle can escape from the potential well.

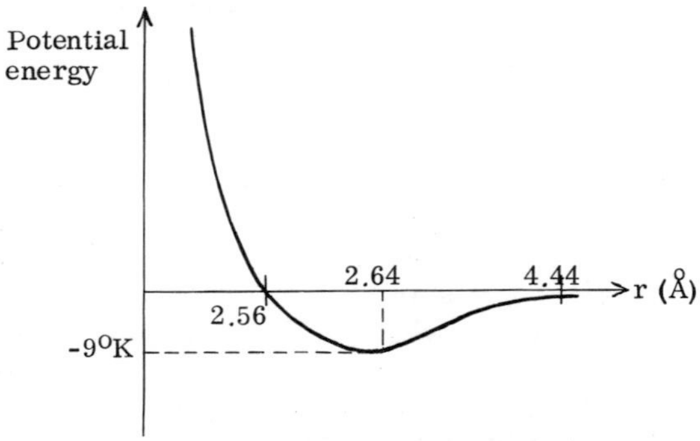


Fig. 4. Potential energy of He-atoms.

It may be added that argon solidifies because of its greater mass and hydrogen because of the stronger interaction between the molecules.

$He_I$ has no remarkable properties. In these lectures only $He_{II}$ will be treated.

2. *Two-fluid model of Tisza*

The phenomenological theory of Tisza suggests, that $He_{II}$ consists of two independent components, a normal fluid and a superfluid. The normal fluid behaves as an ordinary liquid; the superfluid has zero entropy and viscosity, and flows without resistance through channels.

The mass-density and mass-velocity are defined by

$$\begin{aligned} \rho &= \rho_n + \rho_s \\ \rho\,\vec{v} &= \rho_n\,\vec{v}_n + \rho_s\,\vec{v}_s \end{aligned} \tag{2}$$

where the subscripts $s$ and $n$ refer to superfluid and normal fluid respectively.

This theory can already explain three strange phenomena:

a. *The mechano-caloric effect*

The vessels A and B in Fig. 5 are connected by a very thin tube, through which only the superfluid can pass. If one exerts a pressure on the liquid in A some superfluid will flow to B. There is no entropy transport and therefore the entropy per unit mass $S$ will increase in A and decrease in B. According to the equation $C_v\,dT = T\,dS$, $T_A$ will increase and $T_B$ will decrease.

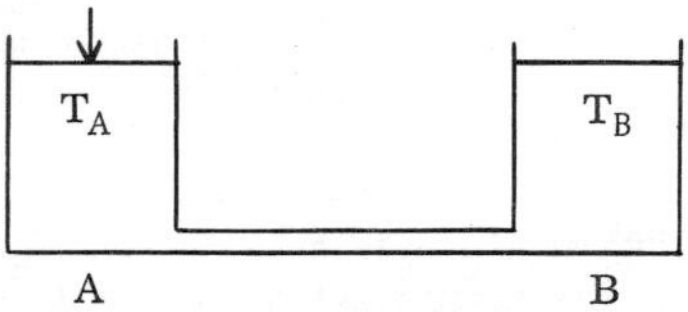


Fig. 5. Mechano-caloric effect.

b. *The fountain effect*

This it the reverse of the mechano-caloric effect. A decrease of $T_B$ or increase of $T_A$ results in a "fountain" of superfluid from A.

c. *Second sound*

Below $T_\lambda$ there exist two kinds of sound waves. The "second sound" is not a pressure wave but rather a temperature or entropy wave. It can be considered as a combination of sound waves in normal fluid and superfluid with opposite phases.

The ratio of mass densities $\rho_n/\rho$ has been determined experimentally by Andronikashvili (see Fig. 6). A pile of disks, spaced 0.2 mm apart and mounted on a shaft, rotates in a helium bath.

The normal fluid rotates between the disks, while the superfluid is supposed to be completely unaffected by the rotating disks. From the moment of inertia one can now determine $\rho_n/\rho$, which has been measured as a function of $T$ along the vapour pressure line. The result is:

$$\frac{\rho_n}{\rho} = \begin{cases} (T/T_\lambda)^{5,6} & \text{for } T < T_\lambda \\ 1 & \text{for } T > T_\lambda \,. \end{cases} \tag{3}$$

In fact the second line of (3) determines the constant of proportionality between the moment of inertia and the ratio $\rho_n/\rho$.

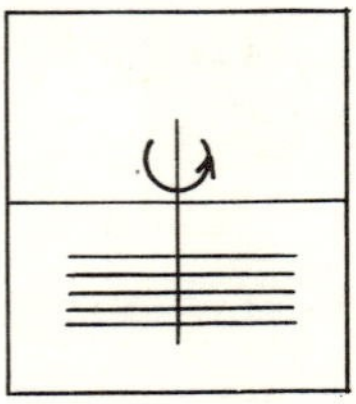

Fig. 6. Experiment of Andronikashvili.

Although Tisza's theory has had some success, it is unsatisfactory, because

1. it has as yet no molecular basis,
2. it is (even phenomenologically) incomplete.

One needs hydrodynamic equations. For instance, the description of the Andronikashvili experiment is not unambiguous.

## 3. *Theory of Landau*

We put $\hbar = 1$ wherever it occurs.

The $T^3$-law for the specific heat near $T = 0$ is the same as in crystals, and may therefore be attributed to the presence of phonons, i.e. quanta of sound waves.

We consider a system of $N$ He-atoms in a box. Landau postulates for the low energy levels:

$$E\{n_k\} = E_0 + \sum_{\vec{k}} \omega_k n_k \qquad n_k = 0, 1, 2, \ldots \,. \tag{4}$$

Although there will be restriction(s) on the numbers $n_k$, they will be treated as independent. This is possible, because for low temperatures, only levels with few excitations are important.

In order to achieve agreement with the experimental curve for the specific heat as a function of $T$, which is the sum of a $T^3$-curve and a exp $(-\Delta/kT)$-curve Landau, postulates furthermore:

$$\omega = \begin{cases} c\,k & \text{for } k \ll k_0 \\ \Delta + \dfrac{(k - k_0)^2}{2\,\sigma} & \text{for } k \approx k_0 \end{cases} \qquad k = |\vec{k}| \tag{5}$$

where $c$, $\Delta$, $k_o$ and $\sigma$ should be adjusted to the experimental curve $C_v = C_v(T)$.

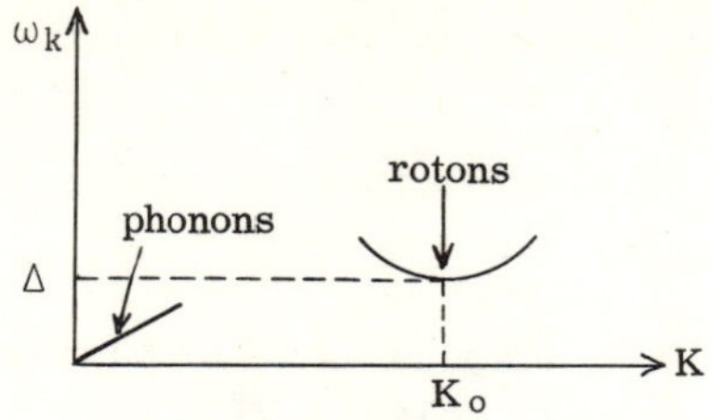


Fig. 7. Spectrum $\omega_k$.

The first line in (5) represents the phonons; $c$ is the velocity of sound. The second line represents rotons, which in some sense may be likened to quantized vortices (spin 1). Applying Bose-statistics, the mean occupation numbers are given by:

$$\langle n_k \rangle = \frac{1}{e^{\beta\omega_k} - 1} . \tag{6}$$

Furthermore:

$$U = E_o + \sum_{\vec{k}} \omega_k \langle n_k \rangle = E_o + 4\,\pi V \int_0^\infty \frac{k^2\,\omega_k\,\mathrm{d}k}{e^{\beta\omega_k} - 1} . \tag{7}$$

Therefore:

$$C_v = \frac{\partial U}{\partial T} \approx (C_v)_{\text{phonon}} + (C_v)_{\text{roton}} \; . \tag{8}$$

One finds:

$$(C_v)_{\text{phonon}} \approx \varkappa\, N \frac{2\,\pi^2\, v\, (\varkappa T)^3}{15\, c^3} \tag{9}$$

$$(C_v)_{\text{roton}} \approx \varkappa\, N \frac{2\,\sigma^{1/2}\, k_o^2\, \Delta^2\, v}{(2\pi)^{3/2}\,(\varkappa T)^{3/2}}\, e^{-\Delta/\varkappa T} \tag{10}$$

where $\varkappa$ is Boltzmann's constant and $v = V/N$ the volume per particle.

From (9) and the experimental $T^3$-law, one finds $c$ = 226 msec$^{-1}$, which is not in bad agreement with the experimental value $c = 239 \pm 2$ msec$^{-1}$.

Fitting the constants to the experimental specific heat, Landau finds

$$\begin{aligned} c &= 226 \text{ msec}^{-1} \\ \Delta/\varkappa &= 9^{\circ}\text{K} \\ k &\cong 2\ \text{Å}^{-1} \\ \sigma/m &\cong .0.3 \end{aligned}$$

Landau derived the spectrum (5) by quantizing the classical equations of hydrodynamics. This procedure is unsatisfactory, because the equations of hydrodynamics should follow from the theory.

## 4. *Theory of Feynman*

All liquids should have sound waves or phonons. But why should these be the sole excitations near $T = 0$?

We shall derive the energy level distribution by determining the wave-functions.

Let us denote the ground state wave-function by $\psi(\vec{r}_1, \ldots, \vec{r}_N)$. The set of vectors $\{\vec{r}_1, \ldots, \vec{r}_N\}$ is termed a configuration.

Properties of $\psi_o$:

1. $\psi_o$ is symmetric in its arguments, because the particles are bosons.
2. $\psi_o$ is real. This is a consequence of the time reversal invariance of the Hamiltonian.
3. $\psi_o \geqslant 0$. Otherwise the kinetic energy would be higher than necessary.

One can picture $\psi_o$ as fairly constant, until atoms touch, when it becomes zero.

We denote the excited state by $\psi(\vec{r}_1, \ldots, \vec{r}_N)$. Properties:

a. $\psi$ is symmetric (Bose statistics).
b. $\psi$ and $\psi_o$ are orthogonal: $(\psi, \psi_o) = 0$.

Real and imaginary parts of $\psi$ must then be positive and negative respectively for half the configuration.

Let $A$ and $B$ be configurations, for which $\psi$ assumes its positive or negative maximum respectively.

To keep the kinetic energy small $A$ and $B$ must be very different. They must certainly differ by more than a simple permutation of the coordinates, since the wave function is invariant under such a permutation.

Fig. 8 gives an idea of $A$ and $B$.

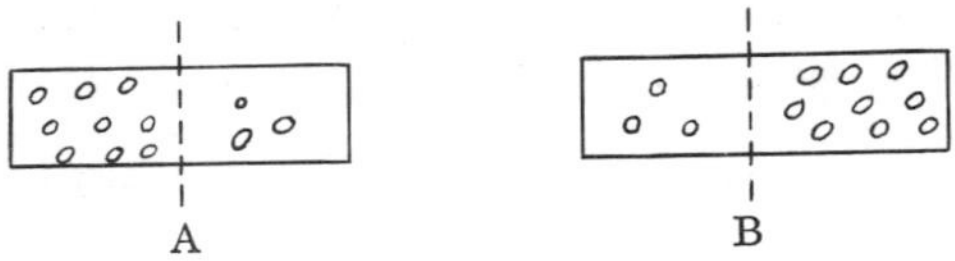

Fig. 8. Configurations for which $|\psi|$ is maximal.

This looks like a sound wave of very long wavelength. Accordingly Feynman postulates:

$$\psi = \sum_{j=1}^{N} f(\vec{r}_j)\, \psi_o . \tag{11}$$

The function $f$ should be determined by minimizing the expectation value of the Hamiltonian.

$$(\psi, H\psi) = \text{minimal} \qquad \text{with the restriction } (\psi,\psi) = 1 \tag{12}$$

$$H = -\frac{1}{2m}\sum_{j=1}^{N} \nabla_j^2 + \sum_{i<j} V_{ij} - E_o . \tag{13}$$

We have shifted the ground state energy to zero, so that

$$H\,\psi_o = 0 .$$

Defining

$$\begin{aligned} \varepsilon &\equiv (\psi, H\psi) \\ \alpha &\equiv (\psi,\psi) \\ F &\equiv \sum_{j=1}^{N} f(\vec{r}_j) \end{aligned} \tag{14}$$

we shall derive the Euler equation corresponding to

$$\delta(\varepsilon - N\omega\alpha) = 0 \tag{15}$$

where $N\omega$ is a Lagrange multiplier. It is easily seen that

$$\begin{aligned} H\psi &= HF\psi_o = [H,F]\,\psi_o + \overbrace{FH\psi_o}^{\overset{O}{\shortparallel}} \\ &= -\frac{1}{2m}\sum_{j=1}^{N}[\nabla_j^2, F]\,\psi_o \\ &= \psi_o/\psi_o\{-\frac{1}{2m}\sum_{j=1}^{N}(\nabla_j^2 F + 2\,\nabla_j F . \nabla_j)\,\psi_o\} \\ &= 1/\psi_o\,(-\frac{1}{2m})\sum_{j=1}^{N}\{(\nabla_j^2 F)\,\psi_o^2 + 2\,(\nabla_j F) \circ \psi_o\,\nabla_j\,\psi_o\} \\ &= 1/\psi_o\,(-\frac{1}{2m})\sum_{j=1}^{N}\nabla_j . (\psi_o^2\,\nabla_j F) \end{aligned}$$

$$\varepsilon = (\psi, H\psi) = -\frac{1}{2m}\sum_{j=1}^{N}\int d^{3N}r\; F^*\,\nabla_j . (\psi_o^2\,\nabla_j F)$$

With suitable boundary conditions partial integration leads to:

$$\begin{aligned} \varepsilon &= \frac{1}{2m}\sum_{j=1}^{N} d^{3N}r\;(\nabla_j F^* . \nabla_j F)\,\psi_o^2 \\ &= \frac{N}{2m}\int d^{3N}r\;\nabla_1 f^*(\vec{r}_1) . \nabla_1 f(\vec{r}_1)\,\psi_o^2(\vec{r}_1,\ldots,\vec{r}_N) . \end{aligned}$$

Defining

$$\rho_o \equiv \int d^3r_2 \ldots d^3r_N\,\psi_o^2(\vec{r}_1,\ldots,\vec{r}_N) \tag{16}$$

which is a constant (translational invariance), we obtain:

$$\varepsilon = \frac{N\rho_o}{2m}\int d^3r\ \nabla f^*(\vec{r})\ .\ \nabla f(\vec{r})\ . \tag{17}$$

Similarly:

$$\alpha = (\psi,\psi) = \int d^{3N}r\ F^*\ F\ \psi_o^2$$

$$= \sum_{i=1}^{N}\sum_{j=1}^{N}\int d^{3N}r\ f^*(\vec{r}_i)f(\vec{r}_j)\ \psi_o^2$$

$$= \sum_{i=1}^{N}\int d^{3N}r\left|f(\vec{r}_i)\right|^2\psi_o^2 + \sum_{i\neq j}\int d^{3N}r\ f^*(\vec{r}_i)f(\vec{r}_j)\ \psi_o^2\ .$$

Or, because $\psi_o$ is symmetric:

$$\alpha = N\int d^{3N}r\ f(\vec{r}_1)^2\ \psi_o^2 + N(N-1)\int d^{3N}r\ f^*(\vec{r}_1)f(\vec{r}_2)\ \psi_o^2\ .$$

Introducing the pair correlation function at absolute zero

$$D(\vec{r}_1 - \vec{r}_2) = \frac{1}{\rho_o}\sum_{i=1}^{N}\sum_{j=1}^{N}\int d^{3N}r'\ \delta(\vec{r}_i' - \vec{r}_1)\ \delta(\vec{r}_j' - \vec{r}_2)\ \psi_o^2 \tag{18}$$

or

$$D = N\delta(\vec{r}_1 - \vec{r}_2) + \frac{N(N-1)}{\rho_o}\int d^3r_3\ldots d^3r_N\ \psi_o^2 \tag{19}$$

we obtain

$$\alpha = \rho_o\int d^3r\ \ d^3r\ f^*(\vec{r}_1)f(\vec{r}_2)\ D(\vec{r}_1 - \vec{r}_2)\ . \tag{20}$$

Inserting (17) and (20) in (15) we see that

$$\delta\left[\frac{N\rho_o}{2m}\int d^3r\ \nabla f^*\ .\ \nabla f - N\omega\rho_o\int d^3r\ \ d^3r\ \ f^*(\vec{r}_1)f(\vec{r}_2)\ D(\vec{r}_1 - \vec{r}_2)\right] = 0\ .$$

The corresponding differential equation reads

$$-\frac{1}{2m}\nabla^2 f(\vec{r}) - \int \omega\ d^3r'\ D(\vec{r}')\ f(\vec{r} + \vec{r}') = 0\ . \tag{21}$$

Assuming a solution of the form

$$f_k(\vec{r}) = e^{i\vec{k}.\vec{r}} \tag{22}$$

we find the relationship

$$\omega_k = \frac{k^2}{2\ m\ S(k)} = E_k - E_o = \frac{\varepsilon}{\alpha} \tag{23}$$

where $S(k)$ is the Fourier transform of the pair correlation function:

$$S(k) = \int d^3r\ e^{i\vec{k}.\vec{r}}\ D(\vec{r})\ . \tag{24}$$

The spectrum $\omega_k$ can be calculated from (23) and (24), because $D(\vec{r})$ can be determined experimentally. The corresponding wave functions are given by

$$\psi_k = \sum_{j=1}^{N} e^{i\vec{k}.\vec{r}_j} \psi_o(\vec{r}_1, \ldots, \vec{r}_N) . \quad (25)$$

The vector $\vec{k}$ is the momentum of the total liquid, because

$$\vec{P}_{op} \psi_k = \frac{1}{i} \sum_{j=1}^{N} \nabla_j \psi_k = \vec{k} \psi_k .$$

Assuming $\vec{P}_{op} \psi_o = 0$ the orthogonality property is proved as follows:

$$\vec{k}(\psi_k, \psi_o) = (\vec{P}_{op} \psi_k, \psi_o) = (\psi_k, \vec{P}_{op} \psi_o) = 0 .$$

Therefore, if $\vec{k} \neq 0$, then $(\psi_k, \psi_o) = 0$ .

The experimental curve $S(k)$ and the corresponding curve $\omega_k$ from (23) are shown in Fig. 9 and 10.

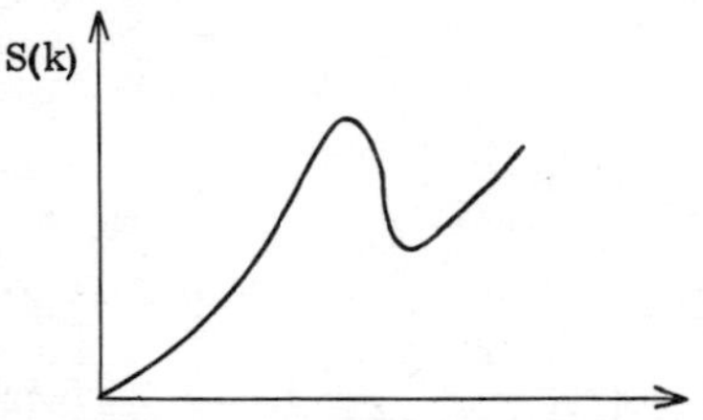


Fig. 9. Experimental curve of the Fourier transform of the pair correlation function.

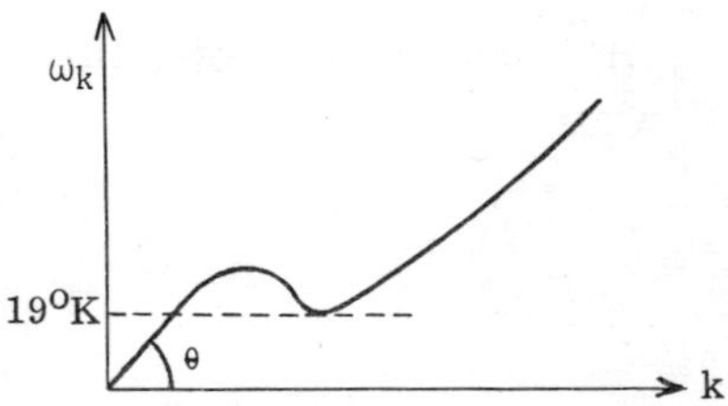


Fig. 10. The spectrum $\omega_k$.

The slope θ agrees with the measured value of the velocity of sound near $T = 0$, but experiment shows that the minimum of (19°K) is too high.

It should be emphasized that Feynman's theory applies only to very low-lying states.

The physical meaning of $\psi_k$ depends crucially on $\psi_o$. If $\psi_o$ were constant, as is the case in an ideal Bose gas, then $\psi_k$ could certainly not represent sound waves. The situation will now be illustrated with a one-dimensional example.

$$Re\ \psi_k = \sum_{j=1}^{N} \cos\ (kx_j)\ \psi_o(x_1, \ldots, x_N) . \quad (26)$$

Suppose first that $kr_o \ll 1$, where $r_o$ is the average distance between particles. The cosines in (26) then tend to enlarge the sum, if the particles bunch together. $\psi_o$, however, becomes zero when particles touch, and therefore favours an extended distribution. The combination of these two opposing effects makes it plausible that

$Re\ \psi_k$ represents sound waves but for this to be so, it is essential that $\psi_o$ should not merely be a constant.

If $kr_o \gg 1$, there exists little correlation between the particles and $\psi_k \approx 0$ independent of $\psi_o$. See Fig. 11 and 12.

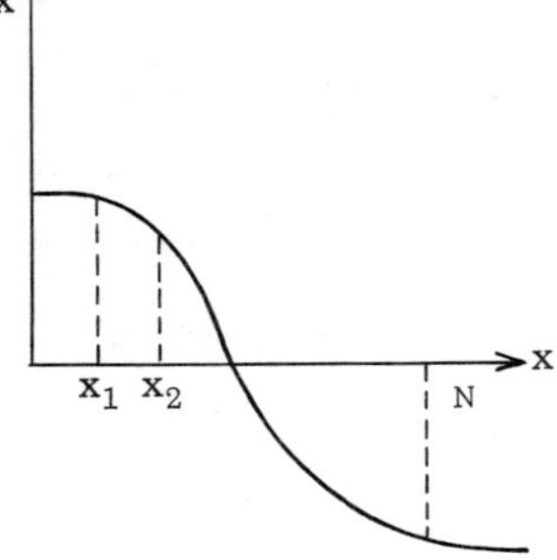


Fig. 11. $kr_o \ll 1$.

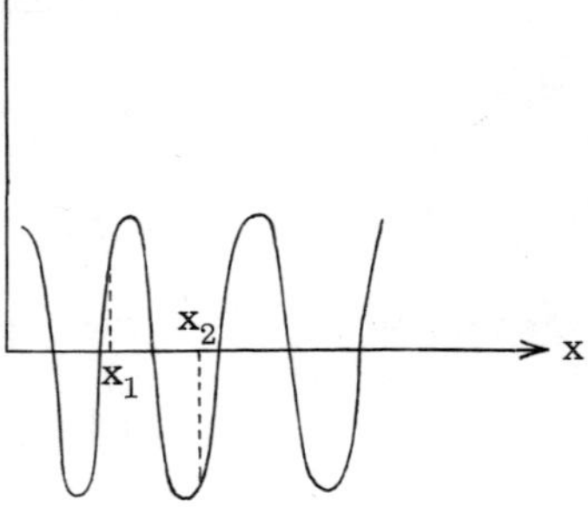


Fig. 12. $kr_o \gg 1$.

*Conclusions*

It is highly plausible, that

$$\psi_k = \sum_{j=1}^{N} e^{i\vec{k}.\vec{r}_j}\ \psi_o \tag{25}$$

$$E_k = E_o + \omega_k \tag{23}$$

$$\left.\begin{aligned} E\{n_k\} &= E_o + \sum_{\vec{k}} n_k\,\omega_k \\ \vec{P}\{n_k\} &= \sum_{\vec{k}} \vec{k}\,n_k \end{aligned}\right\} \quad \begin{aligned} &\text{if } \frac{1}{N}\sum_{\vec{k}} n_k \ll 1 \\ &\text{(few excitations)} \end{aligned} \tag{4}$$

In order to find connections with the two fluid model we start by considering the thermodynamics of the problem near $T = 0$.

## 5. *Equilibrium properties near $T = 0$*

The partition function is given by

$$Z = e^{-\beta A} = \sum_{\{n_k\}} e^{-\beta E\{n_k\}} \tag{28}$$

where $A$ is the free energy and $\beta = 1/\varkappa T$.

At low energies we expect the restrictions on the sets of occupation numbers, if any, to be unimportant. Therefore we take $n_k = 0, 1, 2, \ldots$ independently.

The $\lambda$-transition cannot be described in this way. This is not surprising, however, since the number of particles is indefinite, a fact which also accounts for the failure of the present theory to predict the Bose-Einstein condensation.

Writing

$$Z_k = e^{-\beta\omega_k}$$

it is seen that

$$Z = e^{-\beta E_o} \sum_{\{n_k\}} Z_1^{n_1} \dots Z_k^{n_k} \dots$$

$$= e^{-\beta E_o} \prod_{\vec{k}} \frac{1}{1 - Z_k}$$

$$\ln Z = -\beta E_o + \sum_{\vec{k}} \ln \frac{1}{1 - Z_k}$$

or

$$A = E_o - \frac{1}{\beta} \sum_{\vec{k}} \ln (1 + \langle n_k \rangle) \tag{29}$$

where

$$\langle n_k \rangle = \frac{1}{e^{\beta\omega_k} - 1} . \tag{20}$$

From

$$F = U - TS$$

and

$$U = E_o + \sum_{\vec{k}} \omega_k \langle n_k \rangle$$

we obtain for the entropy:

$$\frac{S}{\varkappa} = \frac{1}{\varkappa T} \sum_{\vec{k}} \omega_k \langle n_k \rangle + \sum_{\vec{k}} \ln (1 + \langle n_k \rangle) . \tag{31}$$

This can be considered to be the entropy of the normal fluid, i.e. the gas of phonons. The superfluid is described by $\psi_o$, its energy is $E_o$ and its entropy zero. The wave function of one phonon is the wave packet

$$\psi_f = \sum_{j=1}^{N} f(\vec{r}_j)\, \psi_o . \tag{32}$$

Two phonons are represented by the double wave packet

$$\psi_{fg} = \sum_{i=1}^{N} \sum_{j=1}^{N} f(\vec{r}_i)\, g(\vec{r}_j)\, \psi_o \tag{33}$$

and so on.

This picture breaks down at somewhat higher temperatures, because then the phonons will overlap.

The next problem is to define quantities such as mass densities, which appear in non-equilibrium phenomena. As a preliminary we need a kinetic theory, which will now be derived with the aid of heuristic arguments.

## 6. *Motion of the superfluid*

In the foregoing we argued that the wave function of the excitations in the system is given by:

$$\Psi_k = \sum_{j=1}^{N} e^{i\vec{k}.\vec{r}_j} \Psi_o \tag{34}$$

(an expression which should hold for every Bose system as we used only Bose statistics as a basis for its derivation). The energy levels are then given by the occupation number $n_k$ of the excitation:

$$E = E_o + \sum \omega_k n_k . \tag{35}$$

The total momentum for such a distribution $n$ reads

$$\vec{P} = \sum_k \vec{k} n_k . \tag{36}$$

The energy momentum, relation $\omega_k$ can be found experimentally by neutron-phonon scattering, see fig. 13.

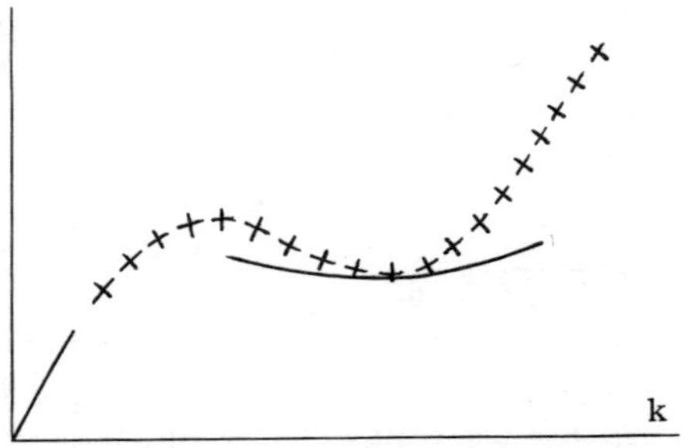


Fig. 13. Energy-momentum relation of the excitations. × experimental, —— Landau.

These excitations have been referred to a background which is at rest. The system also possesses states in which the background (superfluid) has a momentum, as can be seen by considering the following transformation:

$$\begin{cases} \vec{r}_j' = r_j \\ \vec{p}_j' = \vec{p}_j + k_s \end{cases} \quad \text{or} \quad \begin{cases} \vec{r}_j' = e^{-iG} r_j \, e^{+iG} \\ p_j' = e^{-iG} p_j \, e^{+iG} \end{cases} \tag{37}$$

with

$$G = \vec{k}_s . \sum_{j=1}^{N} \vec{r}_j . \tag{38}$$

Now let the hamiltonian $H$ of the system be

$$H = -\frac{1}{2\,m}\sum_{j=1}^{N} \nabla_j^2 + \sum_{i<j} v\left(\left|r_i - r_j\right|\right)$$

and the total momentum

$$\vec{P}_{op} = \frac{1}{i}\sum_{j=1}^{N} \nabla_j \ .$$

Then these operators transform under this canonical transformation as follows:

$$e^{-iG}\, H\, e^{+iG} = H + \frac{1}{m}\,\vec{k}_s \,.\, \vec{P}_{op} + \frac{N\,k_o^2}{2\,m}$$

and
$$e^{-iG}\ P_{op}\ e^{+iG} = \vec{P}_{op} + \vec{N}\,k_s \ .$$

So if $\Psi$ is a simultaneous eigenstate of $H$ and $\vec{P}_{op}$ then

$$\Psi' = e^{+iG}\ \Psi$$

is also a simultaneous eigenstate with the eigenvalues:

$$E(\vec{k}_s) = E_o + \sum_q \left(\omega_q + \frac{1}{m}\,\vec{k}_s \,.\, \vec{q}\right) n_q + \frac{N\,k_s^2}{2\,m} \tag{39}$$

and
$$P(\vec{k}_s) = \sum_q \vec{q}\, n_q + N\,\vec{k}_s \ . \tag{40}$$

So we have the eigenstates extended with new states which are labelled with the quantumnumber $\vec{k}_s$.

Now if $\Psi_o$ is the groundstate at rest (i.e. $\vec{P}_{op}\,\Psi_o = 0$) then

$$e^{+iG}\,\Psi_o$$

is the ground state in a uniformly moving system as follows from:

$$\vec{P}_{op}\,(e^{+iG}\,\Psi_o) = e^{+iG}\,(e^{-iG}\,\vec{P}_{op}\,e^{+iG})\,\psi_o = e^{+iG}\,(\vec{P}_{op} + N\,\vec{k}_s)\,\Psi_o = N\,\vec{k}_s\,e^{iG}\,\Psi_o \ .$$

In the same way we see that if

$$\Psi_q = \sum_{j=1}^{N} e^{\,i\vec{q}.\vec{r}_j}\ \Psi_o$$

is the wave function of an excitation. Then

$$e^{-iG}\,\Psi_q$$

has the momentum $\vec{q} + N\,\vec{k}_s$ since

$$\vec{P}_{op}\,(e^{-iG}\,\Psi_q) = e^{-iG}\,(\vec{P}_{op} + N\,\vec{k}_s)\,\Psi_q = (\vec{q} + N\,\vec{k}_s)\,e^{-iG}\,\Psi_q \ .$$

So for the states $e^{-iG}\Psi_q$, $\vec{q}$ is interpreted as the momentum of the excitation relative to the moving superfluid.

If we take an equilibrium ensemble say the canonical we can prove that

$$\langle k_s \rangle_{ens} = \vec{P}/N \tag{41}$$

where $\vec{P}$ is the (given) total momentum of the system in the ensemble.

This means that at absolute equilibrium the excitations are at rest with respect to the superfluid (which carries the momentum $N\vec{k}_s$).

In our approximation the excitations (phonons) are stable as they have no interaction with each other nor with the superfluid. Thus if we have an ensemble where the average momentum of the phonons is initially finite, it will remain so, at all future times. These situations can be regarded as a quasi-equilibrium. In our approximation we must therefore ignore the relation (41) and admit $k_s$ as an independent parameter of the system. On this basis we set up a kinetic theory wihch has not been derived from first principles up till now.

*Formulation of the problem*. We assume that there is a length $d$ with

a. $d \gg \lambda$ where $\lambda$ is the thermal wavelength, $d \gg r_o$ where $r_o$ is the average intermolecular distance but $d^3$ is a microscopic volume,

b. The thermodynamical functions vary negligibly over $d$. So within $d$ there is quasi-equilibrium (i.e. $\vec{k}_s$ is independent of $\vec{P}/N$).

In order to describe the motion of liquid we need the following quantities:

$\rho$ = mass/volume
$J_\rho$ = mass current
$\varepsilon$ = internal energy/volume
$J_\varepsilon$ = internal energy current
$s$ = entropy/volume
$J_s$ = entropy current.
$\vec{p}$ = momentum/volume
$T_{ij}$ = momentum current.

These densities and currents must obey the conservation laws

$$\begin{gathered}
\frac{\partial \rho}{\partial t} + \nabla . \vec{J}_\rho = 0 \\
\frac{\partial \varepsilon}{\partial t} + \nabla . \vec{J}_\varepsilon = 0 \\
\frac{\partial s}{\partial t} + \overline{\nabla} . \vec{J}_s = 0 \\
\frac{\partial p_i}{\partial t} + \sum_{j=1} \frac{\partial}{\partial x_i} T_{ij} = 0 \qquad i = 1, 2, 3
\end{gathered} \tag{42}$$

Now the problem is to find an expression for the currents. These should be defined with the aid of a distribution function (which has to be solved from a Boltzmann-like equation).

In our case we have only the energy levels and the quasi-equilibrium assumption. We will derive an expression for the currents from quasi-equilibrium thermodynamics.

## 7. *Quasi-equilibrium thermodynamics*

Inside the volumes $d^3$ we use a partition function $Q(\vec{P},\vec{k}_s)$ from which the thermodynamical properties are derived in an ordinary way. $Q(\vec{P},\vec{k}_s)$ is given by:

$$Q(\vec{P},\vec{k}_s) = e^{-\beta E_0 - N\beta k_s^2/2m} \sum_{\{n_q\}}{}' e^{-\beta \sum_q [\omega_q + \frac{\vec{k}_s . \vec{q}}{m}] n_q} \tag{43}$$

where the prime on the summation sign indicates that the summation over the distribution $n_q$ is subject to the condition:

$$\sum_{\vec{q}} \vec{q}\, n_q = \vec{P} - N\vec{k}_s \ . \tag{44}$$

Such a partition function is calculated by means of a generating function and a saddle point integration as follows. The momentum $\vec{P} - N\vec{k}_s$ takes only the values:

$$\vec{P} - N\vec{k}_s = 2\pi\vec{n}/L$$

where $\quad \vec{n} = (n_1, n_2, n_3) \qquad n_j = 0, 1, 2, 3 \ldots$

The generating function is then defined as:

$$J(\vec{t},\vec{k}_s) = \sum_n t_1^{n_1} t_2^{n_2} t_3^{n_3}\, Q(\vec{P},\vec{k}_s)$$

and $Q(\vec{P},\vec{k}_s)$ is obtained from $J(\vec{t},\vec{k}_s)$ by the relation:

$$Q(\vec{P},\vec{k}_s) = \frac{1}{(2\pi i)^3} \oint dt \oint dt \oint dt \ \frac{J(\vec{t},\vec{k}_s)}{t^{n_1+1}\, t^{n_2+1}\, t^{n_3+1}}$$

$$= \frac{\beta^3}{Vi^3} \int dw_1 \int dw_2 \int dw_3\, e^{-\beta\vec{w}(\vec{P}-N\vec{k}_s)}\, J(\vec{w}, k_s) \tag{45}$$

where $\quad t_j = e^{(2\pi\beta/L)w_j}$ .

The generating function $J(\vec{w}, k_s)$ is easily evaluated and yields:

$$\frac{1}{V}\ln J(w,k_s) = -\frac{\beta E_o}{V} - \rho\beta\frac{k_s^2}{2m} - \int\frac{d^3q}{(2\pi)^3}\ln\left[1 - e^{-\beta(\omega_q - \vec{u}.\vec{q})}\right] \quad (46)$$

with $\qquad \vec{u} = \vec{w} - \vec{k}_s/m$ .

The value of $\vec{w}$ (or $\vec{u}$) is determined by the saddle point condition (i.e. the derivative of integrand of (45) = 0)

$$\frac{\vec{P}}{V} - \rho\frac{\vec{k}_s}{m} = \int\frac{d^3q}{(2\pi)^3}\frac{\vec{q}}{e^{\beta(\omega_q - \vec{u}.\vec{q})} - 1} . \quad (47)$$

Then the partition function $Q(\vec{P},k_s)$ is given by:

$$\frac{1}{V}\ln Q(\vec{P},\vec{k}_s) = -\frac{\beta E_o}{V} - \beta\mathcal{E} - \int\frac{d^3q}{(2\pi)^3}\ln\left(1 - e^{-\beta(\omega_q - \vec{u}.\vec{q})}\right) \quad (48)$$

where $\qquad \mathcal{E} = \frac{1}{2}\rho\left(\frac{k_s^2}{m}\right) + \left(\frac{\vec{P}}{V} - \rho\frac{\vec{k}_s}{m}\right)\left(\frac{\vec{k}_s}{m} + \vec{u}\right) \qquad (49)$

and $\vec{u}$ follows from (47).

The average occupation number turns out to be

$$\langle n_q\rangle = \frac{1}{e^{\beta(\omega_q - \vec{u}.\vec{q})} - 1} . \quad (50)$$

The thermodynamic quantities can be expressed in terms of this average occupation number:

1. The pressure:

$$P = P_o + kT\int\frac{d^3q}{(2\pi)^3}\ln\left[1 + \langle n_q\rangle\right]$$

where $P_o$ is zero temperature pressure.

2. The Helmholtz free energy per unit volume $a$

$$a = \varepsilon_o + \mathcal{E} - (P - P_o) \qquad \text{with } \varepsilon_o = E_o/V .$$

3. The Gibbs potential per unit volume $g$

$$g = \varepsilon_o + \mathcal{E} + P_o .$$

4. The internal energy per unit volume $\varepsilon$

$$\varepsilon = \varepsilon_o + \mathcal{E} + \int\frac{d^3q}{(2\pi)^3}\,\omega_q\,\langle n_q\rangle .$$

5. The entropy per unit volume $s$

$$\frac{s}{k} = \frac{\varepsilon - a}{kT} = \int\frac{d^3q}{(2\pi)^3}\langle n_q\rangle\left[\frac{\omega_q}{kT} + \frac{\ln(1 + \langle n_q\rangle)}{\langle n_q\rangle}\right] .$$

The meaning of the vector $\vec{u}$ becomes clear from the following theorem

$$\vec{u} = \frac{\int d^3q \; \langle n_q \rangle \; \nabla_q \; \omega_q}{\int d^3q \; \langle n_q \rangle} = \text{average group velocity of an excitation} \qquad (51)$$

Proof: We use the relation

$$\nabla_q \langle n_q \rangle = - \beta \langle n_q \rangle [1 + n_q ][\nabla_q \; \omega_q - \vec{w}] \; .$$

Multiplying both sides with $\langle n_q \rangle^j$ gives

$$\nabla_q \langle n_q \rangle^{j+1} = - \beta \, (j + 1) \, \langle n_q{}^{j+1} [1 + \langle n_q \rangle ][\nabla_q \; \omega_q - \vec{u}] \; .$$

Integration over $q$ yields then:

$$0 = \int d^3q \; \langle n_q \rangle^{j+1} [1 + \langle n_q \rangle ][\nabla_q \; \omega_q - \vec{u}]$$

for $j = 0, 1 \ldots$

So for every function $f(\langle n_q \rangle)$ expandable in a power series we have

$$0 = \int d^3q \; f(\langle n_q \rangle) [1 + \langle n_q \rangle ][\nabla_q \; \omega_q - \vec{u}] \; .$$

Taking $f(\langle n_q \rangle) = \langle n_q \rangle / (1 + \langle n_q \rangle)$ gives the theorem.

For extremely low temperature we may take for the phonon spectrum the relation

$$\omega_q = c \; q \qquad (52)$$

(where $c$ is the sound velocity at $T = 0$).

Then all integrals can be evaluated explicitly. The saddle point condition becomes:

$$\frac{\vec{P}}{V} - \rho \frac{\vec{k}_s}{m} = \int \frac{d^3q}{(2\pi)^3} \frac{\vec{q}}{\exp \beta \, (cq - \vec{u}.\vec{q}) - 1} \; .$$

As the right hand side must be proportional to $\vec{u}$ we have

$$\frac{\vec{P}}{V} - \rho \frac{\vec{k}_s}{m} = \frac{\vec{u}}{u^2} \int \frac{d^3q}{(2\pi)^3} \frac{\vec{q}.\vec{u}}{\exp \beta \, (cq - \vec{u}.\vec{q}) - 1}$$

$$= \frac{1}{(2\pi)^3} \left(- \frac{\vec{u}}{u^2}\right) \frac{1}{(\beta c)^4} \int_{-1}^{+1} dx \frac{x}{(1 + \frac{u}{c})^4} \int_0^\infty \frac{y^3}{e^y - 1} = \frac{\vec{u}(kT)^4}{[1 - (\frac{u}{c})^2]^3} \frac{2 \, \pi^2}{45 \, c^5} \; .$$

The pressure becomes

$$P = P_o + \frac{\pi^2}{90} \frac{(kT)^4}{c^3} \frac{1}{[1 - (u/c)^2]^2} \; .$$

The internal energy per unit volume

$$\varepsilon = \varepsilon_0 + \mathcal{E} + \frac{\pi^2}{30}\frac{(kT)^4}{c^3}\frac{1+\frac{1}{3}(u/c)^2}{[1-(u/c)^2]^3}$$

The entropy per unit volume

$$\frac{S}{k} = \frac{2\pi^2}{45}\left(\frac{kT}{c}\right)^3 \frac{1+\frac{1}{4}(u/c)^2}{[1-(u/c)^3]^3} \;.$$

For absolute equilibrium ($\vec{u} = 0$) we have

$$P - P_0 = \tfrac{1}{3}(\varepsilon - \varepsilon_0) = \tfrac{4}{9} Ts \;.$$

## 8. *Two fluid model*

For mathematical convenience we introduce:
the velocities $v_s$ and $v_u$

$$v_s = \vec{k}_s/m \qquad \text{(superfluid velocity)} \tag{53}$$

$$v_n = v_s + \frac{\int d^3q \,\langle n_q\rangle\, \nabla_q\, \omega_q}{\int d^3q\, \langle n_q\rangle} = v_s + \vec{u} \qquad \text{(normal velocity)} \tag{54}$$

the mass densities $\rho_s$ and $\rho_n$

$$\rho_s = \rho - \rho_n \qquad \text{(superfluid mass density)} \tag{55}$$

where $\rho_n$ is defined such that

$$\rho_n \vec{u} = \int \frac{d^3q}{(2\pi)^3}\, \vec{q}\, \langle n_q\rangle = \frac{\vec{P} - N\vec{k}_s}{V}$$

which is equivalent to:

$$\rho_n = \frac{1}{u^2}\int \frac{d^3q}{(2\pi)^3}(\vec{u}.\vec{q})\,\langle n_q\rangle \qquad \text{(normal fluid mass density)} \tag{56}$$

Then we have for the total momentum per volume

$$\vec{p} = \frac{\vec{P}}{V} = \rho_n \vec{v}_n + \rho_s \vec{v}_s$$

and for

$$\mathcal{E} = \tfrac{1}{2}\rho_s v_s^2 + \tfrac{1}{2}\rho_n v_n^2 + \tfrac{1}{2}\rho_n |\vec{v}_s - \vec{v}_n|^2 \,.$$

The expression for $\rho_n$ reduces for low temperatures to

$$\rho_n = \frac{2\pi^2}{45\, c^5}\frac{(kT)^4}{1-(\vec{u}/c)^3} \qquad \left(\frac{\rho_n}{\rho} \sim 10^{-6} \text{ at } 0.5^{\circ}\text{K}\right)$$

This set of equations is not unique and not necessary. Moreover one cannot take the two fluid model too literally as we cannot divide the liquid into a set of molecules belonging to the superfluid and a remaining set forming the normal fluid.

Now we postulate the following expressions for the currents:

$$\vec{J}_\rho = \rho_n \vec{v}_n + \rho_s \vec{v}_s \; (= \vec{P}/V) \tag{57}$$

$$\vec{J}_s = s \, \vec{v}_n \tag{58}$$

$$T_{ij} = (\vec{v}_s)_i \, (\rho_s \vec{v}_s)_j + (\vec{v}_n)_i \, (\rho_n v_n)_i + \delta_{ij} \, P \tag{59}$$

The energy current cannot be deduced by this method since the dependence of the energy on the density is not known. We come back to the energy current later. Using these expressions for the currents we have the conservation laws

$$\frac{\partial \rho}{\partial t} + \vec{\nabla} (\rho_n \vec{v}_n + \rho_s \vec{v}_s) = 0 \qquad \text{conservation of mass} \tag{60}$$

$$\frac{\partial s}{\partial t} + \vec{\nabla}.(s \, \vec{v}_n) = 0 \qquad \text{conservation of entropy} \tag{61}$$

$$\frac{\partial}{\partial t}(\rho_n \vec{v}_n + \rho_s \vec{v}_s) + \sum_{i=1} \frac{\partial}{\partial x_i}[(\vec{v}_s)_i \, \rho_s \vec{v}_s + (\vec{v}_n)_i \, \rho_n \vec{v}_n] = - \vec{\nabla} P$$
$$\text{conservation of momentum} \tag{62}$$

The conservation of energy can be written down by using a crude argument due to London.

We take the energy density of the relative motion of the normal fluid with respect to the superfluid as:

$$\mathcal{E}' = \tfrac{1}{2} \rho_n |\vec{v}_n - \vec{v}_s|^2 \; . \tag{63}$$

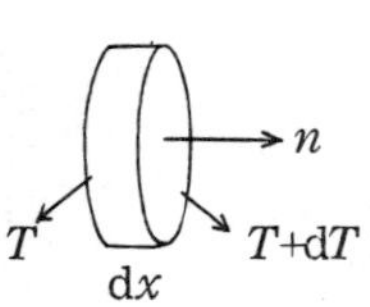


Now we consider the energy balance in a flat slab of the liquid of unit area and of thickness d$x$. During the time d$t$ the kinetic energy of relative motion within the slab decreases by the amount:

$$\mathrm{d}\mathcal{E}' \, \mathrm{d}x$$

Then the heat current $\vec{h}$, which results from the outflow of normal fluid from the slab, is postulated to be

$$\vec{h} = Ts \, (\vec{v}_n - \vec{v}_s) \; . \tag{64}$$

The amount of heat lost by the slab during d$t$ is then

$$\vec{h} . \vec{n} \, \mathrm{d}t$$

where $\vec{n}$ is the normal vector to the flat face of the slab. Since there are no dissipative effects in our model this flow is reversible and the work it performs equals

$$\vec{h}.\vec{n}\,\frac{dT}{T}\,dt$$

where the flat faces of the slab differ in temperature by $dT$.

We now require this work to be supplied by $d\mathcal{E}'\,dx$. Hence

$$-\,d\mathcal{E}'\,dx = \vec{h}.\vec{n}\,\frac{dT}{T}$$

or

$$\frac{d\mathcal{E}'}{dt} + \frac{1}{T}\vec{h}.\nabla T = 0\ .$$

Using the formulae (63) and (64) for $\mathcal{E}$ and $h$ we obtain:

$$(\vec{v}_n - \vec{v}_s)\ \frac{\partial}{\partial}\ \rho_n(\vec{v}_n - \vec{v}_s) + s\ \nabla T\ = 0 \qquad (65)$$

which is the energy conservation law.

We must linearize the conservation equation, as we cannot expect them to hold for large deviations from absolute equilibrium. In absolute equilibrium $\vec{v}_n = \vec{v}_s = 0$ and $\rho_s$ and $\rho_n$ are uniform in space and constant in time. So we shall treat as small:

$$v_n\ ,\ v_s\ ,\ \frac{\partial\rho_n}{\partial t}\ ,\ \frac{\partial\rho_s}{\partial t}\ ,\ \nabla\rho_n\ ,\ \nabla\rho_s\ .$$

We then obtain the following set of linearized conservation laws

$$\frac{\partial\rho}{\partial t} + \rho_n\ \nabla\vec{v}_n + \rho_s\ \nabla\vec{v}_s\ = 0 \qquad (66)$$

$$\frac{\partial s}{\partial t} + s\ \nabla\vec{v}_n = 0 \qquad (67)$$

$$\rho_n\ \frac{\partial v_n}{\partial t} + \rho_s\ \frac{\partial v_s}{\partial t} = -\ \nabla P \qquad (68)$$

$$\rho_n\ \left(\frac{\partial v_n}{\partial t} - \frac{\partial v_s}{\partial t}\right) = -\ s\ \nabla T \qquad (69)$$

Strictly speaking we should add a vector perpendicular to $\vec{v}_n - \vec{v}_s$ to the right hand side. We arbitrarily take this vector to be zero. Subtracting (69) from (68) gives

$$\rho\ \frac{\partial v_s}{\partial t} + \nabla P - s\ \nabla T = 0$$

i.e. for a steady stream $(\partial v_s/\partial t) = 0$ we have

$$\nabla P = s\ \nabla T \qquad (70)$$

which explains the fountain effect.

We can eliminate $v_s$ and $v_n$ from the equations and obtain

$$\frac{\partial^2 \rho}{\partial t} - \nabla^2 P = 0 \tag{71}$$

$$\frac{\partial \rho^2}{\partial t^2} - \frac{\rho}{s}\frac{\partial^2 s}{\partial t^2} + \frac{\rho_s}{\rho_n} s \nabla^2 T = 0 \; . \tag{72}$$

These equations involve $\rho$, $s$, $P$ and $T$ but only two of them are independent. We choose $\rho$ and $T$ and express $S$ and $P$ in terms of them:

$$\frac{\partial^2 s}{\partial t^2} = \left(\frac{\partial s}{\partial T}\right)_\rho \frac{\partial^2 T}{\partial t^2} + \left(\frac{\partial s}{\partial \rho}\right)_T \frac{\partial^2 \rho}{\partial t^2}$$

$$\nabla^2 P = \left(\frac{\partial P}{\partial T}\right)_\rho \nabla^2 T + \left(\frac{\partial P}{\partial \rho}\right)_T \nabla^2 \rho$$

which hold in the same linear approximation. Inserting these relations into (71) and (72) gives

$$\nabla^2 \rho - \frac{1}{c_1^2}\frac{\partial^2 \rho}{\partial t^2} + \gamma_1 \nabla^2 T = 0 \tag{73}$$

$$\nabla^2 T - \frac{1}{c_2}\frac{\partial^2 T}{\partial t^2} + \gamma_2 \frac{\partial \rho}{\partial t^2} = 0 \tag{74}$$

where the coefficients are

$$c_1 = \frac{1}{\left(\frac{\partial P}{\partial \rho}\right)^{\frac{1}{2}}} \qquad c_2 = \left(\frac{s^2 \rho_s}{\rho_n (\partial s/\partial T)}\right)^{\frac{1}{2}}$$

$$\gamma_1 = \frac{(\partial P/\partial T)_\rho}{(\partial P/\partial \rho)_T} = -\left(\frac{\partial \rho}{\partial T}\right)_P \qquad \gamma_2 = \frac{\rho_n}{s \rho_s}\left[1 - \frac{\rho}{s}\left(\frac{\partial s}{\partial \rho}\right)_T\right] \tag{75}$$

When $T \longrightarrow 0$ these constants have the limits

$$c_1 \longrightarrow c \qquad c_2 \longrightarrow c/\sqrt{3}$$
$$\gamma_1 \longrightarrow 0 \qquad \gamma_2 \longrightarrow 0 \tag{76}$$

and at $T = 0$ we have the (uncoupled) equations

$$\nabla^2 \rho - \frac{1}{c^2}\frac{\partial^2 \rho}{\partial t} = 0 \qquad \text{(sound waves)} \tag{77}$$

$$\nabla^2 T - \frac{3}{c^2}\frac{\partial^2 T}{\partial t^2} = 0 \qquad \text{(second sound waves)} \tag{78}$$

In the coupled equation ($T \neq 0$) we can substitute the plane waves:

$$\rho = \rho_o + \rho_1 \exp i\,(\omega t - \vec{k}.\vec{r})$$
$$T = T_o + T_1 \exp i\,(\omega t - \vec{k}.\vec{r}) \tag{79}$$

and obtain for $\omega$ the values

$$\left(\frac{\omega}{k}\right)^2 = \frac{1}{2}\{[c_1^2 + c_2^2 + (c_1\, c_2)^2\, \gamma_1\, \gamma_2] + \pm\sqrt{[(c_1 + c_2)^2 + (c_1\, c_2)^2\, \gamma_1\, \gamma_2]\,[(c_1 - c_2)^2 + (c_1\, c_2)^2\, \gamma_1\, \gamma_2]}\} \tag{80}$$

Lee and Yang have calculated the properties for the hard sphere gas and obtained two curves which start at $c$ and $c/\sqrt{3}$ and cross each other. In higher orders one assumes that this degeneration of the sound velocities is removed (dotted lines). The ratio of first and second sound velocity approaches $1/\sqrt{3}$ in this model. It is, however, difficult to extrapolate the experimental data to $T = 0$.

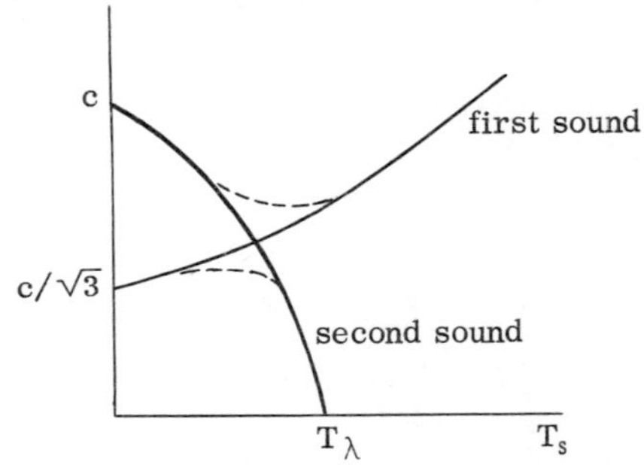


Fig. 14. First and second velocities in a hard sphere gas.

## 9. *Superfluid flow*

We now turn to a somewhat controversial subject.

With the theory developed so far we cannot decide for instance

1. what happens in the Andronikashvili experiment,
2. what the average velocity will be in a tube with a pressure difference between the ends.

In order to discuss these problems we need the concept of viscosity and must introduce boundary conditions at the walls.

In deriving the hydrodynamical equations, we defined a distance $d$ small compared to the macroscopic dimensions of the problem, but much larger than the thermal wavelength $\lambda$. This means that $d$ cannot be arbitrary small. At $T \approx 1^{\circ}$K $\lambda$ is indeed microscopic, but $\lambda = 1$ cm at $T \approx 10^{-20}$ $^{\circ}$K. At these extremely low temperatures, measurements will influence the system strongly. Nevertheless many ideas are based on arguments which in fact apply only to this unrealistic case. For instance:

a. *Landau's explanation* of the fact that external objects can move in $He_{II}$ without encountering any resistance. We denote the velocity of the external object by $\vec{v}_e$. Near $T = 0$ resistance can only be due to excitation of phonons. Assume that an amount of energy $\Delta E$ is transferred to the liquid. Then we have:

$$\Delta E = c \sum_{\vec{k}} |\vec{k}| \; n_k .$$

And the transfer of momentum is given by

$$\Delta \vec{P} = \sum_{\vec{k}} \vec{k} \; n_k .$$

Therefore:

$$|\Delta \vec{P}| \leq \sum_{\vec{k}} |\vec{k}| \; n_k$$

or:

$$c \; |\Delta \vec{P}| \leq \Delta E \; . \tag{81}$$

On the other hand we can write

$$\Delta E = \vec{v}_e \; . \; \Delta \vec{P} \; . \tag{82}$$

We see, that (81) and (82) are inconsistent, unless

$$|\vec{v}_e| \geq c \; . \tag{83}$$

The conclusion is that objects with a velocity less than the velocity of sound in $He_{II}$ (about 230 $msec^{-1}$) do not encounter any resistance.

This argument is probably not relevant for $T \approx 1^oK$, because resistance can then also be due to scattering of phonons. There is no theoretical or experimental evidence that Landau's argument has anything to do with superfluid flow.

Even for $v_e \approx 1$ $msec^{-1}$ strange things happen.

b. *London's proposal*

$$\nabla \times \vec{v}_s = 0 \; . \tag{84}$$

This is based on the idea, that the superfluid is similar to the condensed state of a Bose gas. The superfluid is then a one quantum state changing only adiabatically.

This is correct for $T \approx 10^{-20}$ $^oK$, because then $\lambda$ is of the order of the dimensions of the vessel.

At $T \approx 1^oK$ the theoretical reasons for London's proposal no longer apply. Indeed, even if $\vec{v}_s$ is irrotational in volumes $d^3$, it can nevertheless vary from box to box and have a non-vanishing curl.

There is poor agreement between London's proposal and experiment. Imagine a rotating cylinder with radius $R$ and frequency $\omega$ with $He_{II}$ in it. London's proposal implies that the liquid does not rotate, if $\omega < \hbar/2mR^2 \sim 10^{-4}$ rad/sec for $R$ = 1 cm and rotates in annular layers, if $\omega$ is larger.

The moment of inertia can be calculated on basis of (84) and the result is compared with experiment in Fig. 15.

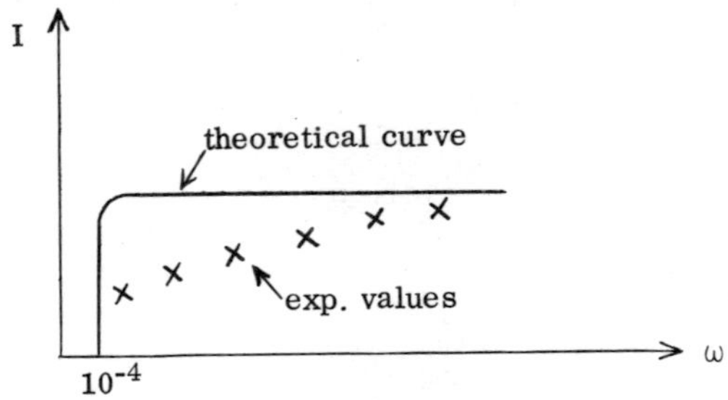


Fig. 15. Moment of inertia of $He_{II}$ in rotating cylinder experiment (R = 1 cm).

c. *Onsager and Feynman* suppose that in addition to the phonons, vortex lines are possible excitations. The wave function should then be such that

$$\frac{\hbar}{2\,m\,i}(\psi^* \,\nabla\psi - \psi\,\nabla\psi^*)$$

behaves as a classical stream.

However, it is very unlikely that at $T \approx 1^{\circ}$K vortex lines of the order of 1 cm exist.

Therefore this argument must also be disregarded.

*Phenomenological theory of Lin*

This theory is not based on first principles but agrees well with experiment.

In the first place Lin introduces viscosity just as in the Navier-Stokes equations.

However, there is no friction between superfluid and normal fluid. The equations now read:

$$\rho_n \frac{\partial \vec{v}_n}{\partial t} + \rho_s \frac{\partial \vec{v}_s}{\partial t} = -\nabla P + [\frac{\rho_n \,\mu_n}{\rho}\nabla^2 \vec{v}_n + \frac{\rho_s\,\mu_s}{\rho}\nabla^2 \vec{v}_s] \tag{85}$$

$$\rho_n \left(\frac{\partial \vec{v}_n}{\partial t} - \frac{\partial \vec{v}_s}{\partial t}\right) = -s\,\nabla T + [\frac{\rho_n \,\mu_n}{\rho}\nabla^2 \vec{v}_n - \frac{\rho_s\,\mu_s}{\rho}\nabla^2 \vec{v}_s] \tag{86}$$

Multiplying (86) by $\rho_s/\rho_n$ and adding one obtains

$$\frac{\partial \vec{v}_n}{\partial t} + \frac{1}{\rho}\nabla P + \frac{\rho_s}{\rho_n} s\,\nabla T = \frac{\mu_n}{\rho}\nabla^2 \vec{v}_n \;. \tag{87}$$

Subtracting (86) from (85) one obtains:

$$\frac{\partial \vec{v}_s}{\partial t} + \frac{1}{\rho}\nabla P - s\,\nabla T = \frac{\mu_s}{\rho}\nabla^2 \vec{v}_s \;. \tag{88}$$

$$\frac{\partial \vec{v}_s}{\partial t} + \frac{1}{\rho} \nabla P - s \nabla T = \frac{\mu_s}{\rho} \nabla^2 \vec{v}_s \, . \tag{88}$$

Denoting the velocity of the wall by $\vec{v}_e$ and tangential components of velocity by the subscript $T$ Lin suggests as boundary conditions:

$$(\vec{v}_n)_T = (\vec{v}_e)_T \tag{89}$$

$$\frac{\partial}{\partial n} (\vec{v}_s)_T = \varkappa \, |\vec{v}_s - \vec{v}_e|^2 \, (\vec{v}_s - \vec{v}_e)_T \tag{90}$$

where $\partial/\partial n$ is the derivative perpendicular to the wall.

Equation (89) is the usual one, whereas (90) assumes a slipping of the superfluid at the boundary and is similar to the boundary condition of Maxwell for a dilute gas. There the accommodation coefficient $\alpha$, the fraction of the gas absorbed by the wall, plays an important role.

Now we are able to discuss the rotating cylinder experiment, cf. Fig. 16.

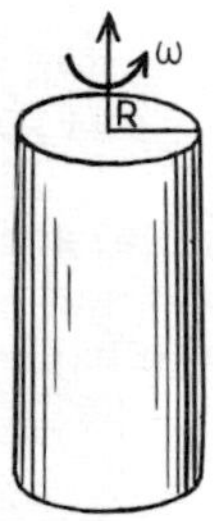


Fig. 16. Rotating cylinder experiment.

We define $L$ as the total angular momentum of the system and $L_c$ as the value $L$ would have if the liquid were classical. Our object is to find $L/L_c$ as a function of $\omega$.

On the circumference of the cylinder application of (89) and (90) gives

$$\begin{aligned} v_n &= \omega R \\ \frac{\partial v_s}{\partial n} &= \varkappa \, (v_s - \omega R)^3 \, . \end{aligned} \tag{91}$$

Due to the viscous terms, the superfluid and normal fluid will eventually rotate as a whole. They flow separately with frequencies, say, $\omega_s$ and $\omega_n$ respectively. Then (91) takes the form

$$\begin{aligned} \omega_n &= \omega \\ \omega_s &= - \varkappa R^3 \, (\omega_s - \omega)^3 \, . \end{aligned} \tag{92}$$

Clearly we can write

$$L = J(\rho_n \omega_n + \rho_s \omega_s) = J(\rho_n \omega + \rho_s \omega_s)$$

$$L_c = J \rho \omega$$

$$\frac{L}{L_c} = \frac{\rho_n}{\rho} + \frac{\rho_s \omega_s}{\rho \omega} . \tag{93}$$

Defining $$x \equiv \rho_n/\rho \qquad y \equiv \omega_s/\omega \tag{94}$$

we obtain from (92)

$$y = - \varkappa R^3 (y - 1)^3 \omega^2$$

or $$\left\{\frac{y}{(1 - y)^3}\right\}^{\frac{1}{2}} = \frac{2\omega}{\omega_0} \qquad \omega_0 \equiv 2 (\varkappa R^3)^{-\frac{1}{2}} . \tag{95}$$

Equation (93) takes the form

$$\frac{L}{L_c} = x + (1 - x) y . \tag{96}$$

Now $y$ can be eliminated from (95) and (96).

If one chooses $\omega_0 = 0.17$ rad sec$^{-1}$ and $\varkappa = 100$ sec$^2$ cm$^{-3}$, the result is in good agreement with experiment, as is illustrated in Fig. 17.

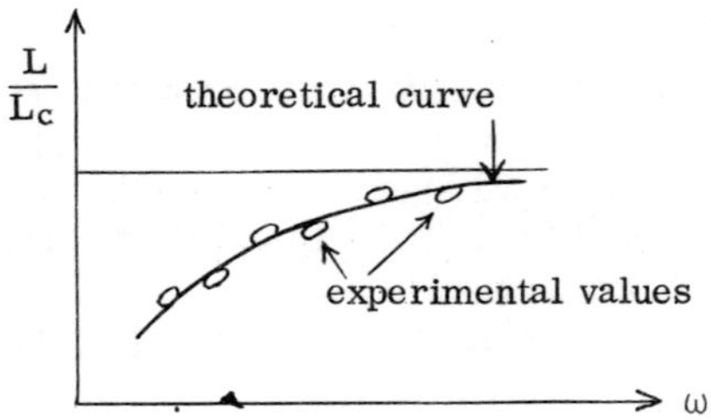


Fig. 17. Rotating cylinder experiment and theory of Lin.

The flow of $He_{II}$ through a channel is another instance in which the theory of Lin is supported by experiment.

*References*

1. F. London, Superfluid II, John Wiley and Sons, New York 1954.
2. K. R. Atkins, Liquid Helium, Cambridge Univ. Press, Cambridge, England 1959.
3. R. P. Feynman, Prog. in low temperature physics (edited by C. J. Gorter), North Holland Publishing Company, Amsterdam, and Phys. Rev. 94 (1954) 262.
4. T. D. Lee and C. N. Yang, Phys. Rev. 113 (1959) 1408.

# MANY PARTICLE ASPECTS OF THE FERMI GAS *

N. M. HUGENHOLTZ
Rijksuniversiteit Groningen, Netherlands

*Introduction*

These lectures contain a discussion of methods, that have been developed in recent years, to derive thermodynamical properties of quantum systems in equilibrium from the Hamiltonian of the system [1, 2, 3, 4, 5, 6, 7]. The methods used by the various authors are similar in many respects. Most authors make use of the formal analogy between many particle systems and field theory. Second quantization is therefore an essential tool, so that the symmetry properties of the particles are introduced right from the start. In this respect the work of Yang and Lee [1] forms an exception. They first consider the Boltzmann gas and symmetrize or antisymmetrize at a later stage.

The presentation given here mainly follows the work of Bloch and De Dominicis [3]. This work was developed as a straightforward extension of earlier work on the ground state of many particle systems [8, 9, 10, 11].

These lectures do not contain a complete treatment, and nothing will really be calculated. An outline will be given of the general method and some difficulties related to the ground state energy will be discussed [12, 13, 14, 15].

## 1. *The ground state of a system of interacting fermions*

Before considering the case of finite temperature we first study the ground state of the system. We shall make use of perturbation theory. We consider a system of a large number $N$ of interacting particles, obeying Fermi-statistics, enclosed in a box of volume $\Omega$. We are in particular interested in properties of such a system in the limiting case that both $\Omega \to \infty$ and $N \to \infty$, while keeping the particle density $n = N/\Omega$ finite.

The hamiltonian is

$$H = H_0 + V \ , \tag{1.1}$$

* These lecture notes were written with the assistance of S. J. Bijl.

where $V$ is the interaction, which will be considered as the perturbation.

The unperturbed ground state $|\varphi_o\rangle$ is characterized by a sphere in momentum space around the origin with radius $k_F$, the Fermi momentum. All single particle states within this sphere are filled, whereas all other states are unoccupied.

The particle density $n$ can be expressed, asymptotically for large systems, in terms of $k_F$:

$$n = \frac{N}{\Omega} \approx \frac{k_F^3}{6\,\pi^2} \,. \tag{1.2}$$

All states of the unperturbed system can be obtained from $|\varphi_o\rangle$ by removing particles from the Fermi sea, thus creating holes, and by adding additional particles outside.

As long as our system is finite we have discrete energy levels and we can apply ordinary time-independent perturbation theory. Accordingly one finds for the energy shift $\Delta E_o = E_o - \varepsilon_o$ of the ground state, up to third order:

$$\Delta E_o^{(1)} = \langle\varphi_o| V |\varphi_o\rangle \tag{1.3}$$

$$\Delta E_o^{(2)} = \sum_{\alpha}{}' \frac{\langle\varphi_o| V |\alpha\rangle\langle\alpha| V |\varphi_o\rangle}{\varepsilon_o - \varepsilon_\alpha} \tag{1.4}$$

$$\Delta E_o^{(3)} = \sum_{\alpha\beta}{}' \frac{\langle\varphi_o| V |\alpha\rangle\langle\alpha| V |\beta\rangle\langle\beta| V |\varphi_o\rangle}{(\varepsilon_o - \varepsilon_\alpha)(\varepsilon_o - \varepsilon_\beta)} \tag{1.5}$$

$$- \langle\varphi_o| V |\varphi_o\rangle \sum_{\alpha} \frac{\langle\varphi_o| V |\alpha\rangle\langle\alpha| V |\varphi_o\rangle}{(\varepsilon_o - \varepsilon_\alpha)^2}$$

(the prime indicates that the summation has to be restricted to states $|\alpha\rangle$ different from the groundstate) where the states $|\alpha\rangle$, $|\beta\rangle$, ... are eigenstates of $H_o$ with eigenvalues $\varepsilon_\alpha$, $\varepsilon_\beta$, ... .

As we see the first term in $\Delta E_o^{(3)}$ has the same structure as $\Delta E_o^{(2)}$, whereas the second term is "irregular". In higher order one always has one "regular" and an increasing number of "irregular" terms. We shall look at the two terms of $\Delta E_o^{(3)}$ somewhat more in detail. We substitute in the first term of $\Delta E_o^{(3)}$ the explicit expression of $V$ in second quantization:

$$V = \frac{1}{4\,\Omega} \sum_{k\,l\,m\,n} (v(k - n) - v(k - m))\, \delta_{kr}(k + l - m - n)\, a_k^* a_l^* a_m a_n \tag{1.6}$$

where the creation and annihilation operators $a_k^*$ and $a_k$ satisfy the commutation relations:

$$\{a_k, a_l\} = 0\,; \qquad \{a_k, a_l^*\} = \delta_{kl}\,; \qquad \{a_k^*, a_l^*\} = 0\,. \tag{1.7}$$

This leads to matrix elements of the form

$$<\varphi_o | V \frac{1}{\varepsilon_o - H_o} V \frac{1}{\varepsilon_o - H_o} V | \varphi_o> . \tag{1.8}$$

The calculation of such ground state expectation values of products of creation and annihilation operators can be carried out with the help of Wick's theorem. To formulate this theorem in a convenient way, we first define a *contraction* of two operators: one takes two operators in the product, brings them next to each other by a number $p$ of permutations, not changing their relative order, and replaces the operators by their ground state expectation value multiplied by $(-1)^p$. Obviously, a contraction differs from zero only if one takes a pair of one creation and one annihilation operator. One can make more than one contraction in a product of operators.

One calls the contraction complete if all the operators in the product have been contracted. Wick's theorem states that the ground state expectation value of a product of creation and annihilation operators is equal to the sum of all different total contractions.

A possible total contraction of (1.8) can be represented schematically by

$$<\varphi_o| a^*_{k_1} a^*_{l_1} a_{m_1} a_{n_1} a^*_{k_2} a^*_{l_2} a_{m_2} a_{n_2} a^*_{k_3} a^*_{l_3} a_{m_3} a_{n_3} |\varphi_o> \tag{1.9}$$

which is equal to

$$<\varphi_o|a^*_{k_1} a_{n_3}|\varphi_o>.<\varphi_o|a^*_{l_1} a_{m_3}|\varphi_o>.<\varphi_o|a_{m_1} a^*_{l_2}|\varphi_o> \dots . \tag{1.10}$$

The proof of the theorem follows from the consideration that every particle or hole that is created has to be removed afterwards. The theorem is, however, not trivial as can be seen in the case that two annihilation operators or two creation operators belong to the same state (e.g. $m_3 = n_3$ and $k_1 = l_1$). The matrix element (1.9) is then zero, whereas (1.10) is not. It can be shown, however, that, if one sums all possible contractions, the errors compensate exactly.

There are two types of contractions in our system:

$$\begin{aligned} &<\varphi_o|a^*_k a_k|\varphi_o> = <\varphi_o|N_k|\varphi_o> \\ &<\varphi_o|a_k a^*_k|\varphi_o> = <\varphi_o|(1 - N_k)|\varphi_o> \end{aligned} \tag{1.11}$$

$$<\varphi_o|a^*_k a_k|\varphi_o> = 1 \quad \text{for } k < k_F , \quad = 0 \quad \text{for } k > k_F$$

$$<\varphi_o|a_k a^*_k|\varphi_o> = 0 \quad \text{for } k < k_F , \quad = 1 \quad \text{for } k > k_F .$$

Following the rule of (1.9) that arrows run from $a^*$ to $a$, we have the following rules for drawing diagrams:

——←—— represents a particle
——→—— represents a hole.

An interaction is represented by a dot (vertex) where four lines come together, two with arrows pointing toward the vertex, corresponding to the $a$'s, and two with arrows pointing away, corresponding to the $a^*$'s.

The contribution to the first term of $\Delta E_0^{(3)}$ of the total contraction given in (1.9) is represented by the diagram of fig. 1.

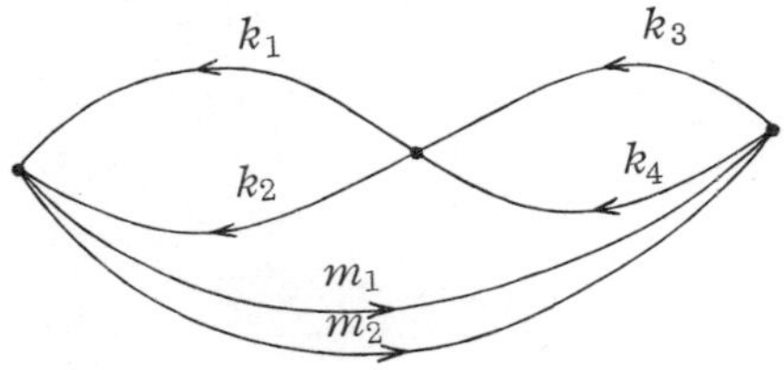


The k's represent particles.
The m's represent holes.

Fig. 1.

The $k_i$ of the particle lines are to be summed over all $k_i > k_F$; the $m_i$ of the hole lines are summed over all $m_i < k_F$.

The energy denominators $(\varepsilon_0 - \varepsilon_\alpha)^{-1}$ can be read immediately from the diagran, by looking at the intermediate states, e.g.

$$\varepsilon_\alpha = \frac{-m_1^2 - m_2^2 + k_1^2 + k_2^2}{2M} . \tag{1.12}$$

We are thus led to a simple set of rules to calculate the contribution of a given diagram:

1. For each vertex we write a factor: $(v(k-n) - v(k-m))\,\delta(k+l-m-n)$.
2. For each intermediate state $|\alpha\rangle$ the energy donominator $(\varepsilon_0 - \varepsilon_\alpha)^{-1}$.
3. Some combinatorial factors (see e.g. ref. 10).
4. We sum particle lines outside the Fermi sphere and hole lines inside.

The diagrams contributing to $\Delta E_0^{(1)}$ and $\Delta E_0^{(2)}$ are shown in fig. 2. Fig. 3 gives some of the diagrams of the regular term in $\Delta E_0^{(3)}$. We notice that all diagrams are connected, except the last one in fig. 3, which is disconnected and consists of two parts.

Let us now for a moment consider the behaviour of various contributions in the limiting case that $\Omega \to \infty$ and $N \to \infty$. We take the example of fig. 1. We have

A. A factor $\Omega^{-3}$ from the three $V$'s.
B. A sum over 6 internal momenta. These momenta are not independent, since momentum is conserved in each interaction. We have three vertices, but we get only two linear relations between the momenta, since then also in the third vertex momentum is conserved. We therefore have $6 - 2 = 4$ independent summations. In the limit of a large $\Omega$ we have:

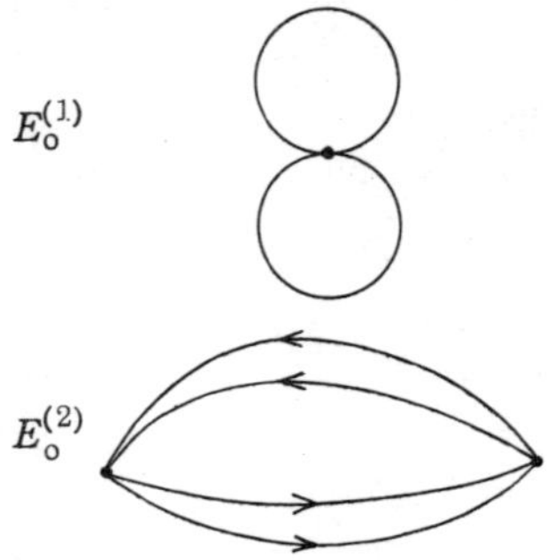


Fig. 2.

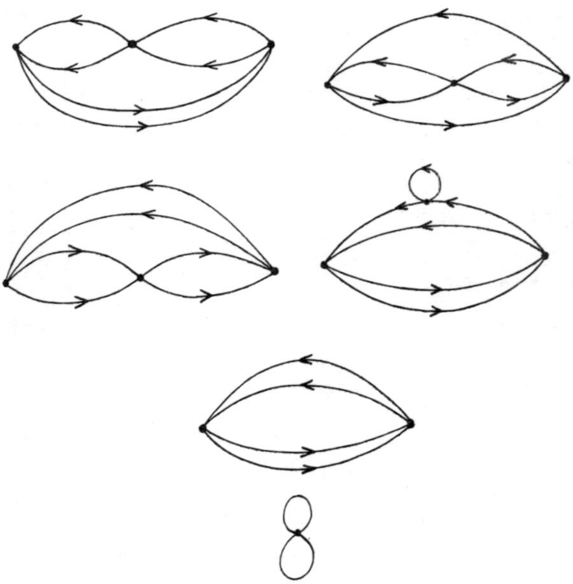

Fig. 3.

$$\frac{(2\pi)^3}{\Omega} \sum_k \rightarrow \int \mathrm{d}^3 k \; . \qquad (1.13)$$

Hence, replacing sums by integrals, we obtain 4 factors $\Omega$.

Combining A and B there remains one factor $\Omega$, which one would expect since $E_o$ is an extensive quantity.

In general one finds that the contribution of any connected diagram is asymptotically $\sim \Omega$. For a disconnected diagram we find as many factors $\Omega$ as connected parts in the diagram. Looking at fig. 3 we see that the "regular" term of $\Delta E_o^{(3)}$ has a contribution of a disconnected diagrams, which consequently is $\sim \Omega^2$. This term therefore must be canceled by some other term. It is immediately seen that this term, in fact, cancels against the "irregular" term in $\Delta E_o^{(3)}$. Brueckner proved that also in fourth order the disconnected contributions to the regular term exactly canceled against the irregular terms. Believing for the moment that this holds true in any order, we are led to the following simple formula for the energy shift:

$$\Delta E_o = \langle\varphi_o| V|\varphi_o\rangle + \langle\varphi_o|[V \frac{1}{\varepsilon_o - H_o} V]_C|\varphi_o\rangle + \dots \tag{1.14}$$

where the subscript $C$ means that one sums over connected diagrams only.

This so-called Brueckner-Goldstone formula was derived first by Goldstone [9]. Before proving (1.14) a preliminary remark should be made on the validity of this formula for large systems. As is well known, for the applicability of perturbation theory in the conventional form one must require that the perturbation is so weak, that the energy shifts are small compared to the level distance. However, for increasingly large systems the levels tend to be more and more closely spaced and the range of applicability of perturbation theory seems to shrink toward weaker and weaker interactions.

It is clear that, to use perturbation methods for large systems, one has to rely on more refined considerations. As we will see, however, extreme caution is necessary.

## 2. *Derivation of the Brueckner-Goldstone formula*

The derivation of the Brueckner-Goldstone formula presented here is due to Cl. Bloch [11]. We consider the resolvent operator

$$R(z) = (H - z)^{-1} , \tag{2.1}$$

where $z$ is a non-real complex number.

We consider the matrix element $\langle\varphi_o|R(z)|\varphi_o\rangle$, for which we can write

$$\langle\varphi_o|R(z)|\varphi_o\rangle = \sum_\alpha \frac{|\langle\varphi_o|\psi_\alpha\rangle|^2}{E_\alpha - z} \tag{2.2}$$

The states $|\psi_\alpha\rangle$ are eigenstates of $H$ with eigenvalues $E_\alpha$. $\langle\varphi_o|R(z)|\varphi_o\rangle$ has clearly a set of poles $E_o, \dots, E_\alpha, \dots$ of which $E_o$ is the smallest.

Instead of $R(z)$ one can consider the operator $e^{-\beta H}$, where $\beta$ is some positive real number. The connection between these operators is a Laplace transformation:

$$e^{-\beta H} = \frac{-1}{2\pi i} \oint dz\, e^{-\beta z} R(z) . \tag{2.3}$$

The integration is to be performed around all singularities of $R(z)$, which are on the real axis. We can also write

$$<\varphi_o|e^{-\beta H}|\varphi_o> = \sum_{\alpha} e^{-\beta E_\alpha}|<\varphi_o|\varphi_\alpha>|^2 \tag{2.4}$$

or $$e^{\beta E_o}<\varphi_o|e^{-\beta H}|\varphi_o> = |<\varphi_o|\psi_o>|^2 + \sum_{\alpha}{}' e^{-\beta(E_\alpha - E_o)}|<\varphi_o|\psi_\alpha>|^2 ,$$

where in the summation the groundstate is excluded. Hence

$$\lim_{\beta\to\infty} e^{\beta E_o}<\varphi_o|e^{-\beta H}|\varphi_o> = |<\varphi_o|\psi_o>|^2 ,$$

which says that for large $\beta$:

$$<\varphi_o|e^{-\beta H}|\varphi_o> \sim |<\varphi_o|\psi_o>|^2 e^{-\beta E_o} . \tag{2.5}$$

We shall now use perturbation theory to calculate these matrix elements. For $R(z)$ we have the identity

$$R(z) = \frac{1}{H_o - z} - \frac{1}{H_o - z} V R(z) . \tag{2.6}$$

This can be iterated, which leads to a power series expansion in $V$

$$R(z) = \frac{1}{H_o - z} - \frac{1}{H_o - z} V \frac{1}{H_o - z} + \ldots . \tag{2.7}$$

Also for $e^{-\beta H}$ one can find a series expansion. This could be obtained from (2.7), making use of (2.3). We do it directly, remarking that $e^{-\beta H}$ is a solution of the Bloch equation:

$$H e^{-\beta H} = -\frac{\partial}{\partial\beta} e^{-\beta H} . \tag{2.8}$$

We define $S(\beta)$ by

$$e^{-\beta H} \equiv e^{-\beta H_o} S(\beta) . \tag{2.9}$$

$S(\beta)$ satisfies the differential equation:

$$\frac{\partial S(\beta)}{\partial\beta} = - V(\beta) S(\beta) \tag{2.10}$$

where

$$V(\beta) = e^{\beta H_o} V e^{-\beta H_o} . \tag{2.11}$$

Eq. (2.10) together with the initial condition

$$S(0) = 1$$

is equivalent to the integral equation

$$S(\beta) = 1 - \int_{o}^{\beta} d\beta_1 \; V(\beta_1)\; S(\beta_1) \;,$$

which by iteration leads to the series

$$S(\beta) = \sum_{n=o}^{\infty} (-1)^n \int_{o}^{\beta} d\beta_1 \int_{o}^{\beta_1} d\beta_2 \ldots \int_{o}^{\beta_{n-1}} d\beta_n \; V(\beta_1) \ldots V(\beta_n) \qquad (2.12)$$

For the matrix element $<\varphi_o|e^{-\beta H}|\varphi_o>$ we have

$$<\varphi_o|e^{-\beta H}|\varphi_o> = e^{-\beta\varepsilon_o} \sum_{n=o}^{\infty} (-1)^n \underset{\beta \geq \beta_1 \geq \ldots \geq \beta_n \geq o}{\int d\beta_1 \ldots \int d\beta_n} <\varphi_o|V(\beta_1)\ldots V(\beta_n)|\varphi_o> \qquad (2.13)$$

The various terms of this expansion can again be represented by diagrams. There is this difference with the previous case that instead of energy denominators one has a $\beta_i$-dependance in each vertex:

$$(v(k-n) - v(k-m))\, e^{\beta_i \frac{k^2+l^2-m^2-n^2}{2M}} \;\delta(k+l-m-n) \;. \qquad (2.14)$$

In this $\beta$-dependent formulation the disentangling of the contribution of disconnected diagrams is very easily possible. Let us consider the diagram of fig. 4. The contribution of this diagram is not a product, due to the region of integration:

$$\beta \geq \beta_1 \geq \beta_1' \geq \beta_2 \geq \beta_2' \;.$$

From the two connected parts of this diagram other disconnected diagrams can be formed by shifting the relative order of the vertices. This is shown schematically in fig. 5. We now take the sum of all contributions. The region of integration is now increased and clearly breaks up into two parts:

$$\left.\begin{array}{l} \beta \geq \beta_1 \geq \beta_2 \geq o \\ \beta \geq \beta_1' \geq \beta_2' \geq o \end{array}\right\} \qquad (2.15)$$

We now obviously have the product of the contributions of the two con-

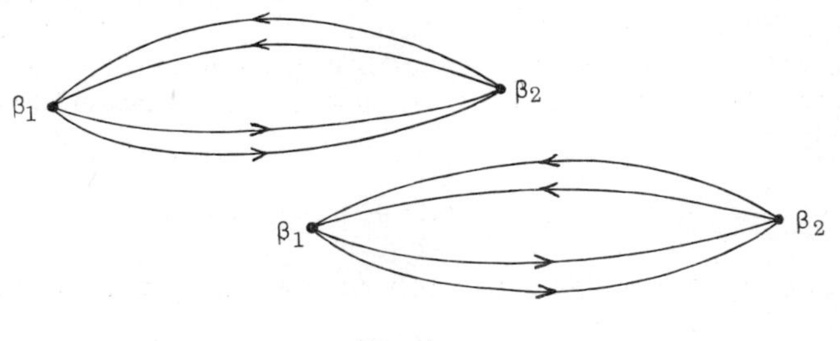


Fig. 4.

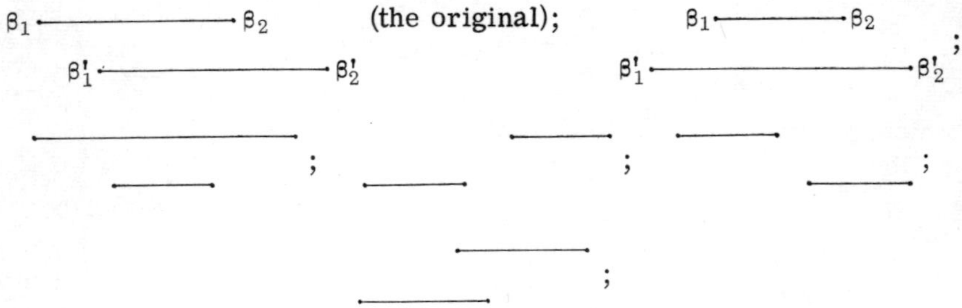


Fig. 5.

nected parts. This result is of course easily generalized to disconnected diagrams of any order and consisting of any number of connected parts.

With the help of this result we derive a simple expression for $\langle\varphi_o|S(\beta)|\varphi_o\rangle$ in terms of connected diagrams only. We write

$$\langle\varphi_o|S(\beta)|\varphi_o\rangle = 1 + \mathrm{I} + \mathrm{II} + \mathrm{III} + \ldots \tag{2.16}$$

where I is the contribution of all connected diagram only. We call this

$$\mathrm{I} = \langle\varphi_o|S_c(\beta)|\varphi_o\rangle . \tag{2.17}$$

II is the contribution of all disconnected diagrams of only two connected parts, etc. According to our previous result this term is equal to

$$\mathrm{II} = \tfrac{1}{2} \langle\varphi_o|S_c(\beta)|\varphi_o\rangle^2 .$$

The factor $\frac{1}{2}$ accounts for the fact that this product gives each diagram twice. Similarly

$$\mathrm{III} = \frac{1}{3!} \langle\varphi_o|S_c(\beta)|\varphi_o\rangle^3 , \quad \text{etc.}$$

Summing this exponential series we get

$$\langle\varphi_o|S(\beta)|\varphi_o\rangle = e^{\langle\varphi_o|S_c(\beta)|\varphi_o\rangle} . \tag{2.18}$$

As we know from (2.5) the energy $E_o$ of the ground state can be obtained from the asymptotic behaviour of $\langle\varphi_o|S(\beta)|\varphi_o\rangle$ for large $\beta$. We therefore must study the asymptotic behaviour of $\langle\varphi_o|S_c(\beta)|\varphi_o\rangle$. We do that by going back to the $z$-dependent formulation. From (2.3) we derive

$$\langle\varphi_o|S(\beta)|\varphi_o\rangle = \frac{-1}{2\pi i} \oint dz\, e^{-\beta z} \langle\varphi_o|R(\varepsilon_o + z)|\varphi_o\rangle \tag{2.19}$$

Since this equation holds separately for each diagram, we can conclude

$$\langle\varphi_o|S_c(\beta)|\varphi_o\rangle = \frac{-1}{2\pi i}\oint dz\, e^{-\beta z}\, \langle\varphi_o|R_C(\varepsilon_o + z)|\varphi_o\rangle \tag{2.20}$$

where

$$\langle\varphi_o|R_C(z)|\varphi_o\rangle = \langle\varphi_o|[-\frac{1}{H_o - z} V \frac{1}{H_o - z} + \dots]_C|\varphi_o\rangle. \tag{2.21}$$

Defining:

$$G_o(z) = \langle\varphi_o|[-V + V\frac{1}{H_o - z}V - \dots]_C|\varphi_o\rangle, \tag{2.22}$$

one has

$$\langle\varphi_o|S_c(\beta)|\varphi_o\rangle = \frac{-1}{2\pi i}\oint dz \frac{e^{-\beta z}}{z^2} G_o(\varepsilon_o + z). \tag{2.23}$$

To derive the asymptotic behaviour of the left hand side of this equation for large $\beta$ one has to know something about $G_o(z)$. From its definition (2.22) we see, since all intermediate states have an energy larger than $\varepsilon_o$, that all singularities of $G_o(z)$ are on the real axis to the right of the point $\varepsilon_o$. For increasingly large systems the poles form a denser and denser set on the real axis. In the limit of $\Omega \to \infty$ and $N \to \infty$ one has a cut along the real axis from $\varepsilon_o$ to $+\infty$. Hence the function is no longer analytical in the point $z = \varepsilon_o$. If one studies the analytical behaviour of the various terms in the expansion one finds, however, that the discontinuity along the cut decreases sufficiently fast, if one approaches the branchpoint $\varepsilon_o$, so that both the function itself and its derivative have a well-defined value in $z = \varepsilon_o$. We then make the assumption that the series converges which implies the existence of $G_o(\varepsilon_o)$ and of $G_o'(\varepsilon_o)$.

The integrand in (2.23) has therefore a pole at $z = o$ and other singularities along the positive real axis. In the limit $\beta \to \infty$ only the contribution from the pole at $z = o$ remains, and we find, for large $\beta$,

$$\langle\varphi_o|S_c(\beta)|\varphi_o\rangle \approx -G_o'(\varepsilon_o) + \beta\, G_o(\varepsilon_o). \tag{2.24}$$

Hence

$$\langle\varphi_o|e^{-\beta H}|\varphi_o\rangle \approx e^{-G_o'(\varepsilon_o)} \cdot e^{-\beta(\varepsilon_o - G_o(\varepsilon_o))}. \tag{2.25}$$

Comparing this with (2.5) one concludes

$$|\langle\varphi_o|\psi_o\rangle|^2 = e^{-G_o'(\varepsilon_o)} \tag{2.26}$$

and

$$\Delta E_o = -G_o(\varepsilon_o). \tag{2.27}$$

The second of these equations is the Brueckner-Goldstone expansion. The first gives the probability of finding $|\psi_o\rangle$ in the unperturbed state $|\varphi_o\rangle$. Since $G_o'(\varepsilon_o)$ like $G_o(\varepsilon_o)$ is asymptotically proportional to the volume $\Omega$, we see that for large systems this probability decreases exponentially with the size of the system. The formulae are, however, exact also for finite systems.

Let us finally make a remark on the validity of these results. The derivation was essentially based on the validity of perturbation theory. If the series in (2.22) does not converge for $z = \varepsilon_0$ one cannot conclude anything.

## 3. *The grandcanonical ensemble*

In quantum statistical mechanics an ensemble is determined by a density operator $\rho$, which in case of a stationary ensemble is a function of the operators $H$ and $N$, the energy and the particle number.

If $F$ is an operator, then the statistical average is given by

$$\bar{F} = \frac{1}{M} \operatorname{Tr} (\rho F) , \tag{3.1}$$

where $M$ is the number of systems in the ensemble. Since $\bar{1} = 1$, we must have

$$\frac{1}{M} \operatorname{Tr} \rho = 1 \qquad \text{or} \qquad \operatorname{Tr} \rho = M . \tag{3.2}$$

Hence

$$\bar{F} = \frac{\operatorname{Tr} (\rho R)}{\operatorname{Tr} \rho} . \tag{3.3}$$

A grand canonical ensemble is given by the density operator

$$\rho = M e^{\beta \chi} . e^{-\beta(H-\mu N)} , \tag{3.4}$$

where the normalizing factor has to be determined in such a way that (3.2) is fulfilled. This gives the equation

$$e^{-\beta \chi} = \operatorname{Tr} e^{-\beta(H-\mu N)} . \tag{3.5}$$

From (3.3) one finds for the average $\bar{F}$ in this case

$$\bar{F} = \frac{\operatorname{Tr} (e^{-\beta(H-\mu N)} F)}{\operatorname{Tr} e^{-\beta(H-\mu N)}} . \tag{3.6}$$

From the Gibbs-potential $\chi$ various thermodynamical quantities such as the pressure $p$, the entropy $S$, the average particle number $\bar{N}$, the free energy $F$ and the internal energy $U$, can be derived:

$$p = - \left(\frac{\partial \chi}{\partial \Omega}\right)_{T,\mu} .$$

For homogeneous systems $\chi \sim \Omega$, so that

$$p\Omega = - \chi .$$

Further

$$S = - \left(\frac{\partial \chi}{\partial T}\right)_{\Omega,\mu} ; \qquad \bar{N} = - \left(\frac{\partial \chi}{\partial \mu}\right)_{\Omega,T} . \tag{3.7}$$

Also $$\chi = F - \mu\bar{N} = U - TS - \mu\bar{N} . \tag{3.8}$$

In the limit $T \to o$ one has:

$$\lim \chi = \lim (U - \mu\bar{N}) . \tag{3.9}$$

Since $\lim\limits_{T\to o} U = E_o$, one can calculate the ground state energy from $\chi$.

We now want to calculate $\chi$ from (3.5) making use of perturbation theory. As in the ground state case we write

$$e^{-\beta H} = e^{-\beta H_o} S(\beta) ,$$

Hence $$e^{-\beta\chi} = \text{Tr}\,(e^{-\beta(H_o - \mu N)} S(\beta)) . \tag{3.10}$$

We now define the unperturbed average of an operator $F$:

$$\langle F\rangle = \frac{\text{Tr}\,[e^{-\beta(H_o-\mu N)} F]}{\text{Tr}\, e^{-\beta(H_o-\mu N)}} . \tag{3.11}$$

Similarly there is an unperturbed Gibbs potential $\chi_o$, defined by

$$e^{-\beta\chi_o} = \text{Tr}\, e^{-\beta(H_o-\mu N)} . \tag{3.12}$$

Combining (3.10), (3.11) and (3.12) one has

$$e^{-\beta(\chi-\chi_o)} = \langle S(\beta)\rangle , \tag{3.13}$$

which, according to (2.12), can be expanded:

$$e^{-\beta(\chi-\chi_o)} = \sum_{n=o}^{\infty} (-1)^n \int\limits_{\beta\geq\beta_1\ldots\geq\beta_n\geq o} d\beta_1 \ldots d\beta_n \langle V(\beta_1) \ldots V(\beta_n)\rangle . \tag{3.14}$$

Instead of ground state expectation values, we now have expressions of the form

$$\langle a^*_{k_1} a^*_{l_1} a_{m_1} a_{n_1} a^*_{k_2} a^*_{l_2} a_{m_2} a_{n_2} a^*_{k_3} a^*_{l_3} a_{m_3} a_{n_3} \rangle$$

To calculate such averages one makes use of a generalization of Wick's theorem, which is due to Bloch and De Dominicis [3]. According to this theorem one contracts the expression in all possible ways. The sum of all different total contractions is equal to the average of the product. We have again two types of contractions:

$$\langle a^*_k a_k\rangle \quad \text{and} \quad \langle a_k a^*_k\rangle$$

The proof of this generalized theorem of Wick will not be given here. We shall only calculate the contractions.

$$\langle a^*_k a_k\rangle = \frac{\text{Tr}\,[e^{-\beta(H_o-\mu N)} N_k]}{\text{Tr}\, e^{-\beta(H_o-\mu N)}} .$$

We calculate this trace in the representation of unperturbed plane wave states. An arbitrary state is then represented by

$$|N_1\, N_2 \ldots N_i \ldots > ,$$

where $N_i$ is the number of particles in the state with momentum $k_i$. We find

$$<a_k^* \, a_k> = \frac{\sum\limits_{N_1, N_2, \ldots} \exp\{-\beta \sum\limits_{l} N_l (\frac{l^2}{2M} - \mu)\} \, . N_k}{\sum\limits_{N_1, N_2, \ldots} \exp\{-\beta \sum\limits_{l} N_l (\frac{l^2}{2M} - \mu)\}} .$$

Both the numerator and the denominator factor out completely; all factors cancel except for momentum $k$. Hence

$$<a_k^* \, a_k> = \frac{\sum\limits_{N_k} e^{-\beta N_k (\frac{k^2}{2M} - \mu)} N_k}{\sum\limits_{N_k} e^{-\beta N_k (\frac{k^2}{2M} - \mu)}} .$$

Now a difference arises between Bose and Fermi statistics. In the first case $N_k$ can take all values from zero to infinity, whereas in the latter case $N_k$ is either zero or one. We find the following result:

A. Bose statistics:

$$<a_k^* \, a_k> = \frac{1}{e^{\beta(\frac{k^2}{2M} - \mu)} - 1} .$$

B. Fermi statistics:

$$<a_k^* \, a_k> = \frac{1}{e^{\beta(\frac{k^2}{2M} - \mu)} + 1} \equiv f_k . \tag{3.15}$$

This is the well-known expression for the average occupation number. We limit ourselves to Fermi particles. We then have for the contractions

$$<a_k^* \, a_k> = f \tag{3.16}$$

and

$$<a_k \, a_k^*> = 1 - f_k = e^{\beta(\frac{k^2}{2M} - \mu)} . f_k ,$$

with $f_k$ given by (3.15).

Exactly like in the ground state case, the various terms of the expansion (3.14) can be calculated by means of diagrams. The only difference is the following:

In the ground state case the integration over the momenta is restricted to the region outside or inside the Fermi sphere for particles or holes respectively. That means that the integrand contains a factor $(1 - f_k^o)$ or $f_k^o$ for each particle line or hole line, where the stepfunction $f_k^o$ is defined by

$$f_k^o = 1 \qquad \text{for} \qquad \frac{k^2}{2M} < \frac{k_F^2}{2M}$$

$$= 0 \qquad \text{for} \qquad \frac{k^2}{2M} > \frac{k_F^2}{2M}.$$

In the present case of finite temperature, the stepfunction $f_k^o$ is replaced by the continuous function $f_k$, which depends on $\beta$.

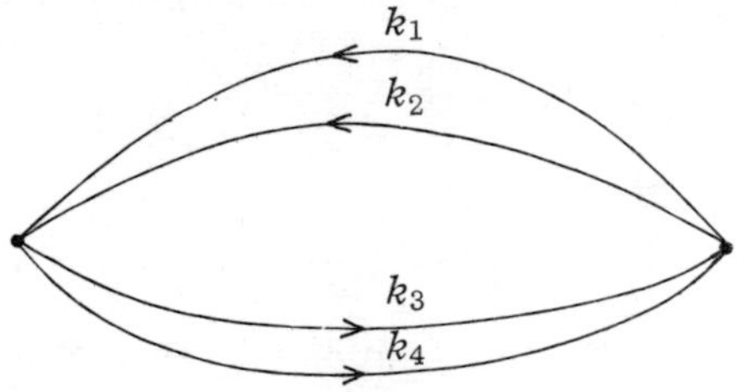

Fig. 6.

As an example we can write down the contribution of the second order diagram of fig. 6. We find the following integral:

$$\frac{1}{2\,\Omega^2} \int\limits_{\beta \geq \beta_1 \geq \beta_2 \geq 0} d\beta_1 \int d\beta_2 \sum_{k_1 k_2 k_3 k_4} \left[ v(k_1 - k_4)^2 - v(k_1 - k_3)\, v(k_1 - k_4) \right]$$

$$(1 - f_{k_1})(1 - f_{k_2})\, f_{k_3} f_{k_4} \cdot \delta_{kr}(k_1 + k_2 - k_3 - k_4)\, e^{\beta_1 \frac{k_3^2 + k_4^2 - k_1^2 - k_2^2}{2M}}\, e^{\beta_2 \frac{k_1^2 + k_2^2 - k_3^2 - k_4^2}{2M}}$$

The fact that in this case the stepfunctions are replaced by continuous functions does not change our previous argument on connected and disconnected diagrams. In complete analogy with equation (2.18) we now have

$$\langle S(\beta) \rangle = e^{\langle S_c(\beta) \rangle}, \qquad (3.17)$$

where $\langle S_c(\beta) \rangle$ is the sum of contributions of connected diagrams only. From (3.13) and (3.17) we obtain

$$\chi - \chi_o = -\beta^{-1} \langle S_c(\beta) \rangle,$$

or
$$\chi - \chi_o = \beta^{-1} \sum_{n=1}^{\infty} (-1)^{n-1} \int\limits_{\beta \geq \beta_1 \geq \dots \geq \beta_n \geq 0} d\beta_1 \dots \int d\beta_n \langle V(\beta_1) \dots V(\beta_n) \rangle_c . \qquad (3.18)$$

This expression of the Gibbs potential in terms of connected diagrams is similar, though not equal, to the Mayer cluster expansion.

For our later discussion on the validity of the Brueckner-Goldstone formula (1.14), we shall write (3.18) in a different form. We do this by making use of results obtained in section 2. Equating both sides of (2.19) term by term one finds

$$\int\limits_{\beta \geq \beta_1 \geq \ldots \geq \beta_n \geq 0} d\beta_1 \ldots \int d\beta_n <\varphi_o| V(\beta_1) \ldots V(\beta_n) |\varphi_o> =$$

$$\frac{-1}{2\pi i} \oint dz\, e^{-\beta z} <\varphi_o| \frac{1}{H_o - \varepsilon_o - z} V \frac{1}{H_o - \varepsilon_o - z} \ldots |\varphi_o> .$$

We know how to relate both sides to diagrams. If we now would replace in both sides of the equation the stepfunctions $f_k^o$ by the continuous functions $f_k$, the equation of course still holds. Hence

$$\int\limits_{\beta \geq \beta_1 \geq \ldots \geq \beta_n \geq 0} d\beta_1 \ldots \int d\beta_n <V(\beta_1) \ldots V(\beta_n)> =$$

$$-\frac{1}{2\pi i} \oint dz\, e^{-\beta z} < \frac{1}{H_o - \varepsilon_o - z} V \frac{1}{H_o - \varepsilon_o - z} \ldots V \frac{1}{H_o - \varepsilon_o - z} >_{f_k} .$$

We know the meaning of the brackets on the lefthand side: the statistical average for the unperturbed system, as defined by (3.11). The bracket on the right, with subscript $f_k$, does not have the meaning of an average. It is calculated by means of diagrams, and it differs from the ground state expectation value only through the functions $f_k$.

Since this equation is valid for each separate diagram we conclude from (3.18)

$$X - X_o = \beta^{-1} \sum_{n=1}^{\infty} (-1)^{n-1} \frac{-1}{2\pi i} \oint dz\, e^{-\beta z} < \frac{1}{H_o - \varepsilon_o - z} V \ldots \frac{1}{H_o - \varepsilon_o - z} >_{f_k, C} . \tag{3.19}$$

To reduce this formula further we make use of invariance under cyclic permutation of the vertices of a diagram. Consider the expression

$$\frac{-1}{2\pi i} \oint dz\, e^{-\beta z} <(\frac{1}{H_o - \varepsilon_o - z} V)^n>_{f_k, C} ,$$

where we notice that one factor $(H_o - \varepsilon_o - z)^{-1}$ is missing at the end. We show now, that the contribution to this expression of a given diagram does not change if we move the vertex at the extreme left to the extreme right, as is shown in fig. 7.

Two effects arise from such a cyclic permutation:

1. All energy denominators are decreased by the amount $\dfrac{k_1^2 + k_2^2 - k_3^2 - k_4^2}{2M}$. This can immediately be read off the diagram. This change of energy in the denominator can be taken care of by

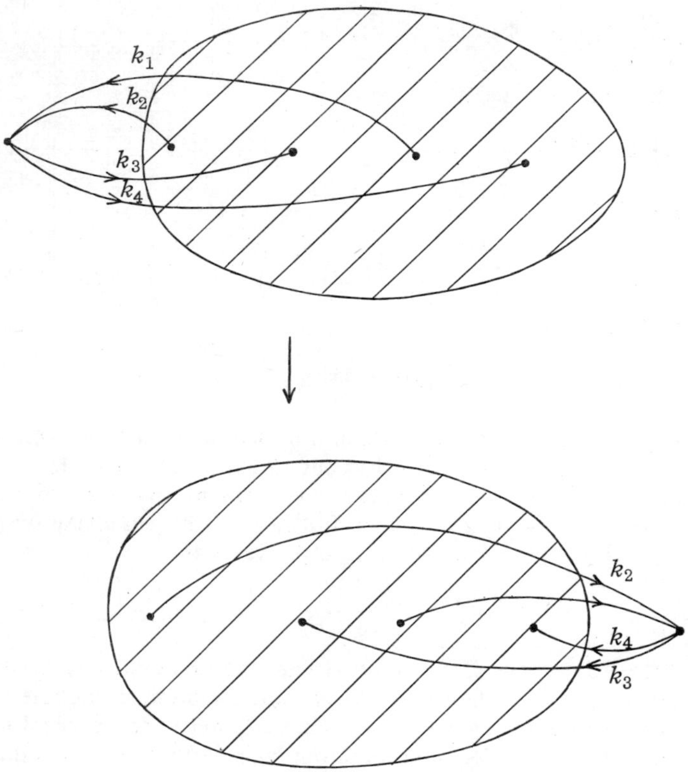


Fig. 7.

a shift of the variable $z$ in the integral. This gives rise to a factor $e^{\beta\frac{k_1^2+k_2^2-k_3^2-k_4^2}{2M}}$

2. The four lines connected to the first vertex change their direction.

This means that $1 - f_{k_1}$ changes into $f_{k_1} = e^{-\beta(\frac{k_1^2}{2M}-\mu)} \cdot (1 - f_{k_1})$, etc.,

which leads to a factor $e^{-\beta\frac{k_1^2+k_2^2-k_3^2-k_4^2}{2M}}$

The two effects cancel.

We make use of this cyclic invariance in the following way. Consider

$$\frac{-1}{2\pi i} \oint dz \, e^{-\beta z} \frac{d}{dz} < \frac{1}{H_o - \varepsilon_o - z} V \ldots \frac{1}{H_o - \varepsilon_o - z} V>_{f_k, C} .$$

Due to the cyclic invariance we can take $n$ times the derivative of the first energy denominator and we obtain

$$n \cdot \frac{-1}{2\pi i} \oint dz\, e^{-\beta z} < \frac{d}{dz} \left(\frac{1}{H_o - \varepsilon_o - z}\right) V \ldots \frac{1}{H_o - \varepsilon_o - z} V >_{f_k, C} ,$$

which, in turn, is equal to

$$n \cdot \frac{-1}{2\pi i} \oint dz\, e^{-\beta z} < \frac{1}{H_o - \varepsilon_o - z} V \cdot \quad \frac{1}{H_o - \varepsilon_o - z} V \frac{1}{H_o - \varepsilon_o - z} >_{f_k, C} .$$

Hence from (3.19)

$$X - X_o = \beta^{-1} \sum_{n=1}^{\infty} \frac{(-1)^{n-1}}{n} \frac{-1}{2\pi i} \oint dz\, e^{-\beta z} \frac{d}{dz} <\left(\frac{1}{H_o - \varepsilon_o - z} V\right)^n>_{f_k, C} .$$

or, integrating by parts,

$$X - X_o = \sum_{n=1}^{\infty} \frac{(-1)^{n-1}}{n} \frac{-1}{2\pi i} \oint dz\, e^{-\beta z} <\left(\frac{1}{H_o - \varepsilon_o - z} V\right)^n>_{f_k, C} . \tag{3.20}$$

If the energies in the different denominators are all different, the contour integration can be carried out immediately. Due to the cyclic invariance we could restrict ourselves to taking the residue only at the first pole, being at $z = o$, and multiplying the resulting expression by $n$. This would lead to the simple formula

$$X - X_o = <[V + V \frac{1}{\varepsilon_o - H_o} V + \ldots]>_{f_k, C} , \tag{3.21}$$

which differs from the Brueckner-Goldstone formula only by the functions $f_k$. One can, however, very easily convince oneself that (3.21) cannot be a correct formula. In contrast to the ground state case, the energies in the denominators can be less than or equal to $\varepsilon_o$. This is due to the fact that in our case "particle" lines can have momenta less than $k_F$ and "hole" lines can have momenta larger than $k_F$. The integrals in (3.21) are therefore no longer well-defined. The difficulties with respect to (3.21) become even more pronounced if one considers the so-called anomalous diagrams, examples of which are given in fig. 8a and b. In such diagrams occur "particle" lines and "hole" lines of the same momentum. In the ground state case

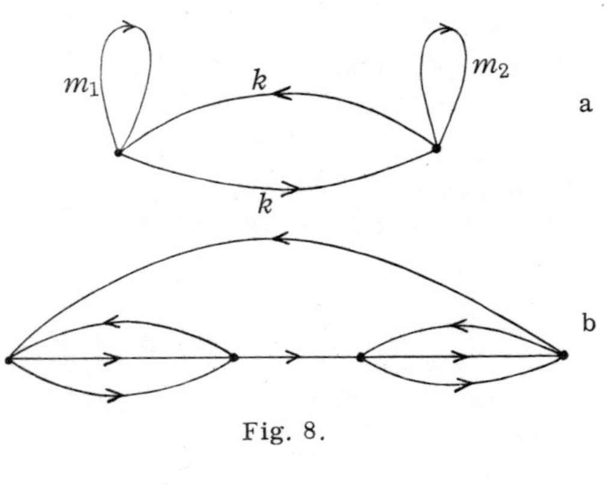


Fig. 8.

such diagrams are not considered, since the particle lines have momenta outside the Fermi sphere, and hole lines inside. In the present case such diagrams should not be excluded. Since these diagrams have an intermediate energy equal to $\varepsilon_o$ the contribution to (3.21) is infinite. We also see clearly why the step from (3.20) to (3.21) was not permissible. In all these cases the pole at $z = o$ is not a single pole, as was assumed.

## 4. *The limit* $T \to o$

In this last section we investigate the limiting case of $T = o$ to obtain an expression for the energy of the ground state. We shall then compare this result with the Brueckner-Goldstone formula.

In this limit the functions $f_k$ change into stepfunctions, i.e.,

$$\lim_{\beta\to\infty} f_k = 1 \qquad \text{for} \qquad \frac{k^2}{2\,M} < \mu$$

$$= o \qquad \text{for} \qquad \frac{k^2}{2\,M} > \mu\ .$$

Note, however, that this stepfunction is not the same as $f_k^o$, since

$$\frac{k_F^2}{2\,M} = \mu_o \neq \mu\ .$$

In this limit (3.21) is correct, if we consider contributions of regular diagrams. Let us see what happens to the contributions of anomalous diagrams in this limit. We calculate the contributions of the second order anomalous diagram of fig. 8a. We make use of the correct formula (3.21):

$$< \frac{1}{z - H_o + \varepsilon_o} V \frac{1}{z - H_o + \varepsilon_o} V >_{f_k} =$$

$$= \frac{1}{z^2} \frac{\Omega}{(2\pi)^9} \int d^3k\, f_k(1 - f_k)\, [\int d^3m\, f_m(v(o) - v(k - m))]^2\ .$$

In the limit $T \to o$ the expression between the brackets becomes

$$F(k) \equiv \int_o^{k_F} d^3m\, (v(o) - v(k - m))\ . \tag{4.1}$$

For the other factor in the integrand we write

$$f_k(1 - f_k) = \frac{e^{\beta(\frac{k^2}{2M} - \mu)}}{(e^{\beta(\frac{k^2}{2M} - \mu)} + 1)^2} = \frac{1}{\beta}\frac{d}{d\mu}\ \frac{1}{e^{\beta(\frac{k^2}{2M} - \mu)} + 1}$$

and we obtain

$$\frac{1}{z^2}\frac{\Omega}{(2\pi)^9}\int d^3k \frac{d}{d\mu} f_k \cdot F(k)^2 .$$

The contribution of this anomalous diagram to $\chi - \chi_o$ is then

$$-\tfrac{1}{2}\frac{\Omega}{(2\pi)^9}\int d^3k \frac{d}{d\mu} f_k \cdot F(k)^2 . \tag{4.2}$$

As long as we have a discrete spectrum, then $k^2/2M - \mu$ is either positive or negative. That means that for each $k$ either $f_k = o$ or $1 - f_k = o$, in the limit $T \to o$. Hence, as long as $\Omega$ is finite,

$$\lim_{T\to o} f_k(1 - f_k) = o .$$

One must, however, perform the two limits $\Omega \to \infty$ and $T \to o$ in the opposite order, i.e., one first takes the limit $\Omega \to \infty$ and afterwards $T \to o$.

In that limit one has

$$\lim_{T\to o} f_k = \Theta\left(\mu - \frac{k^2}{2M}\right) \tag{4.3}$$

where

$$\Theta(x) = o \quad \text{for} \quad x < o ,$$
$$= 1 \quad \text{for} \quad x > o ,$$

and so

$$\frac{d}{d\mu} f_k = \delta\left(\mu - \frac{k^2}{2M}\right) . \tag{4.4}$$

Substituting (4.4) in (4.2) one finds for $T = o$:

$$-\tfrac{1}{2}\frac{\Omega}{(2\pi)^9}\int d^3k\, \delta\left(\mu - \frac{k^2}{2M}\right) \cdot F(k)^2 . \tag{4.5}$$

We see from this example that the anomalous diagrams still give a finite contribution in the limit $T \to o$.

For $\chi - \chi_o$ in the limit $T \to o$ we have finally

$$\lim_{T\to o} (\chi - \chi_o) = G(\mu) + A(\mu) .$$

Here $G(\mu)$ is the sum of all regular diagrams calculated in the ordinary way, however, with the stepfunction $f_k(o) = \Theta(\mu - k^2/2M)$ in contrast to $f_k^o = \Theta(\mu_o - k^2/2M)$ in the Brueckner-Goldstone formula. $A(\mu)$ is the total contribution of the anomalous diagrams, calculated for $T = o$.

For the ground state energy we pbtain

$$E_o(\mu) = \mu\bar{N} + \chi_o(\mu) + G(\mu) + A(\mu) . \tag{4.6}$$

The Brueckner-Goldstone expansion gives, however,

$$E_o^{B.G.}(\mu) = \varepsilon_o(\mu_o) + G(\mu_o) = \mu_o\bar{N} + \chi_o(\mu_o) + G(\mu_o) . \tag{4.7}$$

These two expressions look very different. They have been compared to arbitrary order by Luttinger and Ward [13], and it has been found that in general (4.6) and (4.7) are not equal. Only in the special case of spherical forces between the particles the Brueckner-Goldstone formula gives the correct ground state energy. We shall restrict ourselves to a second order calculation of the difference between (4.6) and (4.7)

$$E_0(\mu) - E_0^{B.G.}(\mu) = \overline{N}(\mu - \mu_0) + \chi_0(\mu) - \chi_0(\mu_0) + G(\mu) - G(\mu_0) + A(\mu) \quad (4.8)$$

We expand $\mu$ into powers of the interaction $V$ and write

$$\mu = \mu_0 + \mu^{(1)} + \mu^{(2)} + \dots$$

where the various terms can be calculated from

$$\overline{N} = -\frac{\partial\chi(\mu)}{\partial\mu}. \quad (4.9)$$

Similarly we have

$$\overline{N} = -\frac{\partial\chi_0(\mu_0)}{\partial\mu_0}. \quad (4.10)$$

In expanding (4.9) into powers of $V$, we have to consider $N$ as a number independent of $V$:

$$\overline{N} = -\frac{\partial\chi_0(\mu)}{\partial\mu} - \frac{\partial\chi_1(\mu)}{\partial\mu} - \dots$$

$$= -\left(\frac{\partial\chi_0(\mu_0)}{\partial\mu_0} + (\mu - \mu_0)\frac{\partial^2\chi_0(\mu_0)}{\partial\mu_0} + \dots\right) - \left(\frac{\partial\chi_1(\mu_0)}{\partial\mu_0} + \dots\right).$$

Equating the various expansion terms on both sides we find first of all (4.10). Secondly one finds

$$\mu_1 = -\frac{\partial\chi_1}{\partial\mu_0} \Big/ \frac{\partial^2\chi_0}{\partial\mu_0^2}. \quad (4.11)$$

Expanding now (4.8) into powers of $V$ we find in lowest, i.e., second order

$$E_0(\mu) - E_0^{B.G.}(\mu) = \tfrac{1}{2}\,\mu^{(1)2}\frac{\partial^2\chi_0(\mu_0)}{\partial\mu_0^2} + \mu^{(1)}\frac{\partial}{\partial\mu_0}G^{(1)}(\mu_0) + A^{(2)}(\mu_0).$$

Since no anomalous diagrams exist in first order $\chi_1 = G^{(1)}$, and so

$$E_0(\mu) - E_0^{B.G.}(\mu) = -\tfrac{1}{2}\frac{\left[\frac{\partial}{\partial\mu_0}G^{(1)}(\mu_0)\right]^2}{\frac{\partial^2}{\partial\mu_0^2}\chi_0(\mu_0)} + A^{(2)}(\mu_0). \quad (4.12)$$

$A^{(2)}(\mu_0)$ has been calculated before and is given by (4.5), if one replaces there $\mu$ by $\mu_0$. $G^{(1)}(\mu_0)$ is the first term of the B.-G. expansion. One finds easily

$$G^{(1)}(\mu_o) = \tfrac{1}{2}\,\frac{\Omega}{(2\pi)^6}\int d^3m_1 \int d^3m_2\, f^o_{m_1} f^o_{m_2}\,(v(o) - v(m_1 - m_2))\ .$$

Hence

$$\frac{\partial}{\partial\mu_o} G^{(1)}(\mu_o) = \frac{\Omega}{(2\pi)^6}\int d^3m_1\ \delta\left(\frac{m_1^2}{2\,M} - \mu_o\right)\int_o^{k_F} d^3m_2 (v(o) - v(m_1 - m_2))$$

$$= \frac{\Omega}{(2\pi)^6}\int d^3k\ \delta\left(\frac{k^2}{2\,M} - \mu_o\right) F(k)\ , \tag{4.13}$$

where $F(k)$ is given by (4.1).

We further have

$$\frac{\partial^2}{\partial\mu_o^2}\,\chi_o(\mu_o) = -\frac{\Omega}{(2\pi)^3}\int d^3k\ \delta\left(\frac{k^2}{2\,M} - \mu_o\right)\ . \tag{4.14}$$

Substituting (4.13), (4.14) and (4.5) in (4.12) one gets

$$E_o - E_o^{B.G.} = \tfrac{1}{2}\,\frac{\left[\frac{\Omega}{(2\pi)^6}\int d^3k\ \delta\left(\frac{k^2}{2\,M} - \mu_o\right) F(k)\right]^2}{\frac{\Omega}{(2\pi)^3}\int d^3k\ \delta\left(\frac{k^2}{2\,M} - \mu_o\right)}$$

$$- \tfrac{1}{2}\,\frac{\Omega}{(2\pi)^9}\int d^3k\ \delta\left(\frac{k^2}{2\,M} - \mu_o\right) F(k)^2\ .$$

This can be put into a more compact form if we define the average of a function $f(k)$ over the unperturbed Fermi-surface:

$$\bar{f} = \frac{\int d^3k\ \delta\left(\frac{k^2}{2\,M} - \mu_o\right) f(k)}{\int d^3k\ \delta\left(\frac{k^2}{2\,M} - \mu_o\right)}\ .$$

We then have

$$E_o - E_o^{B.G.} = \tfrac{1}{2}\,\frac{\Omega}{(2\pi)^9}\int d^3k\ \delta\left(\frac{k^2}{2\,M} - \mu_o\right)\left[\bar{F}^2 - \overline{F^2}\right]\ .$$

We conclude first of all that

$$E_o - E_o^{B.G.} \leq o\ .$$

Secondly we see that $E_o^{B.G.} = E_o$ in the case that $F(k)$ is a function of $|k|$ only. This is clearly the case provided the interaction $v(x_1 - x_2)$ depends only on the distance between the particles. In the more general case where the interaction also depends on the direction of the vector $\vec{x}_1 - \vec{x}_2$ the Fermi surface is deformed by the interaction and the energy shift is not correctly given by the Brueckner-Goldstone formula.

This situation can be illustrated with a simple model, due to Kohn and Luttinger [12]. Take the hamiltonian

$$H = \sum_{k} \frac{k^2}{2\,M}\, a_k^* \, a_k + \sum_{k} u(\vec{k})\; a_k^* \, a_k \, .$$

Clearly the perturbation does not give an interaction between the particles, but merely an energy shift $u(\vec{k})$ to each particle.

We can write

$$H = \sum_{k} E(\vec{k})\; a_k^* \, a_k \, ,$$

where

$$E(\vec{k}) = \frac{k^2}{2\,M} + u(\vec{k}) \, .$$

We can immediately write down the ground state energy,

$$E_o = \sum_{E(k)<\mu} E(\vec{k}) \tag{4.15}$$

where $\mu$ is determined by the requirement

$$N = \sum_{E(k)<\mu} \, . \tag{4.16}$$

If we would perform ordinary perturbation theory we would find:

$$\varepsilon_o = \sum_{\frac{k^2}{2M}<\mu_o} \frac{k^2}{2\,M} \, , \tag{4.17}$$

where $\mu_o$ is determined by

$$N = \sum_{\frac{k^2}{2M}<\mu_o} \, . \tag{4.18}$$

The Brueckner-Goldstone expansion gives only one term, represented by the diagram of fig. 9. Its contribution is

$$\Delta E_o = \sum_{\frac{k^2}{2M}<\mu_o} u(\vec{k}) \, ,$$

and thus

$$E_o^{\text{B.G.}} = \sum_{\frac{k^2}{2M}<\mu_o} E(\vec{k}) \, . \tag{4.19}$$

Comparing (4.19) and (4.15) we see that $E(\vec{k})$ is summed over an equal number of states, but in general not over the same states. If $u(\vec{k})$ and also $E(\vec{k})$ depend on the direction of $\vec{k}$, then the perturbed Fermi surface determined by (4.16) will not be a sphere, whereas the unperturbed Fermi surface is spherical.

Without proving this here, we remark that the temperature dependent approach gives the correct answer. One then has to take

Fig. 9.

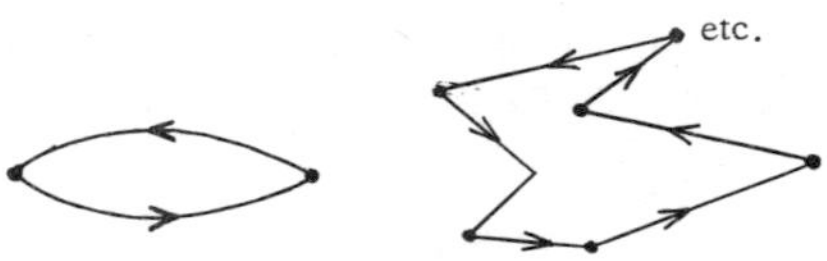


Fig. 10.

into account the contribution of an infinite number of anomalous diagrams as shown in fig. 10.

In this simple example one can see precisely what happens. Due to the anisotropy of the perturbation $u(\vec{k})$ the energy levels will cross. If one makes use of ordinary perturbation theory one starts with the state which for interaction zero has the lowest energy, i.e. the state with a Fermi sphere. If one now switches on the interaction all levels will shift and for a given strength of $V$ another level will cross and become the ground state. In perturbation theory one stays on the same level and the energy one calculated is still an eigenstate, but no longer the ground state.

That statistical mechanics always provides the ground state is not surprising, since in the limit $T \to o$ it automatically chooses the state of lowest energy. For our example one can also see what went wrong in the derivation of the Brueckner Goldstone formula; the unperturbed and perturbed ground state are orthogonal, i.e., $\langle \varphi_o | \psi_o \rangle = 0$.

In a more realistic case the situation is more complicated. There two levels do not in general cross but come close together for a certain interaction strength. This means that the perturbation series will not converge beyond the first nearly crossing of levels. This leads to a radius of convergence which is very small and approaches zero for an infinitely large system.

## *References*


1. C.N. Yang and T.D. Lee, Phys. Rev. 113 (1959) 1165 and subsequent papers.
2. T. Matsubara, Progr. Theor. Phys. 14 (1955) 351.
   The first application of many body methods to thermodynamics.
3. C. Bloch and C. De Dominicis, Nuclear Physics 7 (1958) 459,
   and related to it
4. A.E. Glassgold, W. Heckrotte and K.M. Watson, Physical Review 115 (1959) 1374.
5. E.W. Montroll, J.C. Ward, Physics of Fluids 1 (1958) 55.
6. E.S. Fradkin, Nuclear Physics 12 (1959) 449.

7. P. C. Martin and J. Schwinger, Physical Review 115 (1959) 1342-1373.
8. See, e.g., K. A. Brueckner and C. A. Levinson, Physical Review 97 (1955) 1344 and subsequent articles.
9. J. Goldstone, Proc. Royal Society A 239 (1957) 267.
10. N. M. Hugenholtz, Physica 23 (1957) 481; Boulder Lectures II (1959).
11. C. Bloch, Nuclear Physics 7 (1958) 451.
12. W. Kohn and J. M. Luttinger, Phys. Review 118 (1960) 41.
13. J. M. Luttinger and J. C. Ward, Phys. Review 118 (1960) 1417.
14. R. Brout and F. Englert, Phys. Review 120 (1960) 1519.
15. R. Balian and C. De Dominicis, Nucl. Physics 16 (1960) 502.

# THE BOLTZMANN EQUATION AND ITS GENERALIZATION TO HIGHER DENSITIES

E. G. D. COHEN
Instituut voor Theoretische Fysica, Universiteit van Amsterdam, Amsterdam, Nederland

## *Introduction*

A description of the properties of a classical monatomic not too dense gas that is not in equilibrium can be based on the Boltzmann equation or its generalization to higher densities. In particular a computation of the transport coefficients, that govern the approach to equilibrium, as functions of temperature and density can be achieved in this way. In these lectures we would like to give a survey of the basic ideas that are involved and the main results that have been obtained in this field so far. First we shall give in section 1 a discussion of the description of a gas not in equilibrium based on the Boltzmann equation. As this equation includes only the influence of binary collisions, this treatment is only applicable to *dilute gases* (pressure $p \leq 1$ atm). The method of solving this equation, essentially due to Hilbert and Enskog, will be discussed and the expressions for the transport coefficients that follow from this theory will be given. In order to allow a treatment of the properties of a *moderately dense gas* ($p > 1$ atm) the Boltzmann equation has to be generalized so as to include the effect of triple- and higher order collisions. Thereto in section 2 a more general description of a gas that is not in equilibrium is introduced, based on the Liouville equation. The so-called B-B-G-K-Y hierarchy is derived and the special solution of this hierarchy leading to the Boltzmann equation as a first approximation, is discussed. In section 3 the derivation and generalization of the Boltzmann equation is actually performed. Also a discussion of this derivation is given especially in connection with the introduction of irreversibility. In section 4 finally the results obtained by solving this generalized Boltzmann equation are given and in particular the ensuing expressions for the transport coefficients are discussed.

## 1. *The Boltzmann equation and its solution*

### a. The Boltzmann equation

In this section we shall give a discussion of the properties of a dilute gas that is not in equilibrium on the basis of the Boltzmann equation.

The Boltzmann equation is an equation for the single particle distribution function $f(\vec{q}\ \vec{v}\ t)$ of a gas. Here $f(\vec{q}\ \vec{v}\ t)\ d\vec{q}\ d\vec{v}$ is the average number of molecules to be found at time $t$ in a volume element $d\vec{q} = dx\ dy\ dz$ around $\vec{q}$ in ordinary space and in the volume element $d\vec{v} = dv_x\ dv_y\ dv_z$ around $\vec{v}$ in velocity space or also in a volume element $d\vec{q}\ d\vec{v}$ around the point $\vec{q}\vec{v}$ in the 6-dimensional $\mu$-space. The equation is an integro-differential equation for $f$ and reads in the absence of external forces:

$$\frac{\partial f(\vec{q}\ \vec{v}\ t)}{\partial t} + \vec{v}.\frac{\partial f(\vec{q}\ \vec{v}\ t)}{\partial \vec{q}} = J(f\,f) = \int\int\int [f(\vec{q}\ \vec{v}'\ t)\, f(\vec{q}\ \vec{v}_1'\ t) - f(\vec{q}\ \vec{v}\ t)\, f(\vec{q}\ \vec{v}_1\ t)]\, g\, b\, db\, d\varphi\, d\vec{v}_1 \quad (1.1)$$

It expresses the change per unit of time of the number of molecules per unit of volume around the point $\vec{q}\vec{v}$ in $\mu$-space ($\partial f/\partial t$) as the sum of a streaming term ($-\vec{v}.\,\partial f/\partial\vec{q}$) and a collision term ($J(ff)$). Thus the number of molecules in such a volume element changes due to the fact that they have a velocity $\vec{v}$ and that in addition to this they suffer (binary) collisions. In (1.1) the change due to collisions is computed on the basis of the so-called *Stosszahlansatz*, which has been expressed mathematically in the right hand side of (1.1). The second term in the integrand gives the number of molecules in the volume element around $\vec{q}\vec{v}$ that collides with another molecule so that its velocity will no longer be $\vec{v}$ and the molecule is lost for the $f(\vec{q}\ \vec{v}\ t)$. This number has been computed on the basis of the hypothesis of *molecular chaos* i.e. the hypothesis that the number of molecules with velocity $\vec{v}$ which collides with molecules with velocity $\vec{v}_1$, is proportional to the product $f(\vec{q}\ \vec{v}\ t)f(\vec{q}\ \vec{v}_1\ t)$. Here it has been assumed that the two colliding molecules are uncorrelated i.e. that the presence of a molecule with velocity $\vec{v}$ does not influence the presence of another molecule with velocity $\vec{v}_1$. Also it is somehow assumed that although the two colliding molecules are *not* at the same place $\vec{q}$, the variation of $f$ with $\vec{q}$ is so small that they may nevertheless be taken both at the *same* place $\vec{q}$.

Similarly the first term on the right hand side of (1.1) gives the number of molecules that arrives in the volume element around $\vec{q}\vec{v}$ through collision. Here $\vec{v}'$ and $\vec{v}_1'$, are the initial velocities of two molecules that yield after collision two molecules with velocities $\vec{v}$ and $\vec{v}_1$ (inverse collision). Each collision is characterised not only by the initial velocities of the two molecules but also by their spatial configuration: the impact parameter $b$ and the azimuth $\varphi$ of the plane in which the collision takes place. The expression $g\, b\, db\, d\varphi$ is the (classical differential) collision cross-section, where $g = |\vec{v} - \vec{v}_1| = |\vec{v}' - \vec{v}_1'|$ is the absolute value of the relative velocity of the two colliding molecules. Of course the connection between $\vec{v}', \vec{v}_1'$ and $\vec{v}, \vec{v}_1$ depends also on the intermolecular potential field. It is through this that the intermolecular potential field enters in the Boltzmann equation.

The main problem is to solve the Boltzmann equation for $f(\vec{q}\,\vec{v}\,t)$. If one succeeds in doing so then all relevant properties of a dilute gas which is not in equilibrium and in particular the transport coefficients can be computed (see below). We shall now proceed to a discussion of the Hilbert-Enskog [1] method to solve the Boltzmann equation.

For a discussion of this method of solution of the Boltzmann equation (1.1) it is necessary to consider the general hydrodynamical equations that can be derived from (1.1) by integration. These five equations hold between the five macroscopic quantities that characterize the gas:

1. local density

$$n(\vec{q}\,t) = \int f(\vec{q}\,\vec{v}\,t)\,d\vec{v} \tag{1.2a}$$

2. local average velocity

$$\vec{u}(\vec{q}\,t) = \int f(\vec{q}\,\vec{v}\,t)\,\vec{v}\,d\vec{v}/n(\vec{q}\,t) \tag{1.2b}$$

3. local average kinetic energy (local temperature)

$$\tfrac{3}{2}\,kT(\vec{q}\,t) = \int f(\vec{q}\,\vec{v}\,t)\,\tfrac{1}{2}\,mV^2(\vec{q}\,t)\,d\vec{v}/n(\vec{q}\,t)\,. \tag{1.2c}$$

Here $\vec{V}(\vec{q}\,t) = \vec{v}(\vec{q}\,t) - \vec{u}(\vec{q}\,t)$ is the *peculiar velocity* of a molecule, its velocity relative to the local average velocity $\vec{u}(\vec{q}\,t)$ at $\vec{q}$ and $t$; $V(\vec{q}\,t) = |\vec{V}(\vec{q}\,t)|$, $m$ is the mass of a molecule and $k$ is Boltzmann's constant.

The time- and space derivatives of these quantities are connected by the hydrodynamical equations that can be derived with (1.1) from the definitions (1.2) without solving (1.1) for $f(\vec{q}\,\vec{v}\,t)$. Multiplying the Boltzmann equation (1.1) with $1, m\vec{v}$ and $\tfrac{1}{2}\,mV^2$ respectively, integrating over $\vec{v}$ and using the conservation of the number of particles, momentum and energy during a collision, so that the right hand side of (1.1) does not contribute, the following *hydrodynamical equations* are obtained:

1. Equation of continuity:

$$\frac{Dn}{Dt} = -\,n\,\frac{\partial u_\alpha}{\partial q_\alpha}\,. \tag{1.3a}$$

2. Equation of motion:

$$n\,m\,\frac{Du_i}{Dt} = -\,\frac{\partial P^K_{i\alpha}}{\partial q_\alpha} \qquad (i = x, y, z)\,. \tag{1.3b}$$

3. Energy equation

$$\tfrac{3}{2}\, n\, k \frac{DT}{Dt} = - P^{K}_{\alpha\beta}\, D_{\alpha\beta} - \frac{\partial J^{K}_{\alpha}}{\partial q_{\alpha}}. \tag{1.3c}$$

Here the pressure tensor $P^{K}_{ij}(\vec{q}\ t)$ has been defined as:

$$P^{K}_{ij} = \int f\, m\, V_i\ V_j\ d\vec{v} \qquad (i,j = x, y, z) \tag{1.4a}$$

the heat flux vector $J^{K}_{i}(\vec{q}\, t)$ as:

$$J^{K}_{i} = \int f\, \tfrac{1}{2}\, m\, V^2\, V_i\ d\vec{v} \qquad (i = x, y, z) \tag{1.4b}$$

the rate of strain (deformation) tensor $D_{ij}(\vec{q}\ t)$ by:

$$D_{ij} = \tfrac{1}{2}\left(\frac{\partial u_i}{\partial q_j} + \frac{\partial u_j}{\partial q_i}\right) \qquad (i, j = x, y, z) \tag{1.5}$$

and the material fluxes by $D/Dt \equiv \partial/\partial t + u_{\alpha} \cdot \partial/\partial q_{\alpha}$. Greek characters, as always in these lectures, indicate that summation convention has been assumed so that e.g.

$$u_{\alpha}\frac{\partial}{\partial q_{\alpha}} = u_x \frac{\partial}{\partial x} + u_y \frac{\partial}{\partial y} + u_z \frac{\partial}{\partial z}$$

etc. The superscript K in $P^{K}_{ij}$ and $J^{K}_{i}$ indicates that only contributions due to the translational motion of the molecules (i.e. due to their kinetic energy) have been considered.

The presentation of the Hilbert-Enskog method of the solution of the Boltzmann equation that follows differs somewhat from the usual ones [2,3,4]. It is an adaptation of the presentation given by Uhlenbeck [5] in his Higgins Lectures, so as to stress the analogy with the Bogolubov method of solving the hierarchy in the kinetic stage, treated in the next section.

In the Hilbert-Enskog method for solving the Boltzmann equation, one looks for *special* solutions (the so-called normal solutions) of this equation where the time dependence of $f(\vec{q}\ \vec{v}\ t)$ is only via the five macroscopic quantities (1.2):

$$f(\vec{q}\ \vec{v}\ t) \rightarrow f(\vec{q}\ \vec{v}\,|\,n(\vec{q}\ T), \vec{u}(\vec{q}\ t), T(\vec{q}\ t)) \tag{$\alpha$}$$

where the vertical bar indicates that the $f$ is a functional of $n(\vec{q}\ t)$, $\vec{u}(\vec{q}\ t)$, $T(\vec{q}\ t)$ as far as its time dependence is concerned. At the present time we confine ourselves to an outline of the results obtained on the basis of this assumption. A discussion will be given in section 2 sub c. In order to find solutions $f$ of the Boltzmann equation of the form ($\alpha$), one expands $f$ in a series in a uniformity parameter $\mu$ which is proportional to the gradients in the system and vanishes therefore in equilibrium. The parameter $\mu$ only serves to classify terms according to their order of magnitude: in all final results $\mu$ will be set equal to 1. Thus:

$$f(\vec{q}\ \vec{v}\,|\,n\ \vec{u}\ T) = f_0(\vec{q}\ \vec{v}\,|\,n\ \vec{u}\ T) + \mu f_1(\vec{q}\ \vec{v}\,|\,n\ \vec{u}\ T) + \ldots . \tag{1.6}$$

Introducing (1.6) into the *right* hand side of equation (1.1) we obtain:

$$J(f\,f) = J(f_o\;f_o) + \mu[J(f_o\,f_1) + J(f_1\,f_o)] + \ldots\,. \tag{1.7}$$

Introducing (1.6) into the *left* hand side of equation (1.1), the second term can be expanded immediately in powers of $\mu$, all differentiations with respect to $\vec{q}$ being of the order $\mu$. The first term can then be expanded from:

$$\frac{\partial f(\vec{q}\;\vec{v}\,|n\;\vec{u}\;T)}{\partial t} = \frac{\partial f}{\partial n}\cdot\frac{\partial n}{\partial t} + \frac{\partial f}{\partial u_\alpha}\cdot\frac{\partial u_\alpha}{\partial t} + \frac{\partial f}{\partial T}\cdot\frac{\partial T}{\partial t} \tag{$\alpha'$}$$

in a power series in $\mu$ *only* if also the right hand sides of the hydrodynamical equations (1.3), from which $\partial n/\partial t$, $\partial u_i/\partial t$ and $\partial T/\partial t$ in ($\alpha'$) must be computed, are expanded in powers of $\mu$ *.

Introducing thereto (1.6) into the right hand sides of (1.4) and using then (1.4) in (1.3), one obtains:

$$\frac{\partial n}{\partial t} = \mu\left(\frac{\partial n}{\partial t}\right)_o + \ldots \quad \text{with} \quad \left(\frac{\partial n}{\partial t}\right)_o = -\frac{\partial(nu_\alpha)}{\partial q_\alpha} \tag{1.8a}$$

$$\frac{\partial u_i}{\partial t} = \mu\left(\frac{\partial u_i}{\partial t}\right)_o + \ldots \quad \text{with} \quad \left(\frac{\partial u_i}{\partial t}\right)_o = -u_\alpha\frac{\partial u_i}{\partial q_\alpha} - \frac{1}{n\,m}\frac{\partial P^{K}_{i\alpha,o}}{\partial q_\alpha} \tag{1.8b}$$

$$\frac{\partial T}{\partial t} = \mu\left(\frac{\partial T}{\partial t}\right)_o + \ldots \quad \text{with} \quad \left(\frac{\partial T}{\partial t}\right)_o = -u_\alpha\frac{\partial T}{\partial q_\alpha} - \frac{2}{3\,kn}\left[\frac{\partial J_{\alpha,o}}{\partial q_\alpha} + P^{K}_{\alpha\beta;o}D_{\alpha\beta}\right] \tag{1.8c}$$

In (1.8) the subscript 0 means that the quantity involved has been computed with $f_o$ instead of with $f$. The right hand sides of the equations (1.8) are all of the order $\mu$, because they all involve the gradients of the macroscopic quantities.

Using now ($\alpha'$) and (1.8), also $\partial f/\partial t$ can be expanded in a power series in $\mu$. Equating then the coefficients of equal powers of $\mu$ on both sides of the equation (1.1), because it must hold for all values of $\mu$, one obtains the following set of integral equations for the functions $f_o$, $f_1$, ...:

$$J(f_o\;f_o) = 0 \tag{1.9a}$$

$$J(f_o\;f_1) + J(f_1\;f_o) = \vec{v}\cdot\frac{\partial f_o}{\partial\vec{q}} + \frac{\partial f_o}{\partial n}\left(\frac{\partial n}{\partial t}\right)_o + \frac{\partial f_o}{\partial u_\alpha}\left(\frac{\partial u_\alpha}{\partial t}\right)_o + \frac{\partial f_o}{\partial T}\left(\frac{\partial T}{\partial t}\right)_o\,. \tag{1.9b}$$

The solution of the equation (1.9a) for $f_o$ of the form ($\alpha$) is a local Maxwell-Boltzmann distribution function:

$$f_o(\vec{q}\;\vec{v}\;t) = n(\vec{q}\,t)\,[m/2\pi kT(\vec{q}\,t)]^{\frac{3}{2}}\exp\left[-\,mV^2(\vec{q}\,t)/2kT(\vec{q}\,t)\right] \tag{1.10}$$

* Although one should use in ($\alpha'$) the more general notation as employed in ($\alpha$) in the next section, ordinary partial derivatives ($\partial/\partial n$, $\partial/\partial u_i$, $\partial/\partial T$) suffice for the purpose of this section.

where $n$, $\vec{u}$ and $T$ are the five local macroscopic quantities defined by equation (1.2). Therefore $n$, $\vec{u}$ and $T$ can be computed with $f_o$ as well as with $f$ and this implies that we can require from all higher approximations $f_l$ $(l \geq 1)$ that:

$$\int f_l \, d\vec{v} = 0 \qquad (l \geq 1) \qquad (1.11a)$$

$$\int f_l \, V_i \, d\vec{v} = 0 \quad (i = x, y, z) \qquad (l \geq 1) \qquad (1.11b)$$

$$\int f_l \, \tfrac{1}{2} m \, V^2 \, d\vec{v} = 0 \qquad (l \geq 1) \, . \qquad (1.11c)$$

The equation (1.9b) is an inhomogeneous integral equation for $f_1(\vec{q}\;\vec{v}|n\;\vec{u}\;T)$. Using (1.8) and (1.10) in the right hand side and setting:

$$f_1(\vec{q}\;\vec{v}|n\;\vec{u}\;T) = f_o(\vec{q}\;\vec{v}|n\;\vec{u}\;T)\;\Phi(\vec{q}\;\vec{v}|n\;\vec{u}\;T) \qquad (1.12)$$

the equation (1.9b) can be written in the form:

$$I(\Phi) = \int\int\int f_o(\vec{v})\, f_o(\vec{v}_1)\, [\Phi(\vec{v}') + \Phi(\vec{v}_1') - \Phi(\vec{v}) - \Phi(\vec{v}_1)]\; g\; b\; db\; d\varphi\; d\vec{v}_1 =$$

$$= f_o(\vec{v})\, [\frac{m}{kT}(V_\alpha\; V_\beta - \tfrac{1}{3}\,\delta_{\alpha\beta}\; V^2)\; D_{\alpha\beta} + (\frac{mV^2}{2\;kT} - \tfrac{5}{2})\; V_\alpha\; \frac{\partial \ln T}{\partial q_\alpha}] \qquad (1.13)$$

where we have only indicated explicitly the $\vec{v}$-dependence of the $f_o$- and $\Phi$-functions.

It can be proved that $I$ is a *linear* operator on $\Phi$, so that the integral equation (1.13) is a linear inhomogeneous integral equation for $\Phi$. This integral equation is soluble if the inhomogeneous part (i.e. the right hand side) is orthogonal to the solutions of the homogeneous equation. As the only solutions of the homogeneous equation (i.e. the left hand side of (1.13) = 0) for $\Phi$ are the five conserved quantities 1, $m\vec{V}$, $\frac{1}{2}\,mV^2$ and as one easily sees that the orthogonality conditions are then fulfilled, the equation (1.13) is soluble. The solutions of (1.13) are fixed apart from a linear combination of the five solutions of the homogeneous equations. The solution can be made definite just by requiring that it fulfills the five conditions (1.11) for $l = 1$. Having thus established the existence of a unique solution of the equation (1.9b), we shall now indicate its general form. As the left hand side of equation (1.13) is linear in $\Phi$ and the right hand side is linear in the gradients, $\Phi$ can be shown to be of the form:

$$\Phi(\vec{q}\;\vec{v}|n\;\vec{u}\;T) = -\,A_\alpha(\vec{V})\,\frac{\partial \ln T}{\partial q_\alpha} - B_{\alpha\beta}(\vec{V})\; D_{\alpha\beta} \qquad (1.14)$$

where the vector $A_i(\vec{V})$ and the tensor $B_{ij}(\vec{V})$ follow from the following linear integral equations:

$$I(A_i) = -\frac{1}{n}\,(\frac{mV^2}{2\;kT} - \tfrac{5}{2})\; V_i\; f_o(V) \qquad (1.15a)$$

$$I(B_{ij}) = -\frac{1}{n}\,\frac{m}{kT}\,(V_i\;V_j - \tfrac{1}{3}\,V^2\;\delta_{ij})\; f_o(V) \, . \qquad (1.15b)$$

From (1.14), (1.13), (1.12) and (1.10) two important properties of $f_1$ follow:

1. $f_1$ is proportional to the gradients;
2. $f_1$ is *in*dependent of the density $n$.

In order to determine the vector function $A_i(\vec{V})$ and the tensor function $B_{ij}(\vec{V})$ one proceeds as follows. It can be shown from the equations (1.15) and the special character (isotropy) of the operator $I$ that the $A_i$ and $B_{ij}$ must be of the general form:

$$\begin{aligned} A_i(\vec{V}) &= V_i \, A(V) \\ B_{ij}(\vec{V}) &= (V_i \, V_j - \tfrac{1}{3} V^2 \delta_{ij}) \, B(V) \, . \end{aligned} \tag{1.16}$$

Introduction of $f_1$ from (1.12), (1.14) and (1.16) into the conditions (1.11) for $l = 1$, only leads to a condition on the function $A(V)$:

$$\int f_o(V) \, A(V) \, V^2 \, d\vec{v} = 0 \tag{1.17}$$

because the $B_{ij}(V)$ automatically satisfy the equations (1.11).

The functions $A(V)$ and $B(V)$ can finally be determined by solving the equations (1.15). This can be done by introducing (1.16) into (1.15) and by expanding $A(V)$ and $B(V)$ in terms of a convenient set of orthogonal polynomials in $V$: the Sonine polynomials *. It turns out that already the first polynomial alone leads to very good results for the transport coefficients.

b. The hydrodynamical equations and the transport coefficients

With the solutions for $f_o$ and $f_1$ from the equations (1.10) and (1.14) respectively the hydrodynamical equations can be computed up to the order $\mu$ and $\mu^2$ respectively. This also implies that expressions for the pressure tensor $P^K_{ij}$ and for the heat flux vector $J^K_i$ are found up to the order of $\mu$.

1. With $f_o$ from (1.10) the *Euler-equations* are obtained:

$$\begin{aligned} \frac{Dn}{Dt} &= -n \frac{\partial u_\alpha}{\partial q_\alpha} \\ n \, m \frac{Du_i}{Dt} &= -\frac{\partial p^K}{\partial q_i} \\ \tfrac{3}{2} n \, k \frac{DT}{Dt} &= -p^K \frac{\partial u_\alpha}{\partial q_\alpha} \end{aligned} \tag{1.18}$$

where we have used from (1.4) that

$$P^K_{ij,0} = p^K \delta_{ij} = n \, kT \, \delta_{ij} \qquad (i, j = x, y, z) \tag{1.19}$$

* The Sonine polynomials are the eigenfunctions of the linear operator $I$, for the case that the intermolecular force (entering through the $\Phi(\vec{v}')$, $\Phi(\vec{v}_1')$) is an inverse fifth law of the intermolecular distance (Maxwell molecules).

and $$J^{K}_{i,0} = 0 \qquad (i = x, y, z) .$$

Here $p^K = n\,kT$ is the hydrostatic pressure for an ideal gas.
2. With $f_0 + f_1$ from (1.10), (1.12) and (1.14), one obtains the *Navier-Stokes equations:*

$$\begin{aligned} \frac{Dn}{Dt} &= -\,n\,\frac{\partial u_\alpha}{\partial q_\alpha} \\ n\,m\,\frac{Du_i}{Dt} &= -\,\frac{\partial p^K}{\partial q_i} - \frac{\partial P^K_{i\alpha,1}}{\partial q_\alpha} \\ \tfrac{3}{2}\,n\,k\,\frac{DT}{Dt} &= -\,(p^K\,\delta_{\alpha\beta} + P_{\alpha\beta,1})\,D_{\alpha\beta} - \frac{\partial J^K_{\alpha,1}}{\partial q_\alpha} \end{aligned} \tag{1.20}$$

where it can be shown from (1.4), (1.12), (1.14) and (1.19) that $P^K_{ij}$ is of the form:

$$P^K_{ij} = P^K_{ij,0} + P^K_{ij,1} = p^K\,\delta_{ij} - 2\,\eta\,(D_{ij} - \tfrac{1}{3}\,D_{\alpha\alpha}\,\delta_{ij})$$

and that $J^K_i$ is of the form:

$$J^K_i = J^K_{i,0} + J^K_{i,1} = -\,\lambda\,\frac{\partial T}{\partial q_i} \tag{1.21}$$

where the subscript 1 means that the quantity involved has been computed with $f_1$ and therefore is of the order $\mu$. The equations (1.21) express the proportionality of the off-diagonal elements of the pressure tensor $P^K_{ij}$ and of the $J^K_i$ with the gradients. This is a direct consequence of the proportionality of $f_1$ itself with the gradients. It will also be appreciated that the coefficients of proportionality between the fluxes $(P_{ij,1}\,;\,J_{i,1})$ and the forces $(D_{ij} - \tfrac{1}{3}\,D_{\alpha\alpha}\,\delta_{ij}\,;\,\partial T/\partial q_i)$ i.e. *the transportcoefficients* $\eta$ (viscosity) *and* $\lambda$ (heat conductivity), *are independent of the density* because $f_1$ itself is independent of the density.

Explicit expressions for $\eta$ and $\lambda$ are obtained by using the expressions (1.12), (1.14) and (1.16) for $f_1$. Using the fact that averages of odd functions of $\vec{V}$ vanish when evaluated with the distribution function $f_0(V)$, which is symmetrical in $\vec{V}$, one obtains for the viscosity $\eta$ the formula:

$$\eta(T) = \frac{m^2}{30\,kT}\int B(V)\,V^4 f_0(V)\,d\vec{v} \tag{1.22a}$$

and for the heat conductivity $\lambda$ the formula:

$$\lambda(T) = \frac{k}{3}\sqrt{\frac{m}{2\,kT}}\int A(V)\left(\frac{mV^2}{2\,kT} - \tfrac{5}{2}\right) V^2 f_0(V)\,d\vec{v}\,. \tag{1.22b}$$

Using one Sonine polynomial in the computation of $A(V)$ and $B(V)$ leads then to the following formulae for $\eta$ and $\lambda$ as a function of temperature:

$$\eta(T) = \frac{5\ kT}{8\ \Omega^{(2,2)}(T)} \tag{1.23a}$$

and

$$\lambda(T) = \frac{25\ kT}{16\ \Omega^{(2,2)}(T)}\ C_V = \tfrac{5}{2}\ C_V\ \mu(T) \tag{1.23b}$$

where $\Omega^{(2,2)}(T)$ is defined by:

$$\Omega^{(2,2)}(T) = \frac{\pi^{\frac{1}{2}}}{2}\left(\frac{m}{4\ kT}\right)^{\frac{7}{2}} \int_0^\infty \exp\left(-\ m\ g^2/4\ kT\right)\ g^7\ Q^{(2)}(g)\ \mathrm{d}g \tag{1.24a}$$

as a temperature average over the transport cross-section $Q^{(2)}(g)$ defined by:

$$Q^{(2)}(g) = \frac{2\ \pi}{g} \int_0^\infty (1\ -\ \cos^2\chi)\ g\ b\ \mathrm{d}b\ . \tag{1.24b}$$

Here $g$ is the absolute value of the relative velocity of two colliding molecules: $Q^{(2)}(g)$ the cross-section important for viscosity i.e. a weighted average (with $(1 - \cos^2\chi)$) of the differential cross-section $g\ b\ \mathrm{d}b\ \mathrm{d}\varphi$. The angle of deflection $\chi = \chi(b\ g)$ depends, apart from on $b$ and $g$, also on the intermolecular potential field. $\Omega^{(2,2)}(T)$ is a weighted (with $g^7$) temperature average over $Q^{(2)}(g)$ with the Maxwell-Boltzmann distribution function. $C_V$ is the specific heat per unit mass at constant volume, which is equal to $3\ k/2\ m$.

The formulae (1.23) for $\eta$ and $\lambda$, the viscosity and heat conductivity of a dilute monatomic gas as a function of temperature are in good agreement with experiment over a very wide range of temperature, if a realistic intermolecular potential field (f.i. a 12-6 Lennard-Jones potential field) is used [6].

This can be considered as an important check on the validity of the Boltzmann equation and the Hilbert-Enskog solution method for a description of the transport properties of a dilute gas.

This must conclude our discussion of the Boltzmann equation as a basis for the description of the properties of a dilute gas. In the next section we shall lay the foundations for the derivation of a generalized Boltzmann equation which can serve as a basis for the treatment of a moderately dense gas.

## 2. *The Bogolubov solution of the hierarchy in the kinetic stage*

### a. Introduction

In the previous section a discussion was given, based on the Boltzmann equation, of the non-equilibrium state of a dilute gas where only binary collisions between molecules are important. In particular expressions for the transport coefficients in such a gas were derived. If one wants to generalize this discussion to moderately dense gases where f.i. also triple collisions between molecules are of importance, the problem of generalizing the Boltzmann equation itself so as to include the effect of triple collisions presents itself. If one would succeed in doing so, then one would also be able to give the corrections due to triple- and higher order collisions to the expressions for the transport coefficients, which have been derived in the previous section from the Boltzmann equation under the hypothesis of binary collisions alone.

In this and in the next section we shall discuss a *systematic* generalization of the Boltzmann equation to higher densities due to Bogolubov starting from the Liouville equation. By an expansion in powers of the density we will obtain finally as a first approximation the Boltzmann equation (binary collision only) whereas in the higher approximations the contributions of higher order collisions will be systematically taken into account. The problem of the systematic generalization of the Boltzmann equation is therefore identical with the problem of the systematic derivation of the Boltzmann equation. That one is content to regain the Boltzmann equation in first approximation must be deemed reasonable on the ground of the good agreement that is obtained for the transport coefficients computed on the basis of this equation (see section 1) with experiment. The presentation we shall give of the Bogolubov theory is based essentially on the work of Uhlenbeck and Choh. The basic ideas, the theory of the kinetic stage and an outline of the theory of the hydrodynamical stage were given by Bogolubov in 1946 [7]. Important elucidations and comments were given by Uhlenbeck [5,8] in a number of lectures. A more or less complete theory, upon which these lectures are largely based, was presented in 1958 by Choh and Uhlenbeck [9].

### b. Liouville equation and hierarchy

The derivation and generalization of the Boltzmann equation given by Bogolubov is restricted to systems where the intermolecular potential is additive, central, monotonically *repulsive* and of finite range $\sigma$. No external force, except that due to the walls of the container, is assumed to be present.

We assume that the system of $N$ particles, which is not in equilibrium, is in a container of volume $V$ and is described at time $t$ by an ensemble in the 6 N-dimensional $\Gamma$-space with a probability density $D_N(x_1\, x_2 \,\ldots . x_N;\, t)$. Here $x_i = \vec{q}_i, \vec{p}_i$ $(i = 1, 2, \ldots, N)$ is the phase of the $i$-th particle at the position $\vec{q}_i$ with momentum $\vec{p}_i$, so that $D_N(x_1 \ldots x_N;\, t)$ is the probability density to find the system at time $t$ in a phase where the particle $i$ has the position $\vec{q}_i$ and the momentum $\vec{p}_i$ $(i = 1, \ldots, N)$. The $D_N$ is normalized to unity at every time $t$:

$$\int D_N(x_1 \ldots x_N;\, t)\, dx_1 \ldots dx_N = 1 \,. \tag{2.1}$$

The change of $D_N$ with time is given by *the Liouville equation:*

$$\frac{\partial D_N(x_1 \ldots x_N;\, t)}{\partial t} = \{H_N,\; D_N\} = \sum_{i=1}^{N} \left(\frac{\partial H_N}{\partial \vec{q}_i}\cdot\frac{\partial D_N}{\partial \vec{p}_i} - \frac{\partial H_N}{\partial \vec{p}_i}\cdot\frac{\partial D_N}{\partial \vec{q}_i}\right) \tag{2.2}$$

where the hamilton function $H_N$ of the $N$-particle system is given by:

$$H_N = \sum_{i=1}^{N} \frac{\vec{p}_i^{\,2}}{2\,m} + \sum_{\substack{i<j\\1}}^{N}\sum \varphi(r_{ij}) \tag{2.3}$$

with $\varphi(r_{ij})$ the above mentioned central monotomic repulsive intermolecular potential between the molecules $i$ and $j$ that are at a distance $r_{ij} = |\vec{r}_{ij}| = |\vec{q}_i - \vec{q}_j|$. The { } denote Poisson brackets. It is convenient to introduce instead of them an operator, which is defined for the case of $s$-particles as:

$$\mathcal{H}_s(x_1 \ldots x_s) = \sum_{i=1}^{s} \frac{\vec{p}_i}{m}\cdot\frac{\partial}{\partial \vec{q}_i} - \sum_{\substack{i<j\\1}}^{s}\sum \theta_{ij} \tag{2.4}$$

with
$$\theta_{ij} = \frac{\partial \varphi(r_{ij})}{\partial \vec{q}_i}\cdot\frac{\partial}{\partial \vec{p}_i} + \frac{\partial \varphi(r_{ij})}{\partial \vec{q}_j}\cdot\frac{\partial}{\partial \vec{p}_j}\,.$$

With the help of this operator the Liouville equation can also be written in the form:

$$\frac{\partial D_N(x_1 \ldots x_N;\, t)}{\partial t} = -\,\mathcal{H}_N\, D_N(x_1 \ldots x_N;\, t)\,. \tag{2.5}$$

Solution of the Liouville equation for a given initial condition $D_N(x_1 \ldots x_N;\, 0)$ at $t = 0$ would yield $D_N(x_1 \ldots x_N;\, t)$ for all $t$ from which all properties of the system at all $t$ could be computed. This would, however, imply a solution of the $N$-particle problem, characterized by the hamilton function (2.3). This is not feasible but fortunately also not necessary because the macroscopic properties of the system in which we are interested can all be expressed in terms of the one- and two-particle distribution functions. For a dilute gas already the one particle distribution function alone suffices, as can be seen from the previous section 1.

These lower distribution functions are defined in general by:

$$F_s(x_1 \ldots x_s;\, t) = V^s \int \ldots \int D_N(x_1 \ldots x_N;\, t)\, dx_{s+1} \ldots dx_N \tag{2.6}$$

where $F_s(x_1 \ldots x_s;\, t)$ is the probability density to find the $s$ particles $1 \ldots s$ in the phases $x_1 \ldots x_s$ respectively at time $t$ averaged over all possible phases of the other $(N - s)$ molecules. The single particle (or first-) distribution function $F_1(x_1;\, t)$ and the pair (or second-) distribution function $F_2(x_1\, x_2;\, t)$ follow from (2.6) for $s = 1$ and $s = 2$ respectively.

For the $F_s(x_1 \ldots x_s;\, t)$ one obtains from the Liouville equation by integration over the phases of $(N - s)$ molecules the following set of coupled equations:

It is assumed that $D_N$ is symmetrical under permutations of its arguments.

$$\frac{\partial F_s(x_1 \ldots x_s; t)}{\partial t} + \mathcal{H}_s F_s(x_1 \ldots x_s; t) =$$

$$= \frac{N-s}{V} \int dx_{s+1} \sum_{i=1}^{s} \theta_{i,s+1} F_{s+1}(x_1 \ldots x_{s+1}; t) \,. \quad (2.7)$$

This set of equations is often referred to as the B-B-G-K-Y hierarchy when it is named after a number of authors (Bogolubov, Born, Green, Kirkwood, Yvon) by which it was, independently, derived [7,10].

This hierarchy of equations can be derived in the following way. Multiplying the Liouville equation (2.5) with $V^s$ and integrating over the phases $x_{s+1} \ldots x_N$ one obtains:

$$\frac{\partial F_s}{\partial t} + \mathcal{H}_s F_s = V^s \int \ldots \int dx_{s+1} \ldots dx_N \Big[ - \sum_{i=s+1}^{N} \frac{\vec{p}_i}{m} \cdot \frac{\partial}{\partial \vec{q}_i} + \sum\sum_{k<l} \theta_{kl} + \sum_{i=1}^{s} \sum_{j=s+1}^{N} \theta_{ij} \Big] D_N(x_1 \ldots x_N; t) \,.$$

If we consider only molecules $1 \ldots s$ far from the walls of the container then the effect of the wall potential may be replaced by the assumption that $D_N(x_1 \ldots x_N; t)$ (and with it all $F_s(x_1 \ldots x_s; t)$; $s = 1, 2, \ldots, N$) vanish for large values of $|\vec{q}_i|$, so that

$$\int \ldots \int dx_{s+1} \ldots dx_N \frac{\vec{p}_i}{m} \cdot \frac{\partial}{\partial \vec{q}_i} D_N(x_1 \ldots x_N; t) = 0 \qquad (i \geq s+1)$$

and the first term on the right hand side of the equation for $\partial F_s / \partial t$ vanishes.

Furthermore, one also assumes that particles with infinite momentum do not occur. Therefore one has with (2.4):

$$\int \ldots \int dx_{s+1} \ldots dx_N \, \theta_{kl} \, D_N(x_1 \ldots x_N; t) = 0 \qquad (l > k \geq s+1)$$

so that the second term on the right hand side of the above equation also vanishes.

In the third term on the right hand side one has that for fixed $i$ all the $(N - s)$ terms $\theta_{ij}$ with $j = s+1, \ldots N$ give the same contribution, so that the equation (2.7) is obtained.

In the limit of an *infinite system* at a fixed density i.e. if we take $N \to \infty$, $V \to \infty$, $N/V = 1/v$ = constant, the hierarchy (2.7) reduces to:

$$\frac{\partial F_s}{\partial t} + \mathcal{H}_s F_s = \frac{1}{v} \int dx_{s+1} \sum_{i=1}^{s+1} \theta_{i,s+1} F_{s+1} \qquad (s = 1, 2, \ldots) \,. \quad (2.8)$$

We shall be especially interested in the equations for $F_1 (s = 1)$ and $F_2 (s = 2)$ which read:

$$\frac{\partial F_1(x_1; t)}{\partial t} + \frac{\vec{p}_1}{m} \cdot \frac{\partial F_1(x_1; t)}{\partial \vec{q}_1} = \frac{1}{v} \int dx_2 \, \theta_{12} \, F_2(x_1 x_2; t) \quad (2.9a)$$

$$\frac{\partial F_2(x_1 x_2; t)}{\partial t} + \mathcal{H}_2(x_1 x_2) F_2(x_1 x_2; t) = \frac{1}{v} \int dx_3 (\theta_{13} + \theta_{23}) F_3(x_1 x_2 x_3; t) \quad (2.9b)$$

where

$$\mathcal{H}_2(x_1\, x_2) = \frac{\vec{p}_1}{m}\cdot\frac{\partial}{\partial \vec{q}_1} + \frac{\vec{p}_2}{m}\cdot\frac{\partial}{\partial \vec{q}_2} - \theta_{12}\ .$$

The equations (2.8), (2.9) express the rate of change of $F_s$ as a sum of a streaming term (second term on the left hand side) and a "collision" term (right hand side). The streaming term gives the change due to the motion of the $s$-particles $1 \ldots s$ in their $6s$-dimensional phase space when they only interact with each other i.e. governed by the hamilton function $H_s$ of the $s$ particles. The "collision" term gives the change of $F_s$ due to the interaction of the $s$-particles $1 \ldots s$ with the $(N - s)$ other particles. It should be stressed that this term contains a factor $1/v$ in front. It will also be clear that the hierarchy (2.8) constitutes an infinite set of *coupled* equations, where $F_s$ is expressed in terms of $F_{s+1}$ etc. The Bogolubov method for solving the hierarchy (2.8), that is discussed in this section, makes use of the $1/v$ dependence of the right hand side of (2.8) in order to obtain a set of *uncoupled* equations (2.15) from which the $F_s$ can be found in successive approximation.

### c. Principle of Bogolubov method for solving hierarchy

Like Hilbert and Enskog in their solution of the Boltzmann equation, Bogolubov looks for special solutions of the hierarchy. In fact he seeks solutions of the hierarchy where the $F_s$ for $s \geq 2$ are functionals of the $F_1$ as far as their time dependence is concerned i.e. where the pair- and higher distribution functions depend on time only through the single particle distribution function $F_1$. In order to understand why one should want such solutions of the hierarchy as well as why one should look for the Hilbert-Enskog solution of the Boltzmann equation, Bogolubov argues as follows.

One can distinguish three relevant lengths in the system: the range of the intermolecular forces $\sigma$, the mean free path $l$ and the length of the vessel $L$. One can also distinguish three relevant times in the system if one divides these lengths by an average velocity of the molecules $\bar{v}$. One then obtains: the time of a collision $\tau_0$, the time between two successive collisions $t_0$ and the time to cross the vessel $T_0$. At an average velocity of $\bar{v} = 300$ m/sec, and at a pressure of 1 atm these lengths and times are widely spaced as can be seen from the table 1.

Bogolubov now assumes that coupled with these three widely spaced relaxation times there are three important *stages* in the temporal development of a not too dense gas for *a large class of initial conditions:*

1. *an initial stage* $(0 < t < \tau_0)$ which strongly depends on the precise initial condition at $t = 0$, but which after a time of the order of $\tau_0$ will go over into

Table 1
Relevant lengths and relaxation times for a dilute gas

| | $\sigma$ | $l$ | $L$ |
|---|---|---|---|
| cm | $10^{-8}$ | $10^{-5}$ | 1 |
| | $\tau_o$ | $t_o$ | $T_o$ |
| sec | $10^{-12}$ | $10^{-9}$ | $10^{-4}$ |

2. *a kinetic stage* ($\tau_o < t < t_o$) where, whatever was the initial condition, the $F_s\,(s \geq 2)$ have become functionals of $F_1$ as far as their time dependence is concerned. Again, after a time of the order of $t_o$, this stage will go over into
3. *a hydrodynamical stage* ($t_o < t < T_o$) where, whatever was the initial condition, the $F_1$ (and therefore all the $F_s\,(s \geq 1)$) depends only on its first five moments, the five macroscopic quantities $n$, $\vec{u}$, $T$ defined previously in section 1, equation (1.2).

The last assumption (3), is the same as that which lies at the basis of the Hilbert-Enskog method to solve the Boltzmann equation, the second assumption (2), is a generalization thereof. It will be appreciated that the above mentioned assumptions incorporate the possibility of a great simplification in the description of the system after the lapse of certain relaxation times. They not only make it possible to avoid the exact solution of the difficult initial value problem but also make such a solution superfluous.

Although the existence of the three stages for a large class of initial conditions has never been proved mathematically in any generality, one might still try to understand why, as time proceeds, fewer and fewer variables will be necessary to characterize the system (contraction of the description as Uhlenbeck calls it). One then perhaps best looks first to the assumption (3) concerning the existence of *the hydrodynamical stage*.

During a time of the order of $t_o$ a few collisions will occur during which the number of particles, their momentum and their energy will in first approximation be concerved *. This implies that $n$, $\vec{u}$ and $T$ will be almost constant during such a time and that therefore the other, much faster varying, quantities, in particular the $F_1$, will adjust themselves to them, i.e. will become functionally dependent upon them as to their time dependence. In this way one might visualize that after a time of the order of $t_o$ the $F_1$ (and all $F_s$) will have become functionals of $n$, $\vec{u}$ and $T$ as far as their time dependence is concerned. This argument finds some support in the work of Grad [11)] who showed that if the time development of $F_1$ is determined by thirteen moments instead of by the first five moments, the eight extra moments will have, after a time of the order of $t_o$, practically the same values as they would have had if they had been calculated via an $F_1$ determined by the first five moments only. This suggests that these eight extra moments are *irrelevant* for the time development of $F_1$ after a time of the order of $t_o$.

* Due to the presence of gradients they will change slightly over a few mean free paths, however.

In a similar way one might also try to understand the existence of *a kinetic stage* (assumption (2)). We shall restrict the discussion to a special class of initial conditions and to the pair distribution function of a spatially homogeneous not too dense gas where $F_1(x_1; t)$ does not depend on $\vec{q}_1$ but on $\vec{p}_1$ and $t$ alone. For more details and for a generalization to spatially inhomogeneous systems we refer to elsewhere [12,13,14]. At $t = 0$ a number of molecules is just colliding or will just have collided so that they can be called correlated. Suppose that the initial condition is such that molecules can be taken to be correlated at $t = 0$ only if they are within a distance of the order of the range of the intermolecular forces $\sigma$. If one now waits a time $t \gg \tau_o$, where $\tau_o$ is the duration of a collision, then all molecules that were correlated at $t = 0$ will have moved far apart due to the repulsive character of the intermolecular potential. Furthermore the molecules that collide between $t = 0$ and $t = t$ will have been far apart i.e. uncorrelated, at $t = 0$. Now at not too high densities the $F_2$ will in first approximation change through streaming alone, because the right hand side of the equation (2.9b) for $F_2$ (containing a factor $1/v$ in front) can then be neglected. Thus one has then in first approximation:

$$F_2(x_1\, x_2; t) = F_2(x_1^o\, x_2^o; 0)$$

where $x_1^o$, $x_2^o$ are the phases of the molecules 1 and 2 at $t = 0$. Therefore, if one is only interested in the $F_2(x_1\, x_2; t)$ *for intermolecular distances that are not too large*, then molecules 1 and 2 will have been far apart at $t = 0$ i.e. $r_{12}^o = |\vec{q}_1^{\,o} - \vec{q}_2^{\,o}| > \sigma$ so that they were at that time uncorrelated (see fig. 1).

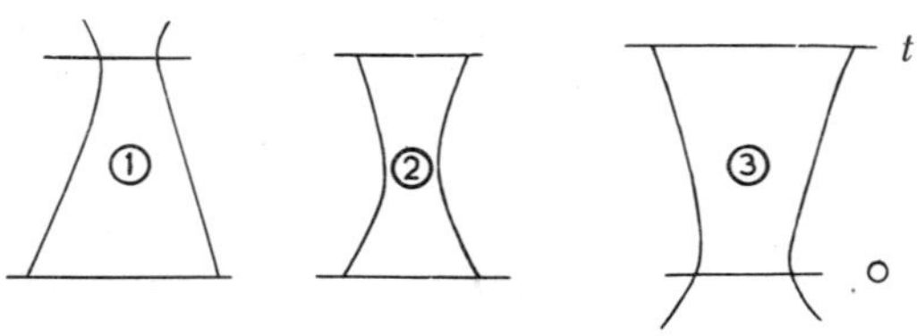


Fig. 1. 1, 2: Molecules that are not far apart at $t \gg \tau_o$ were far apart (uncorrelated) at $t = 0$.
3: Molecules that were not far apart (correlated) at $t = 0$, are far apart at $t \gg \tau_o$.

If one now assumes as an initial condition that the $F_2(x_1^o\, x_2^o; 0)$ at $t = 0$ factorizes into a product of $F_1$ as soon as $r_{12}^o > \sigma$, then one obtains for not too large intermolecular distances at time $t \gg \tau_o$:

$$F_2(x_1\, x_2; t) = F_2(x_1^o\, x_2^o; 0) = F_1(x_1^o; 0)\, F_1(x_2^o; 0)\ .$$

Now for a spatially homogeneous not too dense system one has, with equation (2.9a), *that $F_1$ does not change with time in first approximation*. This is so because one can neglect the right hand side, which contains a factor $1/v$ in front, while also $\partial F_1/\partial\vec{q}_1 = 0$ for such a system. Thus one then has in first approximation:

$$F_1(x_1^o; 0) = F_1(x_1; t)$$

so that one can write:

$$F_2(x_1\, x_2;\, t) = F_1(x_1^o;\, 0)\; F_1(x_2^o;\, 0) = F_1(x_1^o; t)\; F_1(x_2^o;\, t) = F_1(\vec{p}_1^{\,o};\, t)\; F_1(\vec{p}_2^{\,o};\, t)$$

where $\vec{p}_1^{\,o}$ and $\vec{p}_2^{\,o}$ are the time-independent asymptotic initial momenta at $t = 0$ that lead after collision to the $\vec{p}_1$, $\vec{p}_2$ at time $t$. From the last formula one sees that in first approximation for our initial condition and for not too large intermolecular distances at time $t >> \tau_o$ the $F_2$ depends on time only through the $F_1$ i.e. has become a functional of $F_1$ as far as its time dependence is concerned. One has to wait a time $t >> \tau_o$ because after such a lapse of time the *initial correlations* at $t = 0$ will have spread through the system and will have become *unimportant* at not too large intermolecular distances at time $t$. The restriction of the argument to not too large intermolecular distances is not serious because for a derivation and generalization of the Boltzmann equation one is only interested in such distances (see below).

We shall try to find a solution for $F_1$, $F_2$, ... from the hierarchy (2.8) using the above mentioned simplifying assumptions. This will be carried out in two steps, reflecting the existence of the two stages: the kinetic- and the hydrodynamical stage. First we shall try to find a solution for the $F_s$ $(s \geq 2)$ in the kinetic stage as functionals of $F_1$. Then we shall try to find, in the hydrodynamical stage, $F_1$ as a functional of $n$, $\vec{u}$ and $T$ from the hierarchy equation (2.9a) for $F_1$. In the course of this program we shall also, from this equation, derive and generalize the Boltzmann equation.

## d. Hierarchy for $F_s(|F_1)(s \geq 2)$

We are looking for special solutions of the hierarchy equations (2.9b) ... for $F_2$, $F_3$, ... where these distribution functions are functionals of $F_1$ as far as their time dependence is concerned i.e. solutions of the form:

$$F_s(x_1 \dots x_s;\, t) \rightarrow F_s(x_1 \dots x_s |\, F_1(;\, t)) \qquad (\beta)$$

where the vertical bar indicates the functional dependence upon $F_1$. In order to find $F_s$ of the form ($\beta$) one expands the $F_s(x_1 \dots x_s \,|\, F_1)$ in a series in powers of $1/v$:

$$F_s(x_1 \dots x_s\, |F_1) = F_s^{(0)}(x_1 \dots x_s\, |F_1) + \frac{1}{v}\, F_s^{(1)}(x_1 \dots x_s\, |F_1) + \dots \qquad (2.10)$$

The expansion parameter in (2.10) really is the dimensionless quantity $\sigma^3/v$ which reduces with $\sim 1/n\sigma^2$ to $\sigma/l$ or $\tau_o/t_o$ [8], which is a small quantity for a not toc dense gas (see table 1). This procedure is similar to that adopted in section 1, where in the case of the Hilbert-Enskog solution of the Boltzmann equation, a solution of a special form ($\alpha$) was found by expanding in terms of a parameter $\mu$ (equation (2.6)). However, where we were concerned in section 1 with solutions of the *Boltzmann equation* of the special form $f(|n\, \vec{u}\, T)$ with the help of the *hydrodynamical equations for* $n$, $\vec{u}$, $T$, we are concerned here with special solutions of the set of *hierarchy equations* for $F_s$ (with $s \geq 2$) of the form $F_s(|\, F_1)$, with the help of the equation for $F_1$.

We shall now use the expansion (2.10) in the equations (2.8b, c..) for the $F_s$ to obtain a set of equations for the functionals $F_s^{(0)}(|F_1)$, $F_s^{(1)}(|F_1)$ etc.

Introducing thereto (2.10) into the *right* hand side of the equations (2.8b, c..), we obtain immediately an expansion in powers of $1/v$.

Introducing (2.10) into the *left* hand side of the equations (2.8b, c..), the second term can be expanded immediately in powers of $1/v$. The first term can then be expanded from

$$\frac{\partial F_s(|F_1)}{\partial t} = \left[ \frac{\delta F_s(|F_1)}{\delta F_1} , \frac{\partial F_1}{\partial t} \right] \qquad (\beta')$$

in a power series in $1/v$, if also the right hand side of the equation (2.9a) for $\partial F_1/\partial t$ is expanded with (2.10) in powers of $1/v$. In ($\beta'$) the square brackets and the symbol $\delta/\delta F_1$ indicate that the derivative of $F_s(|F_1)$ with respect to time is obtained by adding together the derivatives of $F_s$ with respect to all functions $F_1$ that occur in the functional $F_s(|F_1)$, where each of these derivatives is multiplied by the corresponding derivative of $F_1$ with respect to time. It should be noted that these functions $F_1$ are all taken at the same instant of time $t$, but will depend on different variables $X = \vec{Q}, \vec{P}$. An example will be found in this section under g, equation (2.30a).

An expansion of the right hand side of the equation (2.9a) for $\partial F_1/\partial t$ in powers of $1/v$ can be obtained by introducing the expansion for $F_2$ from (2.10) for $s = 2$ into the right side of equation (2.9a):

$$\frac{\partial F_1}{\partial t} + \frac{\vec{p}_1}{m}\cdot\frac{\partial F_1}{\partial \vec{q}_1} = \frac{1}{v}\int dx_2\, \theta_{12}\left[F_2^{(0)}(x_1\, x_2|F_1) + \frac{1}{v}\, F_2^{(1)}(x_1\, x_2|F_1) + \dots\right] \qquad (2.11a)$$

Writing then for convenience:

$$\frac{\partial F_1(x_1;t)}{\partial t} = A(x_1|F_1) \qquad (2.11)$$

and expanding the functional $A(x_1|F_1)$ also in powers of $1/v$:

$$A(x_1|F_1) = A^{(0)}(x_1|F_1) + \frac{1}{v}A^{(1)}(x_1|F_1) + \frac{1}{v^2}A^{(2)}(x_1|F_1) + \dots \qquad (2.12)$$

one obtains, on comparing (2.11a) with (2.12):

$$A^{(0)}(x_1|F_1) = -\frac{\vec{p}_1}{m}\cdot\frac{\partial F_1}{\partial \vec{q}_1} \qquad (2.13a)$$

$$A^{(1)}(x_1|F_1) = \int dx_2\,\theta_{12}\; F_2^{(0)}(x_1\, x_2|F_1) \qquad (2.13b)$$

$$A^{(2)}(x_1|F_1) = \int dx_2\,\theta_{12}\; F_2^{(1)}(x_1\, x_2|F_1) \qquad (2.13c)$$

Therefore, using (2.11), one can replace ($\beta'$) by:

$$\frac{\partial F_s(|F_1)}{\partial t} = [\frac{\delta F_s(|F_1)}{\delta F_1}, A(|F_1)] \qquad (\beta'')$$

where the right hand side of ($\beta''$) can now be expanded in powers of $1/v$ by inserting the expansion (2.10) for $F_s(|F_1)$ as well as the expansion (2.12) for $A(|F_1)$.

The result of this expansion can be written in the form:

$$\frac{\partial F_s(|F_1)}{\partial t} = D^{(o)} F_s^{(o)}(|F_1) + \frac{1}{v}[D^{(o)} F_s^{(1)}(|F_1) + D^{(1)} F_s^{(o)}(|F_1)] + \ldots \qquad (2.14)$$

where the operators $D^{(l)}$ act on any functional $\chi(|F_1)$ of $F_1$ as:

$$D^{(l)} \chi(|F_1) = [\frac{\delta\chi(|F_1)}{\delta F_1}, A^{(l)}(|F_1)] \qquad (l \geq 0) \qquad (\beta''')$$

(2.14) gives finally the desired expansion of $\partial F_s/\partial t$ in powers of $1/v$, so that now both sides of the equations (2.8b, c..) have been expanded in powers of $1/v$. Equating then the coefficients of equal powers of $1/v$ on both sides of the equations, one obtains the following set of (uncoupled!) functional equations for the functionals $F_s^{(o)}(|F_1)$, $F_s^{(1)}(|F_1)$, ... :

$$D^{(o)} F_s^{(o)}(|F_1) + \mathcal{H}_s F_s^{(o)}(|F_1) = 0 \qquad (2.15a)$$

$$D^{(o)} F_s^{(1)}(|F\ ) + \mathcal{H}_s F_s^{(1)}(|F_1) = \psi_s^{(1)}(|F_1) \qquad (2.15b)$$

$$= - D^{(1)} F_s^{(o)}(|F_1) + \int dx_{s+1} \sum_{i=1}^{s} \theta_{i,s+1} F_{s+1}^{(o)}(|F_1) \qquad (2.15c)$$

- - - - - - - - - - - - - - - - - - - - - - - - - - - - - - - - - - - -

$$D^{(o)} F_s^{(l)}(|F_1) + \mathcal{H}_s F_s^{(l)}(|F_1) = \psi_s^{(l)}(|F_1) =$$

$$= - \sum_{k=1}^{l} D^{(k)} F_s^{(l-k)} + \int dx_{s+1} \sum \theta_{i,s+1} F_{s+1}^{(l-1)}. \qquad (2.15\,l)$$

The solution of the set of equations (2.15) can in principle be effected as follows. One first solves equation (2.15a) for $F_s^{(o)}(|F_1)$ ($s \geq 2$). Then in particular $F_2^{(o)}(|F_1)$ enables one to compute $A^{(1)}(|F_1)$ from (2.13b), so that the right hand side of equation (2.15b) i.e. $\psi_s^{(1)}(|F_1)$ is known. Then one solves equation (2.15b) for $F_s^{(1)}(|F_1)$ ($s \geq 2$) and determines from $F_2^{(1)}(|F_1)$ with equation (2.13c) $A^{(2)}(|F_1)$ etc. or schematically:

$F_2^{(o)}, F_3^{(o)}, \ldots, F_s^{(o)}, \ldots$

↓

$A^{(1)}$

↓

$F_2^{(1)}, F_3^{(1)}, \ldots, F_s^{(1)}, \ldots$

↓

$A^{(2)}$

- - - - - - - - - - - - - - - -

### e. Boundary conditions for $F_s(|F_1)$

It should be emphasized that the set of equations (2.15) are functional equations for the $F_s^{(l)}(|F_1)$ ($s \geq 2$; $l \geq 0$) as functionals of $F_1$, containing derivatives with respect to $F_1$. Therefore just as in the case of ordinary differential equations containing ordinary derivatives, one imposes some boundary or asymptotic conditions on the functional form of the $F_s(|F_1)$ or $F_s^{(l)}(|F_1)$ as functionals of $F_1$. These boundary conditions can be formulated conveniently with the help of *streaming operators* that will be introduced now.

The streaming operator $S^{(l)}(x_1 \ldots x_l)$ of $l$ particles over a time $\tau$ is defined by:

$$S_\tau^{(l)}(x_1 \ldots x_l) = \exp \tau\, \mathcal{H}_l(x_1 \ldots x_l)\ . \tag{2.16}$$

It is the time displacement operator over a time $\tau$ for $l$ particles, which transforms the coordinates and momenta of the $l$ particles $1 \ldots l$ to those a time $\tau$ later, if they move under the influence of their mutual interactions, characterized by the hamilton function $H_l$ (see (2.3) for $N = l$) only. For the $S$-operators defined by (2.16) one can prove the following properties

1. $$S_0^{(l)}(x_1 \ldots x_l) = 1 \tag{2.17a}$$

2. $$S_{\tau_1}^{(l)}(x_1 \ldots x_l)\ .\ S_{\tau_2}^{(l)}(x_1 \ldots x_l) = S_{\tau_1+\tau_2}^{(l)}(x_1 \ldots x_l) \tag{2.17b}$$

3. $$\frac{\partial S_\tau^{(l)}(x_1 \ldots x_l)}{\partial \tau} = \mathcal{H}_l(x_1 \ldots x_l)\, S_\tau^{(i)}(x_1 \ldots x_l) = S_\tau^{(l)}(x_1 \ldots x_l)\, \mathcal{H}_l(x_1 \ldots x_l) \tag{2.17c}$$

4. $$S_\tau^{(l)}(x_1 \ldots x_l)\, f(x_1 \ldots x_l; t) = f(S_\tau^{(l)} x_1 \ldots S_\tau^{(l)} x_l; t) \tag{2.17d}$$

for any function $f(x_1 \ldots x_l; t)$ .

In particular:

$$S_\tau^{(1)}(x_1)\, \vec{q}_1 = \vec{q}_1 + \frac{\vec{p}_1}{m}\tau \tag{2.17e1}$$

$$S_\tau^{(1)}(x_1)\, \vec{p}_1 = \vec{p}_1 \tag{2.17e2}$$

so that

$$S_\tau^{(1)}(x_1)\, f(\vec{q}_1\, \vec{p}_1) = f(S_\tau^{(1)}\vec{q}_1\ ,\ S_\tau^{(1)}\vec{p}_1) = f(\vec{q}_1 + \frac{\vec{p}_1}{m}\tau,\ \vec{p}_1)\ . \tag{2.17e3}$$

In terms of these $S$-operators one can formulate now the boundary condition which the functional $F_s(x_1 \ldots x_s|F_1)$ has to fulfill as functional of $F_1$, according to Bogolubov. For the *spatially homogeneous case* where $F_1$ does not depend on $\vec{q}_1$ one can formulate this condition as follows:

$$S_{-\infty}^{(s)}(x_1 \ldots x_s)\, F_s(x_1 \ldots x_s \mid F_1(;t)) = S_{-\infty}^{(s)}(x_1 \ldots x_s) \prod_{i=1}^{s} F_1(x_i; t) \tag{2.18a}$$

$$(s \geq 2)$$

whereas in the *spatially inhomogeneous case* the assumption reads:

$$S^{(s)}_{-\infty}(x_1 \ldots x_s)\, F_s(x_1 \ldots x_s \,|S^{(1)}_{\infty}\, F_1(;t)) =$$
$$= S^{(s)}_{-\infty}(x_1 \ldots x_s) \prod_{i=1}^{s} S^{(1)}_{\infty}(x_i)\, F_1(x_i;t) \,. \qquad (2.18b)$$

The condition (2.18a) is easiest to interprete. The operator $S^{(s)}_{-\infty}(x_1 \ldots x_s)$ will transform the phases $x_1 \ldots x_s$ of the molecules $1 \ldots s$, which are in the interior of the container, into new phases $S^{(s)}_{-\infty}\, x_1 \ldots S^{(s)}_{-\infty}\, x_s$, where the molecules are far apart, due to the repulsive intermolecular potential between them. Equation (2.18a) now assumes that for such large intermolecular distances the functional $F_s(|F_1)$ will factorize into a product of $F_1$. This should hold at every time $t$ and for arbitrary $F_1$. In particular, therefore, it should hold if one takes $S^{(1)}_{\infty}\, F_1$ for $F_1$, which then leads from the condition (2.18a) to the condition (2.18b). Physically the boundary conditions (2.18) express the absence of correlations at every time $t$ between particles that are far apart. They not only serve as a generalization of the Stosszahlansatz discussed in section 1, but they also ensure the proper ergodic behaviour of the hierarchy solutions which they determine, as will be discussed later (see section 3). For a more detailed discussion of the boundary condition (2.18) we refer to elsewhere [15].

Expanding, with (2.10), the left hand side of the equations (2.18) in powers of $1/v$, one obtains the following boundary conditions for the $F^{(l)}_s(|F_1)$ $(l = 0, 1, 2, \ldots; s \geq 2)$:

$$S^{(s)}_{-\infty}(x_1 \ldots x_s)\, F^{(o)}_s(x_1 \ldots x_s \,|\, F_1(;t)) = S^{(s)}_{-\infty}(x_1 \ldots x_s) \prod_{i=1}^{s} F_1(\vec{p}_i;t) \qquad (2.18a1)$$

and $$S^{(s)}_{-\infty}(x_1 \ldots x_s)\, F^{(l)}_s(x_1 \ldots x_s \,|F_1(;t)) = 0 \qquad (l \geq 1) \qquad (2.18a2)$$

in *the spatially homogeneous case* and

$$S^{(s)}_{-\infty}(x_1 \ldots x_s)\, F^{(o)}_s(x_1 \ldots x_s \,|S^{(1)}_{\infty}\, F_1(;t)) =$$
$$= S^{(s)}_{-\infty}(x_1 \ldots x_s) \prod_{i=1}^{s} S^{(1)}_{\infty}(x_i)\, F_1(x_i;t) \qquad (2.18b1)$$
$$= \mathcal{S}^{(s)}(x_1 \ldots x_s) \prod_{i=1}^{s} F_1(x_i;t)$$

and $$S^{(s)}_{-\infty}(x_1 \ldots x_s)\, F^{(l)}_s(x_1 \ldots x_s \,|\, S^{(1)}_{\infty}\, F_1(;t)) = 0 \quad (l \geq 1) \qquad (2.18b2)$$

in the *spatially inhomogeneous case*, where the operator

$$\mathcal{S}^{(s)}(x_1 \ldots x_s) = S^{(s)}_{-\infty}(x_1 \ldots x_s) \prod_{i=1}^{s} S^{(1)}_{\infty}(x_i) \,.$$

We are now in a position to solve the set of equations (2.15) with the boundary condition (2.18) successively for the $F_s^{(o)}(|F_1)$, $F_s^{(1)}(|F_1)$, ... etc. It should be remarked that it is at this moment not certain that solutions of the equations (2.15) can be found that fulfill the condition (2.18)!

### f. Solution for $F_s^{(o)}(|F_1)$

1. Spatially homogeneous case

The equation (2.15a) reduces in this case to:

$$\mathcal{H}_s(x_1 \ldots x_s)\, F_s^{(o)}(x_1 \ldots x_s\,|F_1) = 0 \tag{2.19}$$

because $D^{(o)}\, F_s^{(o)}(|F_1) = 0$ due to the fact that $A^{(o)}\,(|F_1) = 0$ in this case, according to equation (2.13a).

The equation (2.19) is satisfied by any functional that satisfies the equation:

$$F_s^{(o)}(x_1 \ldots x_s\,|F_1) = S_{-\tau}^{(s)}(x_1 \ldots x_s)\, F_s^{(o)}(x_1 \ldots x_s\,|F_1) \tag{2.20}$$

where $\tau$, as always in these lectures, is an arbitrary real number. In particular one can take $\tau \to \infty$ and then use the boundary condition (2.18a1). This leads to:

$$F_s^{(o)}(x_1 \ldots x_s\,|F_1) = S_{-\infty}^{(s)}(x_1 \ldots x_s)\, F_s^{(o)}(x_1 \ldots x_s\,|F_1)$$

$$= S_{-\infty}^{(s)}(x_1 \ldots x_s)\, F_1(\vec{p}_1;t) \ldots F_1(\vec{p}_s;t) \tag{2.21}$$

as a solution of the equation (2.19) which fulfills the boundary (2.18a1). For the special case of $s = 2$, we have:

$$F_2^{(o)}(x_1\, x_2|F_1) = S_{-\infty}^{(2)}(x_1\, x_2)\, F_1(\vec{p}_1;t)\, F_1(\vec{p}_2;t)\ . \tag{2.21a}$$

Introducing this into the equation (2.13b) for $A^{(1)}(x_1\,|F_1)$ gives:

$$A^{(1)}(x_1\,|F_1) = \int dx_2\, \theta_{12}\ S_{-\infty}^{(2)}(x_1\, x_2)\, F_1(\vec{p}_1;t)\, F_1(\vec{p}_2;t)\ . \tag{2.22}$$

It should be stressed that the operator $S_{-\infty}^{(2)}(x_1\, x_2)$ only depends on the momenta $\vec{p}_1$, $\vec{p}_2$ and the relative coordinates $\vec{r}_{21} = \vec{q}_2 - \vec{q}_1$ *but not on* $\vec{q}_1$, so that also $A^{(1)}$ only depends on $\vec{p}_1$ and not on $\vec{q}_1$.

2. Spatially inhomogeneous case

The equation (2.15a) can be solved through a trick by considering $F_s^{(o)}$ as a functional of $S_\tau^{(1)}\, F_1$ instead of $F_1$ i.e. by considering $F_s^{(o)}(|S_\tau^{(1)}\, F_1)$ instead of $F_s^{(o)}(|F_1)$. This enables one to replace the cumbersome equation (2.15a) for $F_s^{(o)}(x_1 \ldots x_s\,|F_1)$, where a derivative with respect to $F_1$ occurs (via the operator $D^{(o)}$), by an equation for $F_s^{(o)}(x_1 \ldots x_s\,|S_\tau^{(1)}\, F_1)$, where an *ordinary* derivative occurs. This equation can be derived as follows. Replacing $F_1$ by $S_\tau^{(1)}\, F_1$, one has from the definition ($\beta'''$) of $D^{(o)}$:

$$D^{(o)}\ F_s^{(o)}(x_1 \ldots x_s\,|S_\tau^{(1)}\, F_1) = \left[\ \frac{\delta F_s^{(o)}(|S_\tau^{(1)} F_1)}{\delta(S_\tau^{(1)} F_1)}\ ,\ A^{(o)}(|S_\tau^{(1)} F_1)\right]\ . \tag{2.23a}$$

Now, contrary to $A^{(o)}(|F_1)$, $A^{(o)}(|S_\tau^{(1)}F_1)$ can be transformed into a very convenient form and this is the reason that one considers $F_s^{(o)}(|S_\tau^{(1)}F_1)$ rather than $F_s^{(o)}(|F_1)$.

If the variable in anyone of the $A^{(o)}(|S_\tau^{(1)}F_1)$ occurring above is $X \equiv \vec{Q}, \vec{P}$, then one has according to equation (2.13a)

$$A^{(o)}(X|S_\tau^{(1)}F_1) = -\frac{\vec{P}}{m}\cdot\frac{\partial}{\partial\vec{Q}}S_\tau^{(1)}(X)\,F_1(X;t) = -\frac{\partial S_\tau^{(1)}(X)}{\partial\tau}F_1(X;t) = -\frac{\partial S_\tau^{(1)}(X)\,F_1(X;t)}{\partial\tau} \tag{2.23b}$$

because $F_1(x;t)$ is independent of $\tau$. Introducing this into (2.23a), yields with ($\beta'''$):

$$D^{(o)}F_s^{(o)}(x_1\ldots x_s|S_\tau^{(1)}F_1) = \left[\frac{\delta F_s^{(o)}(|S_\tau^{(1)}F_1)}{\delta(S_\tau^{(1)}F_1)},\ -\frac{\partial S_\tau^{(1)}F_1}{\partial\tau}\right] \tag{2.23c}$$

$$= -\frac{\partial F_s^{(o)}(|S_\tau^{(1)}F_1)}{\partial\tau}. \tag{2.23c}$$

Replacing $F_1$ in (2.15a) by $S_\tau^{(1)}F_1$ and using (2.23c) leads to the following functional equation for $F_s^{(o)}(|S_\tau^{(1)}F_1)$

$$\frac{\partial F_s^{(o)}(x_1\ldots x_s|S_\tau^{(1)}F_1)}{\partial\tau} - \mathcal{H}_s(x_1\ldots x_s)\,F_s^{(o)}(x_1\ldots x_s|S_\tau^{(1)}F_1) = 0 \tag{2.24}$$

where the derivative with respect to $F_1$ occurring in (2.15a) has been replaced by an ordinary derivative.

The equation (2.24) is satisfied by any functional that satisfies the equation:

$$F_s^{(o)}(x_1\ldots x_s|S_\tau^{(1)}F_1) = S_\tau^{(s)}(x_1\ldots x_s)\,F_s^{(o)}(x_1\ldots x_s|F_1) \tag{2.25}$$

as can easily be verified with the help of equation (2.17c).

Applying the operators $S_{-\tau}^{(s)}(x_1\ldots x_s)$ on both sides of the equation (2.25) and using (2.17b) yields:

$$F_s^{(o)}(x_1\ldots x_s|F_1) = S_{-\tau}^{(s)}(x_1\ldots x_s)\,F_s^{(o)}(x_1\ldots x_s|S_\tau^{(1)}F_1)\ . \tag{2.26}$$

As equation (2.26) holds for arbitrary $\tau$, one can take in particular $\tau\to\infty$ and then use the boundary condition (2.18b1). This leads to:

$$F_s^{(o)}(x_1\ldots x_s|F_1) = S_{-\infty}^{(s)}(x_1\ldots x_s)\,F_s^{(o)}(x_1\ldots x_s|S_\infty^{(1)}F_1) = \mathcal{S}^{(s)}(x_1\ldots x_s)\prod_{i=1}^{s}F_1(x_i;t)\quad (s\geq 2) \tag{2.27}$$

as a solution of the equation (2.15a) which fulfills the boundary condition (2.18b1).

For the special case of $s = 2$ we have:

$$F_2^{(o)}(x_1\, x_2|F_1) = \mathcal{S}^{(2)}(x_1\, x_2)\, F_1(x_1;t)\, F_1(x_2;t)\ . \tag{2.27a}$$

Introducing this into the equation (2.13b) for $A^{(1)}(x_1|F_1)$ gives:

$$A^{(1)}(x_1|F_1) = \int dx_2\ \theta_{12}\ \mathcal{S}^{(2)}(x_1\, x_2)\ F_1(\vec{q}_1\,\vec{p}_1;t)\ F_1(\vec{q}_2\,\vec{p}_2;t) \tag{2.28}$$

where now $A^{(1)}$ depends in general on $\vec{q}_1$ as well as on $\vec{p}_1$.

## g. Solution for $F_s^{(1)}(|F_1)$

We shall only derive here the solution for $F_2^{(1)}(x_1\, x_2|F_1)$ for *the spatially homogeneous case*. Results for the spatially inhomogeneous case as well as for $F_s^{(1)}(|F_1)$ with $s \geq 3$ will only be quoted. For details we refer to elsewhere [5,9,14]. For the spatially homogeneous case, the equation (2.15b) for $F_2^{(1)}(|F_1)$ reads:

$$\mathcal{H}_2(x_1\, x_2)\, F_2^{(1)}(x_1\, x_2|F_1) = \psi_2^{(1)}(x_1\, x_2|F_1) =$$

$$= -D^{(1)}\ F_2^{(o)}(x_1\, x_2|F_1) + \int dx_3\ (\theta_{13} + \theta_{23})\ F_3^{(o)}(x_1\, x_2\, x_3)\ . \tag{2.29}$$

With the help of the expressions for $F_2^{(o)}(|F_1)$, $F_3^{(o)}(|F_1)$ and $A^{(1)}(|F_1)$ given under f), the right hand side of this equation can be written down explicitly. First, using ($\beta'''$) and (2.22) we find:

$$\begin{aligned} D^{(1)}\ F_2^{(o)}(x_1\, x_2|F_1) &= D^{(1)}\ S_{-\infty}^{(2)}(x_1\, x_2)\ F_1(\vec{p}_1;t)\ F_1(\vec{p}_2;t) \\ &= S_{-\infty}^{(2)}(x_1\, x_2)\ F_1(\vec{p}_1;t) \\ &\quad \cdot \int dx_3\ \theta_{23}\ S_{-\infty}^{(2)}(x_2\, x_3)\ F_1(x_2;t)\ F_1(x_3;t) \\ &\quad + S_{-\infty}^{(2)}(x_1\, x_2)\ F_1(\vec{p}_2;t) \\ &\quad \cdot \int dx_3\ \theta_{13}\ S_{-\infty}^{(2)}(x_1\, x_3)\ F_1(x_1;t)\ F_1(x_3;t) = \\ &= S_{-\infty}^{(2)}(x_1\, x_2) \int dx_3\ [\theta_{13}\ S_{-\infty}^{(2)}(x_1\, x_3) + \\ &\quad + \theta_{23}\ S_{-\infty}^{(2)}(x_2\, x_3)]\ F_1(\vec{p}_1;t)\ F_1(\vec{p}_2;t)\ F_1(\vec{p}_3;t) \end{aligned} \tag{2.30a}$$

Then with (2.21) for $s = 3$ we have:

$$F_3^{(o)}(x_1\, x_2\, x_3|F_1) = S_{-\infty}^{(3)}(x_1\, x_2\, x_3)\ F_1(\vec{p}_1;t)\ F_1(\vec{p}_2;t)\ F_1(\vec{p}_3;t)\ . \tag{2.30b}$$

Thus, using (2.30) in (2.29) one obtains the following equation for $F_2^{(1)}(x_1\, x_2|F_1)$:

$$\mathcal{H}_2(x_1\, x_2)\ F_2^{(1)}(x_1\, x_2|F_1) = \psi_2^{(1)}(x_1\, x_2|F_1)$$

$$\begin{aligned} &= \int dx_3\ [(\theta_{13} + \theta_{23})\ S_{-\infty}^{(3)}(x_1\, x_2\, x_3) - S_{-\infty}^{(2)}(x_1\, x_2)\ \theta_{13}\ S_{-\infty}^{(2)}(x_1\, x_3) \\ &\quad - S_{-\infty}^{(2)}(x_1\, x_2)\ \theta_{23}\ S_{-\infty}^{(2)}(x_2\, x_3)]\ F_1(\vec{p}_1;t)\ F_1(\vec{p}_2;t)\ F_1(\vec{p}_3;t)\ . \end{aligned} \tag{2.31}$$

We now have to solve the functional equation (2.31) for $F_2^{(1)}(|F_1)$ with the boundary condition (2.18a2).

We remark that a *necessary* condition for the existence of such a solution is that:

$$S_{-\infty}^{(2)}(x_1\, x_2)\, \psi_2^{(1)}(x_1\, x_2 | F_1) = 0 \tag{2.32}$$

For, applying the operator $S_{-\infty}^{(2)}(x_1\, x_2)$ to both sides of the equation (2.31), the left hand side can be shown to vanish by using the commutability of $\mathcal{H}_2$ and $S_{-\infty}^{(2)}$ (see (2.16) and (2.17c)) and the boundary condition (2.18a2) for $l = 2$. This leads to the equation (2.32).

That the right hand side of equation (2.31) actually fulfills the condition (2.32) can be seen as follows.

Consider phases $x_1$ and $x_2$ where $r_{12}$, the distance between the particles 1 and 2, is much greater than the range of the intermolecular forces $\sigma$ i.e. where $r_{12} >> \sigma$. For such phases of the molecules 1 and 2, molecule 3 will have to be *either* close to molecule 1 (i.e. $r_{13} < \sigma$) *or* close to molecule 2 (i.e. $r_{23} < \sigma$) if the integrand on the right hand side of equation (2.31) is not to vanish. For if molecule 3 is far from both molecule 1 and molecule 2, the integrand vanishes due to the vanishing of both the $\Theta_{13}$ and $\Theta_{23}$ operators. If molecule 3 is close say to molecule 1, then $\Theta_{23} = 0$ and the operator $S_{-\infty}^{(3)}(x_1\, x_2\, x_3)$ will, in the large majority of cases *, factorize into:

$$S_{-\infty}^{(3)}(x_1\, x_2\, x_3) = S_{-\infty}^{(2)}(x_1\, x_3)\, S_{-\infty}^{(1)}(x_2) \qquad (r_{12} >> \sigma)\ .$$

But if such a factorization occurs, then one sees from equation (2.31) that $\psi_2^{(1)}(x_1\, x_2 | F_1)$ vanishes. A similar argument can be used to show that also if molecule 3 is close to molecule 2 and $\Theta_{13} = 0$, $\psi_2^{(1)}(x_1\, x_2 | F_1)$ vanishes. Therefore for all phases of molecules 1 and 2 for which $r_{12} >> \sigma$, $\psi_2^{(1)}(x_1\, x_2 | F_1)$ can be taken to vanish. A direct consequence of this is that in particular the condition (2.32) will be satisfied because the operator $S_{-\infty}^{(2)}(x_1\, x_2)$ will separate the particles 1 and 2 completely due to the assumed repulsive potential between them.

We now proceed to the solution of the equation (2.31). A solution can be written in the form:

$$F_2^{(1)}(x_1\, x_2 | F_1) = \int_0^\infty d\tau\ S_{-\tau}^{(2)}(x_1\, x_2)\, \psi_2^{(1)}(x_1\, x_2 | F_1) \tag{2.33}$$

For, applying the operator $\mathcal{H}_2(x_1\, x_2)$ on both sides of the equation, the right hand side can be shown to be equal to $\psi_2^{(1)}(x_1\, x_2 | F_1)$ by interchanging the $\mathcal{H}_2$ operator and the integration and by using (2.17c) and (2.32):

* This means the following. The operator $S_{-\infty}^{(3)}(x_1\, x_2\, x_3)$ transforms the phases $x_1\, x_2\, x_3$ of the molecules 1, 2 and 3 back into the indefinite past under their mutual interaction. It is now assumed that in the large majority of cases the molecule 2 remains far from the molecules 1 and 3 during this whole transformation, so that no collisions between molecule 2 and the molecules 1, 3 occur if they were originally far apart. This means that it is assumed that if the molecules 1 and 3 are originally far away from molecule 2, they will in the large majority of cases remain so.

$$\mathcal{H}_2(x_1\, x_2)\, F_2^{(1)}(x_1\, x_2|F_1) = \mathcal{H}_2(x_1\, x_2) \int_0^\infty \mathrm{d}\tau\; S_{-\tau}^{(2)}(x_1\, x_2)\, \psi_2^{(1)}(x_1\, x_2|F_1) =$$

$$= \int_0^\infty \mathrm{d}\tau\; \mathcal{H}_2(x_1 x_2)\, S_{-\tau}^{(2)}(x_1\, x_2)\, \psi_2^{(1)}(x_1\, x_2|F_1) =$$

$$= \int_0^\infty \mathrm{d}\tau \frac{\mathrm{d}S_{-\tau}^{(2)}(x_1\, x_2)}{\mathrm{d}\tau}\, \psi_2^{(1)}(x_1\, x_2|F_1) =$$

$$= S_{-\tau}^{(2)}\, \psi_2^{(1)}(x_1\, x_2|F_1)\Big|_0^\infty = \psi_2^{(1)}(x_1\, x_2|F_1)\,.$$

We shall now sketch briefly how the right hand side of the expression (2.33) for $F_2^{(1)}$ can be transformed into a different and more transparent form, where the $\tau$-integration has been carried out and where the $\theta$-operators that occur in the $\psi_2^{(1)}(|F_1)$ have disappeared. Using thereto the following identities:

$$\theta_{13} + \theta_{23} = \mathcal{H}_2(x_1\, x_2) - \mathcal{H}_3(x_1\, x_2\, x_3) + \frac{\vec{p}_3}{m}\cdot\frac{\partial}{\partial\vec{q}_3}$$

$$\theta_{13} = \mathcal{H}_2(x_1\, x_2) - \mathcal{H}_2(x_1\, x_3) - \frac{\vec{p}_2}{m}\cdot\frac{\partial}{\partial\vec{q}_2} + \frac{\vec{p}_3}{m}\cdot\frac{\partial}{\partial\vec{q}_3}$$

$$\theta_{23} = \mathcal{H}_2(x_1\, x_2) - \mathcal{H}_2(x_2\, x_3) - \frac{\vec{p}_1}{m}\cdot\frac{\partial}{\partial\vec{q}_1} + \frac{\vec{p}_3}{m}\cdot\frac{\partial}{\partial\vec{q}_3}$$

in $\psi_2^{(1)}(x_1\, x_2|F_1)$ as given by equation (2.31), equation (2.33) can be written in the form:

$$F_2^{(1)}(x_1\, x_2|F_1) = \int_0^\infty \mathrm{d}\tau\; S_{-\tau}^{(2)}(x_1\, x_2) \int \mathrm{d}x_3\, [\mathcal{H}_2(x_1\, x_2) + \frac{\vec{p}_3}{m}\cdot\frac{\partial}{\partial\vec{q}_3}]$$
$$.\, [S_{-\infty}^{(3)}(x_1\, x_2\, x_3) - S_{-\infty}^{(2)}(x_1\, x_2)\, S_{-\infty}^{(2)}(x_1\, x_3) - S_{-\infty}^{(2)}(x_1\, x_2)\, S_{-\infty}^{(2)}(x_2\, x_3)]\,.$$
$$.\, F_1(\vec{p}_1;\, t)\, F_1(\vec{p}_2;\, t)\, F_1(\vec{p}_3;\, t) \tag{2.34}$$

because $$\frac{\vec{p}_2}{m}\cdot\frac{\partial}{\partial\vec{q}_2} S_{-\infty}^{(2)}(x_1\, x_3)\, F_1(\vec{p}_1;\, t)\, F_1(\vec{p}_2;\, t)\, F_1(\vec{p}_3;\, t) = 0$$

$$\frac{\vec{p}_1}{m}\cdot\frac{\partial}{\partial\vec{q}_1} S_{-\infty}^{(2)}(x_2\, x_3)\, F_1(\vec{p}_1;\, t)\, F_1(\vec{p}_2;\, t)\, F_1(\vec{p}_3;\, t) = 0$$

and $$S_{-\infty}^{(2)}(x_1\, x_2)\, \theta_{12}\, F_1(\vec{p}_1;\, t)\, F_1(\vec{p}_2;\, t)\, F_1(\vec{p}_3;\, t) = 0$$

because $S_{-\infty}^{(2)}(x_1\, x_2)\, \theta_{12} = 0$.

As from equation (2.19) for $s = 2$ follows that

$$\mathcal{H}_2(x_1\, x_2)\, S_{-\infty}^{(2)}(x_1\, x_2)\, F_1(\vec{p}_1;\, t)\, F_1(\vec{p}_2;\, t)\, F_1(\vec{p}_3;\, t) = 0$$

and as also

$$\frac{\vec{p}_3}{m}\cdot\frac{\partial}{\partial\vec{q}_3} S_{-\infty}^{(2)}(x_1\, x_2)\, F_1(\vec{p}_1;\, t)\, F_1(\vec{p}_2;\, t)\, F_1(\vec{p}_3;\, t) = 0$$

one obtains from (2.34)

$$F_2^{(1)}(x_1\, x_2 | F_1) = \int_0^{\infty} d\tau\, S_{-\tau}^{(2)}(x_1\, x_2)\, \mathcal{H}_2(x_1\, x_2) \int dx_3\, [S_{-\infty}^{(3)}(x_1\, x_2\, x_3) - S_{-\infty}^{(2)}(x_1\, x_2)\, S_{-\infty}^{(2)}(x_1\, x_3) - S_{-\infty}^{(2)}(x_1\, x_2)\, S_{-\infty}^{(2)}(x_2\, x_3) + S_{-\infty}^{(2)}(x_1\, x_2)]\, F_1(\vec{p}_1; t)\, F_1(\vec{p}_2; t)\, F_1(\vec{p}_3; t) \tag{2.35}$$

because one can set

$$\int dx_3 \frac{\vec{p}_3}{m}\cdot\frac{\partial}{\partial \vec{q}_3} [S_{-\infty}^{(3)}(x_1\, x_2\, x_3) - S_{-\infty}^{(2)}(x_1\, x_2)\, S_{-\infty}^{(2)}(x_1\, x_3) - S_{-\infty}^{(2)}(x_1\, x_2)\, S_{-\infty}^{(2)}(x_2\, x_3) + S_{-\infty}^{(2)}(x_1\, x_2)]\, F_1(\vec{p}_1; t)\, F_1(\vec{p}_2; t)\, F_1(\vec{p}_3; t) = 0 \tag{2.36}$$

Using now (2.17c) in (2.35), the $\tau$-integration can be carried out and because

$$S_{-\infty}^{(2)}(x_1\, x_2) \int dx_3\, [S_{-\infty}^{(3)}(x_1\, x_2\, x_3) - S_{-\infty}^{(2)}(x_1\, x_2)\, S_{-\infty}^{(2)}(x_1\, x_3) - S_{-\infty}^{(2)}(x_1\, x_2)\, S_{-\infty}^{(2)}(x_2\, x_3) + S_{-\infty}^{(2)}(x_1\, x_2)]\, F_1(\vec{p}_1; t)\, F_1(\vec{p}_2; t)\, F_1(\vec{p}_3; t) = 0 \tag{2.37}$$

the expression (2.35) or (2.33) for $F_2^{(1)}(x_1\, x_2 | F_1)$ can be written in the form:

$$F_2^{(1)}(x_1\, x_2 | F_1) = \int dx_3\, [S_{-\infty}^{(3)}(x_1\, x_2\, x_3) - S_{-\infty}^{(2)}(x_1\, x_2)\, S_{-\infty}^{(2)}(x_1\, x_3) - S_{-\infty}^{(2)}(x_1\, x_2)\, S_{-\infty}^{(2)}(x_2\, x_3) + S_{-\infty}^{(2)}(x_1\, x_2)]\, F_1(\vec{p}_1; t)\, F_1(\vec{p}_2; t)\, F_1(\vec{p}_2; t) \tag{2.38}$$

This expression for $F_2^{(1)}(x_1\, x_2 | F_1)$ in the spatially homogeneous case has also been given by Green [13].

The proof of the properties (2.36) and (2.37) follows if one realizes that in the large majority of cases the integrand vanishes if particle 3 is far from both particles 1 and 2, or if $r_{12} > \sigma$. This is so because in those cases the $S_{-\infty}^{(3)}(x_1\, x_2\, x_3)$-operator factorizes into products of $S_{-\infty}^{(2)}$ and $S_{-\infty}^{(1)}$ operators which makes the integrand vanish by a similar argument as has been used in the justification of the equation (2.32) (see also foonote * there).

The equation (2.38) for $F_2^{(1)}(x_1\, x_2 | F_1)$ can be given a simple interpretation. The integrand on the right hand side of equation (2.38) vanishes *unless* the *three* particles 1, 2 and 3 are close together, i.e. unless with $r_{12} < \sigma$ at least either $r_{13} < \sigma$ or $r_{23} < \sigma$. Therefore, similarly to the third virial coefficient in the theory of the equation of state for a gas in equilibrium, $F_2^{(1)}(|F_1)$ gives the contribution of "triple collisions" ($S_{-\infty}^{(3)}(x_1\, x_2\, x_3)$) *in excess* over the contributions of successive binary collisions ($S_{-\infty}^{(2)}(x_1\, x_2)\, S_{-\infty}^{(2)}(x_1\, x_3)$) and ($S_{-\infty}^{(2)}(x_1\, x_2)\, S_{-\infty}^{(2)}(x_2\, x_3)$) which have been taken into account in $F_2^{(0)}(x_1\, x_2 | F_1)$ already. For a more detailed discussion we refer to elsewhere [12,13,14].

Introducing the expression (2.38) for $F_2^{(1)}(|F_1)$ into the right hand side of equation (2.13c) for $A^{(2)}(x_1 | F_1)$ one finds immediately:

$$A^{(2)}(\vec{p}_1| F_1) = \int dx_2\, \theta_{12} \int dx_3\, [S^{(3)}_{-\infty}(x_1\, x_2\, x_3) - S^{(2)}_{-\infty}(x_1\, x_2)\, S^{(2)}_{-\infty}(x_1\, x_3)$$

$$- S^{(2)}_{-\infty}(x_1\, x_2)\, S^{(2)}_{-\infty}(x_2\, x_3) + S^{(2)}_{-\infty}(x_1\, x_2)]\, F_1(\vec{p}_1; t)\, F_1(\vec{p}_2; t)\, F_1(\vec{p}_3; t) \quad (2.39)$$

where $A^{(2)}$ depends only on $\vec{p}_1$.

It can be shown [14] that the correct generalization of equation (2.38) to the *spatially inhomogeneous case* is given by an expression of the same form as (2.38) where the $S^{(s)}_{-\infty}$-operators are replaced everywhere by $\mathcal{S}^{(s)}$-operators:

$$F_2^{(1)}(x_1\, x_2 | F_1) = \int dx_3\, [\mathcal{S}^{(3)}(x_1\, x_2\, x_3) - \mathcal{S}^{(2)}(x_1\, x_2)\, \mathcal{S}^{(2)}(x_1\, x_3) -$$

$$- \mathcal{S}^{(2)}(x_1\, x_2)\, \mathcal{S}^{(2)}(x_2\, x_3) + \mathcal{S}^{(2)}(x_1\, x_2)]\, F_1(x_1; t)\, F_1(x_2; t)\, F_1(x_3; t) \quad (2.40)$$

Introducing the expression (2.40) into the right hand side of the equation (2.13c) for $A^{(2)}(x_1|F_1)$ leads to:

$$A^{(2)}(x_1 | F_1) = \int dx_2\, \theta_{12} \int dx_3\, [\mathcal{S}^{(3)}(x_1\, x_2\, x_3) - \mathcal{S}^{(2)}(x_1\, x_2)\, \mathcal{S}^{(2)}(x_1\, x_3)$$

$$- \mathcal{S}^{(2)}(x_1\, x_2)\, \mathcal{S}^{(2)}(x_2\, x_3) + \mathcal{S}^{(2)}(x_1\, x_2)]\, F_1(x_1; t)\, F_1(x_2; t)\, F_1(x_3; t) \quad (2.41)$$

for the spatially inhomogeneous case.

## h. Formulae for $F_s^{(l)}(|F_1)$

Before closing this section we shall quote the *general formulae* for the $F_s^{(l)}(|F_1)$, which are the solutions of the equations (2.15) with the boundary conditions (2.18a) in the spatially homogeneous- as well as in the spatially inhomogeneous case. For the *spatially homogeneous case* one has in general:

$$F_s^{(l)}(x_1 \ldots x_s | F_1) = \int_0^\infty d\tau\, S^{(s)}_{-\tau}(x_1 \ldots x_s)\, \psi_s^{(l)}(x_1 \ldots x_s | F_1) \quad (2.42)$$

$$(l \geq 1) \qquad (s \geq 2)$$

whereas the general expression for the *spatially inhomogeneous case* can be obtained from that of the spatially homogeneous case by replacing $\psi_s^{(l)}(|F_1)$ by $\psi_s^{(l)}(|S_\tau^{(1)}F_1)$

$$F_s^{(l)}(x_1 \ldots x_s | F_1) = \int_0^\infty d\tau\, S^{(s)}_{-\tau}(x_1 \ldots x_s)\, \psi_s^{(l)}(x_1 \ldots x_s | S_\tau^{(1)} F_1) \; . \quad (2.43)$$

It can be shown [14] that also the expressions (2.42) and (2.43) for the $F_s^{(l)}(x_1 \ldots x_s | F_1)$, that are generalizations of the equation (2.33) for $F_2^{(1)}(x_1\, x_2 | F_1)$, can be reduced to expressions of a similar structure as (2.38) and (2.40) respectively i.e. expressions where the $\tau$-integration has been carried out and where the $\theta$-operators do not occur anymore explicitly.

Thus we have accomplished the first part of our programme: the solution of the hierarchy in the kinetic stage. We have found explicit expressions for the $F_s$ (with $s \geq 2$) as functionals of $F_1$, from the equations (2.15) which also fulfill the boundary conditions (2.18). The $F_s$ were actually determined in successive approximation by an expansion in powers of $1/v$. An important check on these expressions for the $F_s$ is that upon introduction of the Maxwell-Boltzmann distribution for $F_1$, they reduce to the correct expansion of the $F_s$ in powers of the density $n$, known from equilibrium statistical mechanics [7,9,13,14].

In the following two sections we shall carry out the second part of our programme: the determination of $F_1$. In the next section, section 3, the formulae found for $A^{(1)}(x_1|F_1)$ and for $A^{(2)}(x_1|F_1)$ in this section, will be used to derive a selfcontained equation for $F_1$. This equation will appear to be identical with the Boltzmann equation to lowest order in *two* expansion parameters $\sigma^3/v$ and $\mu$. The next terms in the expansion with respect to these two parameters will also be given. They incorporate a.o. a generalization of the Boltzmann equation to higher densities.

In the last section, section 4, this generalized Boltzmann equation will be solved in the hydrodynamical stage by the same method as was used in section 1.

## 3. *The derivation and generalization of the Boltzmann equation*

### a. Introduction

We shall confine ourselves in these notes to the first two terms of a systematic generalization of the Boltzmann equation. Clearly by taking into account higher order terms in the expansions in powers of $1/v$ and $\mu$, higher order correction terms to the Boltzmann equation can be computed.

In the previous section, expressions have been derived for $A^{(o)}(x_1|F_1)$ (equation (2.13a)), $A^{(1)}(x_1|F_1)$ (equation (2.22)) and $A^{(2)}(x_1|F_1)$ (equation (2.41)). Introduction of these expressions into the right hand side of equation (2.12) yields the first three terms of the expansion of the functional $A(x_1|F_1)$ in powers of $1/v$. Using this expansion for $A(x_1|F_1)$ in equation (2.11a) for $\partial F_1/\partial t$ one obtains the following equation for $F_1$ in the *spatially inhomogeneous case*, valid up to terms of the order $1/v^2$:

$$\frac{\partial F_1(x_1;\,t)}{\partial t} - \frac{\vec{p}_1}{m}\cdot\frac{\partial F_1(x_1;\,t)}{\partial \vec{q}_1} = \frac{1}{v}A^{(1)}(x_1|F_1) + \frac{1}{v^2}A^{(2)}(x_1|F_1) \tag{3.1}$$

where the collision term

$$A^{(1)}(x_1|F_1) = \frac{1}{v}\int dx_2\, \theta_{12}\, F_2^{(o)}(x_1\, x_2|F_1) =$$

$$= \frac{1}{v}\int dx_2\, \theta_{12}\, \mathcal{S}^{(2)}(x_1\, x_2)\, F_1(x_1;\, t)\, F_1(x_2;\, t) \qquad (3.2a)$$

contains the contribution of binary collisions and the collision term

$$A^{(2)}(x_1|F_1) = \frac{1}{v^2}\int dx_2\, \theta_{12}\, F_2^{(1)}(x_1\, x_2|F_1) =$$

$$= \frac{1}{v^2}\int dx_2\, \theta_{12} \int dx_3\, [\mathcal{S}^{(3)}(x_1\, x_2\, x_3) - \mathcal{S}^{(2)}(x_1\, x_2)\, \mathcal{S}^{(2)}(x_1\, x_3) -$$

$$- \mathcal{S}^{(2)}(x_1\, x_2)\, \mathcal{S}^{(2)}(x_2\, x_3) + \mathcal{S}^{(2)}(x_1\, x_2)]\, .$$

$$F_1(x_1;\, t)\, F_1(x_2;\, t)\, F_1(x_3;\, t) \qquad (3.2b)$$

incorporates contributions of triple collisions.

In order to obtain from the equations (3.1), (3.2) the Boltzmann equation and its systematic generalization, one has to expand the right hand side of equation (3.1) or also $F_2^{(o)}(x_1\, x_2|F_1)$ and $F_2^{(1)}(x_1\, x_2|F_1)$ in powers of a uniformity parameter $\mu$ *in addition to* the expansion in powers of $1/v$ that has already been performed.

### b. Expansion of $F_2^{(o)}(x_1\, x_2|F_1)$ and $F_2^{(1)}(x_1\, x_2|F_1)$ in powers of $\mu$

#### b1. Expansion of $F_2^{(o)}(x_1\, x_2|F_1)$

We first expand $F_2^{(o)}(x_1\, x_2|F_1)$, given by equation (2.27a) in powers of a uniformity parameter $\mu$ around the position $\vec{q}_1$. This is possible because due to the presence in the $A^{(1)}$ and $A^{(2)}$ of the operator $\theta_{12}$, which vanishes if $r_{21} = |\vec{q}_2 - \vec{q}_1| > \sigma$, we are only interested in phases $x_1$ and $x_2$ where $r_{21} \leq \sigma$. Introducing then center of mass- and relative coordinates by the equations:

$$2\,\vec{R}_2 = \vec{q}_1 + \vec{q}_2\, ; \qquad \vec{r}_{21} = -\,\vec{r}_{12} = \vec{q}_2 - \vec{q}_1 \qquad (3.3)$$

one has:

$$\mathcal{S}^{(2)}\, \vec{q}_1 = \mathcal{S}^{(2)}(\vec{R}_2 - \frac{\vec{r}_{21}}{2}) = \vec{R}_2 - \mathcal{S}^{(2)}\, \frac{\vec{r}_{21}}{2} = \vec{q}_1 + \frac{\vec{r}_{21}}{2} - \tfrac{1}{2}\mathcal{S}^{(2)}\, \vec{r}_{21} \qquad (3.4a)$$

$$\mathcal{S}^{(2)}\, \vec{q}_2 = \mathcal{S}^{(2)}(\vec{R}_2 + \frac{\vec{r}_{21}}{2}) = \vec{R}_2 + \mathcal{S}^{(2)}\, \frac{\vec{r}_{21}}{2} = \vec{q}_1 + \frac{\vec{r}_{21}}{2} + \tfrac{1}{2}\mathcal{S}^{(2)}\, \vec{r}_{21} \qquad (3.4b)$$

$$\mathcal{S}^{(2)}(x_1\, x_2)\, \vec{p}_i = S_{-\infty}^{(2)}(x_1\, x_2)\, \vec{p}_i = \vec{P}_i^{(2)} \qquad (i = 1, 2) \qquad (3.4c)$$

because of (2.17e2). Here we have used that the $\mathcal{S}^{(2)}(x_1\, x_2)$-operator only operates on the relative coordinates and not on the center of mass coordinates.

Using (3.4) in (2.27a), one obtains the following expansion of $F_2^{(o)}(x_1\, x_2 | F_1)$ around $\vec{q}_1$:

$$F_2^{(o)}(x_1\, x_2| F_1) = F_1(\mathcal{S}^{(2)}\, \vec{q}_1,\ \mathcal{S}^{(2)}\, \vec{p}_1;\ t)\ F_1(\mathcal{S}^{(2)}\, \vec{q}_2,\ \mathcal{S}^{(2)}\, \vec{p}_2;\ t) =$$

$$= F_{2,0}^{(o)}(x_1\, x_2 | F_1) + \mu\, F_{2,1}^{(o)}(x_1\, x_2 | F_1) + \dots \tag{3.5}$$

with
$$F_{2,0}^{(o)}(x_1\, x_2 | F_1) = F_1(\vec{q}_1\, \vec{P}_1^{(2)};\ t)\ F_1(\vec{q}_1\, \vec{P}_2^{(2)};\ t) \tag{3.5a}$$

and
$$\begin{aligned} F_{2,1}^{(o)}(x_1\, x_2| F_1) = {} & \frac{\vec{r}_{21}}{2}\cdot\frac{\partial}{\partial \vec{q}_1}[F_1(\vec{q}_1\, \vec{P}_1^{(2)};\ t)\ F_1(\vec{q}_1\, \vec{P}_2^{(2)};\ t)] - \\ & - \frac{\mathcal{S}^{(2)}\, \vec{r}_{21}}{2}\cdot[\frac{\partial F_1(\vec{q}_1\, \vec{P}_1^{(2)};\ t)}{\partial \vec{q}_1}\, F_1(\vec{q}_1\, \vec{P}_2^{(2)};\ t) - \\ & - F_1(\vec{q}_1\, \vec{P}_1^{(2)};\ t)\, \frac{\partial F_1(\vec{q}_1\, \vec{P}_2^{(2)};\ t)}{\partial \vec{q}_1}] \end{aligned} \tag{3.5b}$$

where $\mathcal{S}^{(2)}\, \vec{r}_{21}$ is of the same order of magnitude as $\vec{r}_{21}$.

Strictly speaking, one can only say that $F_2^{(o)}(x_1\, x_2 | F_1)$ is a sum of terms that are proportional to the gradient of $F_1$ at $\vec{q}_1$. As we shall only be interested in the case that the $F_1$ depends on $\vec{q}_1$ through the macroscopic quantities $n$, $\vec{u}$ and $T$, one can also say, be it at this stage a little prematurely, that $F_{2,1}^{(o)}(| F_1)$ consists of terms which are proportional to the gradients of the macroscopic quantities i.e. which are of the first order in the uniformity parameter $\mu$.

Introducing the expansion (3.5) into the equation (3.2a) for $A^{(1)}(x_1 | F_1)$, leads immediately to an expansion of $A^{(1)}(x_1 | F_1)$ in powers of $\mu$:

$$A^{(1)}(x_1| F_1) = A_o^{(1)}(x_1| F_1) + \mu\, A_1^{(1)}(x_1| F_1) + \dots \tag{3.6}$$

where
$$\begin{aligned} A_o^{(1)}(x_1| F_1) &= \int dx_2\, \theta_{12}\, F_{2,0}^{(o)}(x_1\, x_2| F_1) = \\ &= \int dx_2\, \theta_{12}\, F_1(\vec{q}_1\, \vec{P}_1^{(2)};\ t)\ F_1(\vec{q}_1\, \vec{P}_2^{(2)};\ t) \end{aligned} \tag{3.6a}$$

and

$$\begin{aligned} A_1^{(1)}(x_1 | F_1) &= \int dx_2\, \theta_{12}\, F_{2,1}^{(o)}(x_1\, x_2| F_1) = \\ &= \int dx_2\, \theta_{12} \Big\{ \frac{\vec{r}_{21}}{2}\cdot\frac{\partial}{\partial \vec{q}_1}[F_1(\vec{q}_1\, \vec{P}_1^{(2)};\ t)\ F_1(\vec{q}_1\, \vec{P}_2^{(2)};\ t)] - \\ & \quad - \frac{\mathcal{S}^{(2)}\, \vec{r}_{21}}{2}\cdot[\frac{\partial F_1(\vec{q}_1\, \vec{P}_1^{(2)};\ t)}{\partial \vec{q}_1}\, F_1(\vec{q}_1\, \vec{P}_2^{(2)};\ t) - \\ & \quad - F_1(\vec{q}_1\, \vec{P}_1^{(2)};\ t)\, \frac{\partial F_1(\vec{q}_1\, \vec{P}_2^{(2)};\ t)}{\partial \vec{q}_1}] \Big\}. \end{aligned} \tag{3.6b}$$

b2. Expansion of $F_2^{(1)}(x_1\, x_2 \,| F_1)$

The expansion of $F_2^{(1)}(x_1\, x_2| F_1)$ from (2.38) in powers of $\mu$ can be performed in a similar way as that of $F_2^{(o)}(x_1\, x_2 | F_1)$, by introducing center of mass- and relative coordinates of the three particles 1, 2 and 3 [9]. As will become clear in the following section we only need the first term in this expansion. This term can be obtained from (2.38) by replacing $\mathcal{S}^{(3)}\, \vec{q}_i$ $(i = 1,2,3)$ as well as $\mathcal{S}^{(2)}\, \vec{q}_i$ $(i = 1,2)$ by $\vec{q}_1$, so that one can write:

$$F_2^{(1)}(x_1\, x_2 | F_1) = F_{2,0}^{(1)}(x_1\, x_2 | F_1) + \mu\, F_{2,1}^{(1)}(x_1\, x_2 | F_1) + \dots \tag{3.7}$$

where (with 3.4c):

$$F_{2,0}^{(1)}(x_1\, x_2 | F_1) = \int dx_3\, [S_{-\infty}^{(3)}(x_1\, x_2\, x_3) - S_{-\infty}^{(2)}(x_1\, x_2)\, S_{-\infty}^{(2)}(x_1\, x_3) - S_{-\infty}^{(2)}(x_1\, x_2)\, S_{-\infty}^{(2)}(x_2\, x_3) + S_{-\infty}^{(2)}(x_1\, x_2)]_p \cdot F_1(\vec{q}_1\, \vec{p}_1; t)\, F_1(\vec{q}_1\, \vec{p}_2; t)\, F_1(\vec{q}_1\, \vec{p}_3; t)\,. \tag{3.7a}$$

The subscript $p$ on the last bracket reminds of the fact that the $S_{-\infty}$-operators in (3.7a) *only operate on the momenta* $\vec{p}_i$ $(i = 1,2,3)$ that occur in the $F_1$ and *not* on the $\vec{q}_1$. Furthermore, the expression for $F_{2,1}^{(1)}(x_1\, x_2 | F_1)$ will consist of terms proportional to the gradients of $F_1$, which we shall not need here [9].

Introducing the expansion (3.7) for $F_2^{(1)}(x_1\, x_2 | F_1)$ into the equation (3.2b) for $A^{(2)}(x_1 | F_1)$, leads immediately to an expansion of $A^{(2)}(x_1 | F_1)$ in powers of $\mu$:

$$A^{(2)}(x_1 | F_1) = A_o^{(2)}(x_1 | F_1) + \mu\, A_1^{(2)}(x_1 | F_1) + \dots \tag{3.8}$$

where

$$A_o^{(2)}(x_1 | F_1) = \int dx_2\, \theta_{12} \int dx_3\, [S_{-\infty}^{(3)}(x_1\, x_2\, x_3) - S_{-\infty}^{(2)}(x_1\, x_2)\, S_{-\infty}^{(2)}(x_1\, x_3) - S_{-\infty}^{(2)}(x_1\, x_2)\, S_{-\infty}^{(2)}(x_2\, x_3) + S_{-\infty}^{(2)}(x_1\, x_2)]_p \cdot F_1(\vec{q}_1\, \vec{p}_1; t)\, F_1(\vec{q}_1\, \vec{p}_2; t)\, F_1(\vec{q}_1\, \vec{p}_3; t) \tag{3.8a}$$

and where $A_1^{(2)}(x_1 | F_1)$ follows from $F_{2,1}^{(1)}(x_1\, x_2 | F_1)$ and will not be given here.

We remark that the expressions (3.5a) and (3.7a) for $F_{2,0}^{(o)}(x_1\, x_2 | F_1)$ and $F_{2,0}^{(1)}(x_1\, x_2 | F_1)$ are the same as the expressions (2.21a) and (2.27a) for $F_2^{(o)}(x_1\, x_2 | F_1)$ and $F_2^{(1)}(x_1\, x_2 | F_1)$ for the spatially homogeneous case respectively, but for the fact that the $F_1$ in (3.5a) and (3.7a) depends on $\vec{q}_1$, $\vec{p}_1$ and $t$ instead of on $\vec{p}_1$ and $t$ alone, as in the spatially homogeneous case.

## c. Derivation and generalization of the Boltzmann equation

Introducing the $\mu$-expansions (3.6) for $A^{(1)}(x_1|F_1)$ and (3.8) for $A^{(2)}(x_1|F_1)$ into the equation (3.1) for $F_1$ leads to *the following equation for $F_1$ valid up to terms of the second order in $1/v$ and $\mu$, the two expansion parameters used:*

$$\frac{\partial F_1(x_1;\,t)}{\quad} = \mu\, A^{(o)}(x_1|F_1) + \frac{1}{v}\,[A_o^{(1)}(x_1|F_1) + \mu\, A_1^{(1)}(x_1|F_1)] + + \frac{1}{v^2} A_o^{(2)}(x_1|F_1) \tag{3.9}$$

where $A^{(o)}(x_1|F_1)$ follows from equation (2.13a), $A_o^{(1)}(x_1|F_1)$ and $A_1^{(1)}(x_1|F_1)$ follow from the equations (3.6) and $A_o^{(2)}(x_1|F_1)$ from equations (3.8). Written out in full the equation (3.9) reads:

$$\begin{aligned}\frac{\partial F_1(x_1;\,t)}{\partial t} + \frac{\vec p_1}{m}\cdot\frac{\partial F_1(x_1;\,t)}{\partial \vec q_1} = \frac{1}{v}\Big\{\int dx_2\; \theta_{12}\; F_1(\vec q_1\, \vec P_1^{(2)};\,t)\; F_1(\vec q_1\, \vec P_2^{(2)};\,t) + \\ + \int dx_2\; \theta_{12}\; [\frac{\vec r_{21}}{2}\cdot\frac{\partial}{\partial \vec q_1}(F_1(\vec q_1\, \vec P_1^{(2)};\,t)\; F_1(\vec q_1\, \vec P_2^{(2)};\,t)) - \\ - \frac{\mathcal{S}^{(2)}\,\vec r_{21}}{2}\cdot(\frac{\partial F_1(x_1\, \vec P_1^{(2)};\,t)}{\partial \vec q_1}\, F_1(\vec q_1\, \vec P_2^{(2)};\,t) - F_1(\vec q_1\, \vec P_1^{(2)};\,t)\,\frac{\partial F_1(\vec q_1\, \vec P_2^{(2)};\,t)}{\partial \vec q_1})]\Big\} \\ + \frac{1}{v^2}\int dx_2\; \theta_{12} \int dx_3\; [S_{-\infty}^{(3)}(x_1\, x_2\, x_3) - S_{-\infty}^{(2)}(x_1\, x_2)\, S_{-\infty}^{(2)}(x_1\, x_3) - \\ - S_{-\infty}^{(2)}(x_1\, x_2)\, S_{-\infty}^{(2)}(x_2\, x_3) + S_{-\infty}^{(2)}(x_1\, x_2)]_p\; F_1(\vec q_1\, \vec p_1;\,t)\; F_1(\vec q_1\, \vec p_2;\,t)\; F_1(\vec q_1\, \vec p_3;\,t)\end{aligned} \tag{3.9a}$$

$A^{(o)}(x_1|F_1)$ is of the order $\mu$ because it contains a derivative $\partial F_1/\partial \vec q_1$. We remark that no term of the order $\mu^2$ appears in the equation (3.9).

We shall now show that the equation (3.9) reduces to the Boltzmann equation (1.1) given in section 1, if one takes into account the first two terms on the right hand side only. Once this has been established, it will be clear that *the equation (3.9) constitutes a generalization of the Boltzmann equation not only to higher values of the density (triple collisions) but also to higher values of the gradients that exist in the system.*

*Derivation of the Boltzmann equation.*

To convert the equation

$$\frac{\partial F_1(x_1;\,t)}{\partial t} = A^{(o)}(x_1|F_1) + \frac{1}{v}\,A_o^{(1)}(x_1|F_1) \tag{3.10}$$

into the Boltzmann equation (1.1) we first change from $\Gamma$-space, where $F_1$ is defined, to $\mu$-space where the $f$ of section 1 is defined. We define a function $f$ by the relation:

$$f(\vec{q}\,\vec{p};\,t) = \frac{N}{V}F_1(\vec{q}\,\vec{p};\,t)\,. \tag{3.11}$$

Introducing now, with (3.11), $f$ instead of $F_1$ into (3.10) and using the equations (2.13a) and (3.6a) for $A^{(o)}$ and $A_o^{(1)}$ respectively, we obtain the following equation for $f$:

$$\frac{\partial f(\vec{q}_1\,\vec{p}_1;\,t)}{\partial t} + \frac{\vec{p}_1}{m}\cdot\frac{\partial f(\vec{q}_1\,\vec{p}_1;\,t)}{\partial \vec{q}} = \int d\vec{p}_2 \int d\vec{q}_2\, \theta_{12}\, f(\vec{q}_1\,\vec{P}_1^{(2)};\,t)\, f(\vec{q}_1\,\vec{P}_2^{(2)};\,t)\,. \tag{3.12}$$

The left hand side of this equation is already identical in form with the left hand side of the Boltzmann equation (1.1).

We shall now transform the right hand side of the equation (3.12) into the right hand side of the Boltzmann equation (1.1) i.e. into the familiar Boltzmann collision term. We first remark that the operator $\theta_{12}$ in the right hand side of (3.12) only operates on the momenta $\vec{p}_1$, $\vec{p}_2$ and not on the coordinate $\vec{q}_1$. This means that, as far as the $\theta_{12}$ is concerned, the $\vec{q}_1$ on the right hand side can be ignored and that the equation (2.19) of the spatially homogeneous case can be used to transform the right hand side. Changing then from $\vec{q}_1$ and $\vec{q}_2$ to $\vec{q}_1$ and $\vec{r}_{21} = \vec{q}_2 - \vec{q}_1$ as new coordinates, so that

$$\frac{\vec{p}_1}{m}\cdot\frac{\partial}{\partial\vec{q}_1} + \frac{\vec{p}_2}{m}\cdot\frac{\partial}{\partial\vec{q}_2} = \frac{\vec{p}_1}{m}\cdot\frac{\partial}{\partial\vec{q}_1} + \frac{\vec{p}_2 - \vec{p}_1}{m}\cdot\frac{\partial}{\partial\vec{r}_{21}}$$

and using the independence of the $\vec{P}_i^{(2)}$ from $\vec{q}_1$, one obtains:

$$\theta_{12}\, f(\vec{q}_1\,\vec{P}_1^{(2)};\,t)\, f(\vec{q}_1\,\vec{P}_2^{(2)};\,t) = \frac{\vec{p}_2 - \vec{p}_1}{m}\cdot\frac{\partial}{\partial\vec{r}_{21}} f(\vec{q}_1\,\vec{P}_1^{(2)};\,t)\, f(\vec{q}_1\,\vec{P}_2^{(2)};\,t)\,. \tag{3.13}$$

Introducing equation (3.13) into the right hand side of (3.12), changing the $\vec{q}_2$-integration to an $\vec{r}_{21}$-integration, taking for the $\vec{r}_{21}$-integration cylindrical coordinate with the axis in the direction of the relative velocity $\vec{g}_{21} = (\vec{p}_2 - \vec{p}_1)/m$, denoting the coordinate along this axis $l$ and the polar coordinates perpendicular to this axis $b$, $\varphi$, one can write the right hand side of equation (3.12) in the form:

$$\int d\vec{p}_2\, g \int db\, b \int d\varphi \int_{-\infty}^{+\infty} dl\, \frac{\partial}{\partial l}\left[F_1(\vec{q}_1\,\vec{P}_1^{(2)};\,t)\, F_1(\vec{q}_1\,\vec{P}_2^{(2)};\,t)\right] \tag{3.14}$$

where $g = |\vec{g}_{21}|$. The $l$-integration can be carried out immediately leading to:

$$\int_{-\infty}^{+\infty} dl\, \frac{\partial}{\partial l}\left[F_1(\vec{q}_1\,\vec{P}_1^{(2)};\,t)\, F_1(\vec{q}_1\,\vec{P}_2^{(2)};\,t)\right] = F_1(\vec{q}_1\,\vec{P}_1^{(2)};\,t)\, F_1(\vec{q}_1\,\vec{P}_2^{(2)};\,t)\Big|_{l=-\infty}^{l=+\infty} \tag{3.15}$$

For the evaluation of the upper limit ($l = +\infty$) we change the azimuth $\varphi$ to $\varphi + \pi$. Then we have to find the initial momenta $\vec{P}_1^{(2)}$, $\vec{P}_2^{(2)}$ of the molecules 1 and 2 at $\tau = -\infty$ that lead to the momenta $\vec{p}_1\,\vec{p}_2$ with

$\vec{r}_{21}$ $(l = +\infty,\ b,\ \varphi + \pi)$ at $\tau = 0$. These *initial* momenta are the momenta of the *restituting* collision $\vec{p}_1'$, $\vec{p}_2'$ at $\tau = 0$ *:

$$\lim_{l\to+\infty} \vec{P}_i^{(2)}(\vec{p}_1 \vec{p}_2 \vec{r}_{21}\ (l,\ b,\ \varphi+\pi)) = \vec{p}_i' \qquad (i = 1, 2) \qquad (3.16a)$$

Similarly for the lower limit, we have to find the initial momenta $\vec{P}_1^{(2)}$, $\vec{P}_2^{(2)}$ at $\tau = -\infty$ which lead to the momenta $\vec{p}_1$, $\vec{p}_2$ with $\vec{r}_{21}$ $(l = -\infty,\ b,\ \varphi)$ at $\tau = 0$. These initial momenta must be the $p_1$, $p_2$ themselves, because in this case the relative motion of the two molecules 1 and 2 is such that they will not have a chance to come inside their range of interaction. Thus:

$$\lim_{l\to-\infty} \vec{P}_i^{(2)}(\vec{p}_1 \vec{p}_2 \vec{r}_{21}\ (l,\ b,\ \varphi)) = \vec{p}_i \qquad (i = 1, 2) \qquad (3.16b)$$

Using (3.13) - (3.16) in the right hand side of the equation (3.12), one obtains:

$$\frac{\partial f(\vec{q}_1\ \vec{p}_1;\ t)}{\partial t} + \frac{\vec{p}_1}{m}\cdot\frac{\partial f}{\partial \vec{q}_1} = \int d\vec{p}_2\ g \int db\ b\ [f(\vec{q}_1\ \vec{p}_1';\ t)\ f(\vec{q}_1\ \vec{p}_2';\ t) - f(\vec{q}_1\ \vec{p}_1;\ t)\ f(\vec{q}_1\ \vec{p}_2;\ t)]\ . \qquad (3.17)$$

This equation reduces immediately to the familiar Boltzmann equation (1.1) in the spatially inhomogeneous case if we replace momenta $\vec{p}$ by velocities $\vec{v}$ everywhere, and $\vec{q}_1$, $\vec{v}_1$, $\vec{v}_2$ by $\vec{q}$, $\vec{v}$, $\vec{v}_1$ respectively.

### d. Boltzmann equation and irreversibility

We would now like to make a few remarks concerning the derivation of the Boltzmann equation, which we have just completed.

For that purpose we can restrict ourselves to the spatially homogeneous case, where the $f$ do not depend on $\vec{q}$, so that the $\mu$-expansion becomes superfluous. The Boltzmann equation reads in this case:

$$\frac{\partial f(\vec{v};\ t)}{\partial t} = \int d\vec{v}_1 \int db\ b \int d\varphi\ [f(\vec{v}';\ t)\ f(\vec{v}_1';\ t) - f(\vec{v};\ t)\ f(\vec{v}_1;\ t)\ . \qquad (3.18)$$

As is well-known, one can prove on the basis of this equation that the $H$-function

$$H = \int f(\vec{v};\ t)\ \ln f(\vec{v};\ t)\ d\vec{v} \qquad (3.19)$$

*never increases with time* i.e. that $\partial H/\partial t \le 0$, leading finally for any initial distribution function $f(\vec{v};\ t)$ to a stationary value of $H$ $(\partial H/\partial t = 0)$ for the Maxwell-Boltzmann equilibrium distribution function only.

* These momenta $\vec{p}_1'$, $\vec{p}_2'$ are also the final momenta of the collision with $\vec{p}_1$, $\vec{p}_2$ as initial momenta and with $\vec{r}_{21}$ $(l = -\infty,\ b,\ \varphi)$.

This *irreversible* behaviour of the Boltzmann equation (3.18) in $\mu$-space stands in sharp contrast to the *reversible* behaviour of the Liouville equation in $\Gamma$-space from which it has been derived.

As was pointed out already by Ehrenfest [16], the Liouville equation shows only an irreversible behaviour in a coarse-grained sense. An initial distribution that will tend to spread over the energy surface in $\Gamma$-space will only approach the homogeneous (equilibrium) distribution if one looks coarse-grained i.e. if one looks in small but finite elements of the energy surface. However, for any initial distribution moving *towards* equilibrium, there is also one moving away from equilibrium. One cannot say therefore, even if one looks in a coarse-grained sense, that the Liouville equation shows the same monotonic irreversible behaviour as the Boltzmann equation, where *any* initial distribution approaches monotonically to equilibrium. Thus in the derivation of the Boltzmann equation from the Liouville equation one will have to introduce, in addition to the Ehrenfest coarse-graining, an assumption that leads to the monotonic approach to equilibrium just mentioned.

In order to investigate this we shall now briefly review the essential steps in the derivation of the Boltzmann equation from the Liouville equation.

1. The B-B-G-K-Y hierarchy for an infinite system was derived from the Liouville equation.

2. The second and higher order distribution functions were all considered to be functionals of the first distribution function as far as their time dependence is concerned (assumption ($\beta$)).

3. A boundary condition was used for the asymptotic functional form of the second and higher order distribution functions in terms of the first distribution function (equations 2.18).

4. An expansion of the $F_2(|F_1)$ in powers of $1/v$ was used of which only the first term was kept.

Of these four assumptions only the second and the third seem to be relevant for the introduction of irreversibility. As to the first assumption, one expects the hierarchy to show still an essentially reversible behaviour, although the Poincaré cycles have become infinitely long, due to the infinite size of the system. Also the restriction to low density, as incorporated in the fourth assumption, seems to be of no consequence for the introduction of irreversibility.

The second assumption selects the $F_1$ as the secular variable and leads in principle to a self-contained equation for $F_1$, where $F_1$ is not anymore determined by $F_2$, $F_3$, ... as in the hierarchy (2.7) or (2.8). Now $F_1$ is defined as a multiple integral in $\Gamma$-space and will therefore be the same in finite regions of $\Gamma$-space. This means that looking for $F_1$ is already looking in a coarse-grained sense. Thus the second assumption seems to introduce the Ehrenfest coarse-graining, which is *not* sufficient, however, to ensure the monotonic approach to equilibrium as required by the H-theorem. In order to obtain this, the third assumption i.e. the boundary condition, is needed.

One can also say that the boundary conditions (2.18) select from all possible solutions of the hierarchy of the form $F_s(|F_1)$ those with the proper ergodic behaviour. This can be illustrated as follows [15].

One could have formulated a different, but equally plausible boundary condition for the spatially homogeneous case, by replacing $S_{-\infty}^{(s)}(x_1 \ldots x_s)$ in (2.18) by $S_{+\infty}^{(s)}(x_1 \ldots x_s)$:

$$S_{+\infty}^{(s)}(x_1 \ldots x_s)\, F_s(x_1 \ldots x_s\,|F_1(;t)) = S_{+\infty}^{(s)}(x_1 \ldots x_s) \prod_{i=1}^{s} F_1(x_i;t) \quad (3.20)$$

because also the operator $S_{+\infty}^{(s)}(x_1 \ldots x_s)$ will transform the phases $x_1 \ldots x_s$ into new phases where the molecules are far apart and not correlated.

With the boundary condition (3.20) instead of (2.18a) one derives the following expression for $F_2^{(0)}(x_1\, x_2|F_1)$ in the spatially homogeneous case:

$$F_2^{(0)}(x_1\, x_2|F_1) = S_{+\infty}^{(2)}(x_1\, x_2)\, F_1(\vec{p}_1;t)\, F_1(\vec{p}_2;t) \quad (3.21)$$

as can easily be seen from equation (2.20) by taking $\tau \to -\infty$ and using equation (3.20). Introducing (3.21) for $F_2^{(0)}(x_1\, x_2|F_1)$ into the equation (2.13b) for $A^{(1)}(x_1|F_1)$ and changing from $\Gamma$-space to $\mu$-space one obtains the equation:

$$\frac{\partial f(\vec{p}_1;t)}{\partial t} = \int d\vec{p}_2 \int d\vec{q}_2\, \theta_{12}\, S_{+\infty}^{(2)}(x_1\, x_2)\, f(\vec{p}_1;t)\, f(\vec{p}_2;t)\,. \quad (3.22)$$

Using the equations (3.13) - (3.15), the right hand side of the equation (3.22) can be shown to reduce to the Boltzmann collision term in the spatially homogeneous case *but for the sign*. Therefore the boundary condition (3.20) leads to a Boltzmann equation with the wrong sign before the collision term. This equation is only correct for the time reversed motion.

The reason for this wrong result is clearly the use of the boundary condition (3.20) instead of the boundary condition (2.18a) that was used before, because all other assumptions and approximations mentioned before are the same in both cases. Thus although the boundary condition (3.20) is a priori just as plausible as the boundary condition (2.18a) *it can not be used as a boundary condition for the equations (2.19)*.

This can perhaps be understood in the following way. The equations (2.19) in connection with which both boundary conditions are used, express that the functionals $F_s^{(0)}(|F_1)$ do not change through streaming of the particles $1 \ldots s$ in their $6s$-dimensional space. If one uses the boundary conditions (2.18a), the $F_s^{(0)}(|F_1)$ evolve, so to speak, according to equation (2.20) from a product of $F_1$ "in the past" at $\tau \to \infty$ (as given by the equations (2.18a) for uncorrelated particles) into $F_s^{(0)}(|F_1)$ that cannot in general be written anymore as a product of $F_1$. This is due to the fact that during the streaming the particles $1 \ldots s$ can in general come close enough together so that collisions take place and they become correlated. Thus if one would use, the boundary conditions (3.20) for uncorrelated particles, this would imply that these particles must have been correlated in general in a very special way before collision i.e. "in the past". Such a special correlation is presumably not true for an actual system, so that the boundary condition (3.20) should not be used in connection with the equations (2.19). For a more detailed account we refer to elsewhere [15].

This may conclude our discussion of the derivation of the Boltzmann equation. We shall now return to the generalized Boltzmann equation (3.9) and in the next section try to solve it formally for $F_1$ in the hydrodynamical stage. We shall particularly be interested in how far the results obtained from the generalized Boltzmann equation differ from those obtained on the basis of the ordinary Boltzmann equation treated in section 1.

## 4. *The solution of the generalized Boltzmann equation in the hydrodynamical stage*

### a. Introduction

In order to make the analogy with section 1 as close as possible, we shall first transform the equation (3.9) for $F_1(\vec{q}_1\, \vec{p}_1; t)$ in $\Gamma$-space to an equation for $f(\vec{q}\, \vec{v}; t)$ in $\mu$-space, using the relation (3.11) between $F_1$ and $f$.

For the first two terms on the right hand side of equation (3.9), which are of the first order in $1/v$ and $\mu$, this has already been done in the preceding section 3 sub c. Including also the terms of the second order in $1/v$ and $\mu$, replacing momenta $\vec{p}$ by velocities $\vec{v}$ and then $\vec{q}_1\, \vec{v}_1$ by $\vec{q}\, \vec{v}$ and $\vec{q}_2\, \vec{v}_2$ by $\vec{q}_1\, \vec{v}_1$, the following generalized Boltzmann equation for $f$ is obtained:

$$\frac{\partial f(\vec{q}\, \vec{v}; t)}{\partial t} + \vec{v}\cdot\frac{\partial f(\vec{q}\, \vec{v}; t)}{\partial \vec{q}} = J(ff) + J_1(ff) + K(fff)\ . \qquad (4.1)$$

Here $J(ff)$ is the Boltzmann collision term given in (1.1). $J(ff)$ has been derived from the equation (3.6a) for $A_0^{(1)}(x_1|\, F_1)$ in the previous section sub c. $J_1(ff)$ is a correction term due to spatial inhomogeneity and can be derived directly from the equation (3.6b) for $A_1^{(1)}(x_1\,|\, F_1)$ by replacing $F_1$ by $f$ with (3.11). $K(fff)$ is the correction term containing the contributions of triple collisions and can be derived from the equation (3.8a) for $A_0^{(2)}(x_1\,|\, F_1)$ in the same way.

In the equation (4.1) the explicit $1/v$ dependence has disappeared because of the relation (3.11) between $F_1$ and $f$. We remark that the second terms on both sides of the equation are of the first order in $\mu$.

## b. Formal solution of the generalized Boltzmann equation

In the following we shall briefly outline how the generalized Boltzmann equation (4.1) can be solved formally. For all details we refer to elsewhere [9]. The procedure is completely analogous to the Hilbert-Enskog method, used in section 1 for the Boltzmann equation (1.1). We look for special solutions of the equation (4.1), which are of the form (α):

$$f(\vec{q}\ \vec{v}\,;t) \to f(\vec{q}\ \vec{v}\,|\,n(\vec{q}\,;t)\,\vec{u}(\vec{q}\,;t)\,T(\vec{q}\,;t)) \tag{α}$$

where the $f$ depends on time only through the five macroscopic quantities $n$, $\vec{u}$ and $T$. This means that one looks for those special solutions of the equation (4.1) that should suffice, according to Bogolubov (see section 2), for a description of the system after the lapse of a time of the order of $t_o$, the time between two successive collisions i.e. for a description of the hydrodynamical stage. Again one needs the connection between time- and space derivatives of the macroscopic quantities $n$, $\vec{u}$ and $T$, which can be derived from the generalized Boltzmann equation (4.1), without solving it for $f$. Multiplying the equation (4.1) with $1, m\vec{v}$ and $\frac{1}{2}\,mV^2$ respectively and integrating over $\vec{v}$, the following hydrodynamical equations are obtained:

1. Equation of continuity:

$$\frac{Dn}{Dt} = -n\,\frac{\partial u_\alpha}{\partial q_\alpha} \tag{4.2a}$$

2. Equations of motion:

$$n\,m\,\frac{Du_i}{Dt} = -\,\frac{\partial P_{i\alpha}}{\partial q_\alpha} \qquad (i = x, y, z) \tag{4.2b}$$

3. Kinetic energy equation:

$$\tfrac{3}{2}\,n\,k\,\frac{DT}{Dt} = -\,P_{\alpha\beta}\,D_{\alpha\beta} - \frac{\partial J_\alpha^{\prime}}{\partial q_\alpha} - R \tag{4.2c}$$

where the pressure tensor $P_{ij}(\vec{q},\,t)$ has been defined as:

$$P_{ij} = P_{ij}^{K} + P_{ij}^{\varphi} \tag{4.3a}$$

with $P_{ij}^{K}$ from (1.4a) and $P_{ij}^{\varphi}$ from:

$$P_{ij}^{\varphi} = -\int d\vec{v}\int d\vec{v}_1\int d\vec{r}\,\frac{r_i}{2}\,\frac{\partial\varphi(r)}{\partial r_j}\,F_{2,0}^{(0)}\,(\vec{q}\ \vec{r}\ \vec{v}\ \vec{v}_1|f) \tag{4.3b}$$

the heat flux vector $J_i^{\prime}(\vec{q},t)$ as:

$$J_i^{\prime} = J_i^{K} + J_i^{\varphi 1} \tag{4.3c}$$

with $J_i^{K}$ from (1.4b) and $J_i^{\varphi 1}$ from:

$$J_i^{\varphi 1} = -\tfrac{1}{2}\int d\vec{v}\int d\vec{v}_1\int d\vec{r}\,(V_\alpha + V_{1\alpha})\frac{r_i}{2}\frac{\partial\varphi}{\partial r_\alpha}\,F_{2,0}^{(0)}(\vec{q}\,\vec{r}\,\vec{v}\,\vec{v}_1|f) \tag{4.3d}$$

and where $R$ is given by an expression similar to (4.3d), which shall not be given here.

The equations (4.2) are of the same general form as the equations (1.3). The pressure tensor $P_{ij}$ and the heat flux vector $J_i'$ contain, however, in addition to a contribution from the translational motion of the molecules, $P_{ij}^K$ and $J_i^K$, also contributions from the intermolecular potential energy $\varphi(r)$ of the molecules: $P_{ij}^\varphi$ and $J_i^{\varphi 1}$. The functional $F_{2,0}^{(0)}(\vec{q}\,\vec{r}\,\vec{v}\,\vec{v}_1|f)$ of $f$ is obtained from the functional $F_{2,0}^{(0)}(x_1\,x_2|F_1)$ given by equation (3.5a) of the previous section by replacing, with (3.11), $F_1$ by $f$ and by changing from momenta $\vec{p}_1$, $\vec{p}_2$ to velocities $\vec{v}$, $\vec{v}_1$ respectively and from the coordinates $\vec{q}_1$ and $\vec{q}_2$ to $\vec{q}$ and $\vec{q}_1$ respectively, so that $\vec{r}_{12} = \vec{q}_1 - \vec{q}_2$ changes into $\vec{r} = \vec{q} - \vec{q}_1$. The dependence of $F_{2,0}^{(0)}(\vec{q}\,\vec{r}\,\vec{v}\,\vec{v}_1|f)$ on $\vec{r}$ stems from the dependence of the $\vec{P}_i^{(2)} = S_{-\infty}^{(2)}(x_1\,x_2)\,\vec{p}_i$ $(i = 1,2)$ on $\vec{r}_{12}$.

In addition to the kinetic energy equation (4.2c) which is needed for the computation of the rate of change $\partial T/\partial t$, there exists now also a potential energy equation. This equation can be obtained from the second hierarchy equation (2.9b) for $F_2$ by multiplying with $\varphi(r_{12})$ and integrating over $\vec{p}_1$ and $\vec{x}_2$. After the usual transformations from $\Gamma$-space to $\mu$-space, one then obtains the following potential energy equation:

3') Potential energy equation:

$$n\frac{D\varepsilon^\varphi}{Dt} = -\frac{\partial J_\alpha^{\varphi 2}}{\partial q_\alpha} + R \tag{4.2d}$$

where the potential energy per molecule $\varepsilon^\varphi(\vec{q},t)$ is defined by:

$$n\varepsilon^\varphi(\vec{q}\;t) = \tfrac{1}{2}\int d\vec{v}\int d\vec{v}_1\int d\vec{r}\,\varphi(r)\,F_{2,0}^{(0)}(\vec{q}\,\vec{r}\,\vec{v}\,\vec{v}_1|f) \tag{4.4}$$

and the heat flux vector $J_i^{\varphi 2}(\vec{q},t)$ by:

$$J_i^{\varphi 2}(\vec{q},t) = \tfrac{1}{2}\int d\vec{v}\int d\vec{v}_1\int d\vec{r}\,\varphi(r)\,V_i\,F_{2,0}^{(0)}(\vec{q}\,\vec{r}\,\vec{v}\,\vec{v}_1|f) \tag{4.3e}$$

and where we have replaced in the right hand sides of the equations (4.4) and (4.3e) $F_2$ by $F_{2,0}^{(0)}$ in order to obtain the same degree of approximation as in the equations (4.3b) and (4.3d).

Adding the equations (4.2e) and (4.2d) yields the total energy equation:

4. Total energy equation:

$$n\frac{D\varepsilon}{Dt} = -P_{\alpha\beta}D_{\alpha\beta} - \frac{\partial J_\alpha}{\partial q_\alpha} \tag{4.2e}$$

where the total heat flux vector $J_i(\vec{q},t)$ has been defined by:

$$J_i = J_i^f + J_i^{\varphi 2} = J_i^K + J_i^{\varphi 1} + J_i^{\varphi 2} \tag{4.5}$$

from the equations (1.4b), (4.3c), (4.3d) and (4.3e).

The contributions $P_{ij}^{\varphi}$ and $J_i^{\varphi 1}$, $J_i^{\varphi 2}$ to the pressure tensor and the heat flux vector respectively all come from the correction term $J_1(ff)$ on the right hand side of the generalized Boltzmann equation (4.1), the $K(fff)$ only contributing to $R$. These contributions, which were called *collisional transfer* by Enskog [2)], arise from the transport of momentum or energy from one molecule to another upon collision through the intermolecular potential field. They are unimportant in dilute gases where, due to the rareness of collisions, only the transport of momentum and energy through the motion of the molecules from one point to another gives significant contributions $P_{ij}^K$ and $J_i^K$ to the pressure tensor and the heat flux vector.

We now try to find a solution of the equation (4.1) for $f$ of the form ($\alpha$) by expanding $f$ in a series in a uniformity parameter $\mu$, that is proportional to the gradients in the system. Thus we try:

$$f(\vec{q}\,\vec{v}\,|\,n\,\vec{u}\,T) = f_0(\vec{q}\,\vec{v}\,|\,n\,\vec{u}\,T) + \mu f_1(\vec{q}\,\vec{v}\,|\,n\,\vec{u}\,T) + \dots \tag{4.6}$$

The expansion parameter in (4.6) will appear to be the variation of the macroscopic quantities over a mean free path i.e. $l\nabla$ or $l/L = t_0/T_0$ (compare the expansion (2.10) in the kinetic stage) [2,5,8)].

Introducing (4.6) into the *right* hand side of equation (4.1) we obtain:

$$\begin{aligned} J(ff) + \mu\, J_1(ff) + K(fff) &= [J(f_0 f_0) + K(f_0 f_0 f_0)] + \\ &+ \mu\,[J(f_0 f_1) + J(f_1 f_0) + J_1(f_0 f_0) + K(f_0 f_0 f_1) + \dots + \dots] + \dots \end{aligned} \tag{4.7}$$

Introducing (4.6) into the *left* hand side of equation (4.1), the second term, which is already of the order $\mu$, can be expanded immediately in powers of $\mu$. The first term can then be expanded from:

$$\frac{\partial f(\vec{q}\,\vec{v}\,|\,n\,\vec{u}\,T)}{\partial t} = \left[\frac{\delta f}{\delta n}\frac{\partial n}{\partial t} + \frac{\delta f}{\delta u_\alpha}\frac{\partial u_\alpha}{\partial t} + \frac{\delta f}{\delta T}\frac{\partial T}{\partial t}\right] \tag{$\alpha'$}$$

in a power series in $\mu$ *only* if the right hand sides of the hydrodynamical equations (4.2a-c), from which $\partial n/\partial t$, $\partial \vec{u}/\partial t$ and $\partial T/\partial t$ in ($\alpha'$) must be computed, are expanded in powers of $\mu$. Unlike in section 1, we have used in ($\alpha'$) the more general notation, also employed in ($\beta'$) in section 2.

Introducing (4.6) into the right hand sides of the equations (4.3a-d) and using then (4.3a-d) in (4.2a-c), one obtains:

$$\frac{\partial n}{\partial t} = \mu \left(\frac{\partial n}{\partial t}\right)_o' + \ldots \quad \text{with} \quad \left(\frac{\partial n}{\partial t}\right)_o' = \left(\frac{\partial n}{\partial t}\right)_o = - \frac{\partial (n u_\alpha)}{\partial q_\alpha}$$

$$\frac{\partial u_i}{\partial t} = \mu \left(\frac{\partial u_i}{\partial t}\right)_o' + \ldots \quad \text{with} \quad \left(\frac{\partial u_i}{\partial t}\right)_o' = - u_\alpha \frac{\partial u_i}{\partial q_\alpha} - \frac{1}{n\,m} \frac{\partial P_{i\alpha,0}}{\partial q_\alpha}$$

$$\frac{\partial T}{\partial t} = \mu \left(\frac{\partial T}{\partial t}\right)_o' + \ldots \quad \text{with} \quad \left(\frac{\partial T}{\partial t}\right)_o' = - u_\alpha \frac{\partial T}{\partial q_\alpha} - \frac{2}{3\,kn} P_{\alpha\beta,0}\, D_{\alpha\beta} - \tag{4.8}$$

$$- \frac{2}{3\,kn} \frac{\partial J'_{\alpha,0}}{\partial q_\alpha} + \frac{2}{3\,kn} R_1 \;.$$

These equations are direct generalizations of the equations (1.8), the main difference being the replacement of $P_{ij}^K$ and $J_i^K$ by $P_{ij}$ and $J_i'$ respectively. The $R_1$ term results from the expansion of $R$. It contains $f_1$ in addition to $f_o$, but we shall not write it down here.

Using now ($\alpha'$) and (4.8), also $\partial f/\partial t$ can be expanded in a power series in $\mu$. Equating then the coefficients of equal powers of $\mu$ on both sides of the equation (4.1), because it must hold for all values of $\mu$, one obtains the following set of integral equations for $f_o, f_1, \ldots$

$$J(f_o\, f_o) + K(f_o\, f_o\, f_o) = 0 \tag{4.9a}$$

$$J(f_o\, f_1) + J(f_1\, f_o) + K(f_o\, f_o\, f_1) + \ldots + \ldots = \vec{v}\cdot\frac{\partial f_o}{\partial \vec{q}} - J_1(f_o\, f_o) +$$

$$+ \frac{\partial f_o}{\partial n}\left(\frac{\partial n}{\partial t}\right)_o' + \frac{\partial f_o}{\partial u_\alpha}\left(\frac{\partial u_\alpha}{\partial t}\right)_o' + \frac{\partial f_o}{\partial T}\left(\frac{\partial T}{\partial t}\right)_o' \;. \tag{4.9b}$$

---

One can again show that the local Maxwell-Boltzmann distribution (1.10) is a solution of the equation (4.9a) for $f_o$. This implies that one can require again the conditions (1.11) for the $f_l$ ($l \geq 1$).

The second equation (4.9b) is an inhomogeneous integral equation for $f_1$. It can be solved by expanding $f_1$ into a power series in the local density $n$ i.e. by setting:

$$f_1(\vec{v}) = f_1^{(0)}(\vec{v}) + n\, f_1^{(1)}(\vec{v}) + \ldots \;. \tag{4.10}$$

Introducing (4.10) for $f_1$ into equation (4.9b), expanding both sides of the equation in powers of $n$ with the help of the equations (4.8) and equating the coefficients of equal powers of $n$ on both sides of the equation leads to separate equations for $f_1^{(0)}$ and $f_1^{(1)}$. Setting then

$$f_1^{(0)}(\vec{v}) = f_o(\vec{v})\; \Phi_1^{(0)}(\vec{v}) \tag{4.11}$$

the following linear inhomogeneous integral equation for $\Phi_1^{(0)}(\vec{v})$ is obtained:

$$I(\Phi_1^{(0)}) = \int\int f_o(\vec{v}) f_o(\vec{v}_1) [\Phi_1^{(0)}(\vec{v}') + \Phi_1^{(0)}(\vec{v}_1') - \Phi_1^{(0)}(\vec{v}) - \Phi_1^{(0)}(\vec{v}_1)] \, g \, b \, db \, d\vec{v}_1$$

$$= f_o(\vec{v}) [\frac{m}{kT} (V_\alpha V_\beta - \tfrac{1}{3} \delta_{\alpha\beta} V^2) D_{\alpha\beta} + (\frac{mV^2}{2kT} - \tfrac{5}{2}) V_\alpha \frac{\partial \ln T}{\partial q_\alpha} \quad (4.12)$$

which is identical with the equation (1.13) for $\Phi(\vec{v})$ obtained in the Hilbert-Enskog theory of the Boltzmann equation treated in section 1. As there, the right hand side of this equation contains contributions from $\vec{v}.(\partial f_o/\partial\vec{q})$ and from the kinetic part of the hydrodynamical equations ($P_{ij}^K$, $J_i^K$).

Similarly, setting:

$$f_1^{(1)}(\vec{v}) = f_o(\vec{v}) \, \Phi_1^{(1)}(\vec{v}) \quad (4.13)$$

one obtains a linear inhomogeneous integral equation for $\Phi_1^{(1)}(\vec{v})$ of the form:

$$I(\Phi_1^{(1)}) = \int\int f_o(\vec{v}) f_o(\vec{v}_1) [\Phi_1^{(1)}(\vec{v}') + \Phi_1^{(1)}(\vec{v}_1') - \Phi_1^{(1)}(\vec{v}) - \Phi_1^{(1)}(\vec{v}_1)] \, g \, b \, db \, d\vec{v}_1$$

$$= f_o(\vec{v}) [M_{\alpha\beta}(\vec{V}) D_{\alpha\beta} + L_\alpha(\vec{V}) \frac{\partial \ln T}{\partial q_\alpha}] \quad (4.14)$$

where $M_{ij}(\vec{V})$ is of the form:

$$M_{ij}(\vec{V}) = M_1(\vec{V}) \frac{m}{kT} (V_i V_j - \tfrac{1}{3} V^2 \delta_{ij}) + M_2(V) \, \delta_{ij} \quad (4.15a)$$

and $L_i(\vec{V})$ of the form:

$$L_i(\vec{V}) = L(V) \, V_i \, . \quad (4.14b)$$

The functions $M_1(V)$, $M_2(V)$ and $L(V)$ contain contributions from triple collisions (from $K$) as well as from collisional transfer (from $J_1$). They will not be given here. The linear inhomogeneous integral equations (4.12) and (4.14) *both* contain the same linear operator $I$ which involves the *binary* collision cross-section only. They are both soluble if the inhomogeneous part is orthogonal to the five solutions of the homogeneous equation 1, $m\vec{V}$, $\frac{1}{2} mV^2$. As this can easily be established, the equations must have solutions. These are fixed apart from a linear combination of the five solutions of the homogeneous equation, just mentioned. We make the solutions of the equations (4.12) and (4.14) definite by requiring that $f_1^{(0)}$ as well as $f_1^{(1)}$ each fulfills the equations (1.11) for $l = 1$.

The general form of the solutions of the equations (4.12) and (4.14) for $\Phi_1^{(0)}$ and $\Phi_1^{(1)}$ can be derived in the same way as that of $\Phi$ in section 1. Although the solution of the equation (4.12) has already been given in section 1, we shall, for the sake of comparison with the solution of equation (4.14), repeat some of the main results here.

As the left hand sides of the equations (4.12) and (4.14) are linear in the $\Phi$ and the right hand sides of these equations linear in the gradients, one can derive that the $\Phi$ must be of the form:

$$\Phi_1^{(0)}(\vec{V}) = -A_\alpha^{(0)}(\vec{V})\frac{\partial \ln T}{\partial q_\alpha} - B_{\alpha\beta}^{(0)}(\vec{V})\, D_{\alpha\beta} \tag{4.16}$$

where $A_i^{(0)}(\vec{V})$ and $B_{ij}^{(0)}(\vec{V})$ satisfy the integral equations:

$$I(A_i^{(0)}) = -\frac{1}{n}\left(\frac{mV^2}{2\,kT} - \tfrac{5}{2}\right) V_i\, f_o(V) \tag{4.17a}$$

$$I(B_{ij}^{(0)}) = -\frac{1}{n}\frac{m}{kT}\left(V_i\, V_j - \tfrac{1}{3}\, V^2 \delta_{ij}\right) f_o(V) \tag{4.17b}$$

and

$$\Phi_1^{(1)}(\vec{v}) = -A_\alpha^{(1)}(\vec{V})\frac{\partial \ln T}{\partial q_\alpha} - B_{\alpha\beta}^{(1)}(\vec{V})\, D_{\alpha\beta} \tag{4.18}$$

where $A_i^{(1)}(\vec{V})$ and $B_{ij}^{(1)}(\vec{V})$ satisfy the integral equations:

$$I(A_i^{(1)}) = -L(V)\, V_i\, f_o(V) \tag{4.19a}$$

$$I(B_{ij}^{(1)}) = -M_1(V)\frac{m}{kT}\left(V_i\, V_j - \tfrac{1}{3}\, V^2\, \delta_{ij}\right) f_o(V)$$

$$-M_2(V)\, \delta_{ij}\, f_o(V)\;. \tag{4.19b}$$

From (4.10) - (4.19) the following two important properties of $f_1$ follow:

1. $f_1^{(0)}$, $f_1^{(1)}$ and $f_1$ are proportional to the gradients;
2. $f_1 = f_1^{(0)} + n\, f_1^{(1)}$ is no longer independent of the density, as in section 1, but contains a correction term that is proportional to the density.

The vector functions $A_i^{(0)}(\vec{V})$, $A_i^{(1)}(\vec{V})$ and the tensor functions $B_{ij}^{(0)}(\vec{V})$, $B_{ij}^{(1)}(\vec{V})$ can be determined as follows. First one can show on the basis of the equations (4.17) and (4.19) that they must be of the general form:

$$A_i^{(0)}(\vec{V}) = V_i\, A^{(0)}(V) \tag{4.20a}$$

$$A_i^{(1)}(\vec{V}) = V_i\, A^{(1)}(V) \tag{4.20b}$$

and

$$B_{ij}^{(0)}(\vec{V}) = \left(V_i\, V_j - \tfrac{1}{3}\, V^2 \delta_{ij}\right) B^{(0)}(V) \tag{4.20c}$$

$$B_{ij}^{(1)}(\vec{V}) = \left(V_i\, V_j - \tfrac{1}{3}\, V^2 \delta_{ij}\right) B^{(1)}(V) + \delta_{ij}\, B^{(2)}(V)\;. \tag{4.20d}$$

Introduction of $f_1^{(0)}$ from (4.11), (4.16) and (4.20) and of $f_1^{(1)}$ from (4.13), (4.18) and (4.20) into the conditions (1.11) leads to conditions on the $A^{(0)}(V)$, $A^{(1)}(V)$ and $B^{(2)}(V)$:

$$\int f_o(V)\, V^2 A^{(0)}(V)\, d\vec{v} = 0 \tag{4.21a}$$

$$\int f_o(V)\, V^2 A^{(1)}(V)\, d\vec{v} = 0 \tag{4.21b}$$

$$\int f_o(V)\, B^{(2)}(V)\, d\vec{v} = 0 \tag{4.21c}$$

$$\int f_o(V)\, V^2\, B^{(2)}(V)\, d\vec{v} = 0 \tag{4.21d}$$

of which the first (equation (4.21a)) has been obtained before in section 1 (eq. (1.17)).

c. The hydrodynamical equations and the transport coefficients

With the solutions for $f_o$, and $f_1 = f_1^{(0)} + n\, f_1^{(1)}$, obtained above, one can evaluate the hydrodynamical equations up to terms of the order $\mu$ and $\mu^2$ respectively. Of course, the results obtained with $f_o$ and $f_o + f_1^{(0)}$ will be identical with the equations (1.18) - (1.21) obtained before in section 1 on the basis of the Boltzmann equation.

In order to appreciate the difference of these results with those obtained in this section on the basis of the generalized Boltzmann equation, we shall give both results together (see table 2 on page 154).

In all equations in table 2 a subscript 0 or 1 means that the quantity involved has been calculated with the help of $f_o$ or $f_1$ respectively. The superscript $K$ in the pressure tensor or in the heat flux vector means that only contributions due to translational motion of the molecules (i.e. their kinetic energy) have been included, whereas the superscript $\varphi$ means that contributions to these quantities due to the intermolecular potential field have been included. It should be noted that in the equations (4.25a'), (4.25b') and (4.25b") a superindex $K'$ instead of $K$ has been used. This is due to the fact that in the case of the *ordinary* Boltzmann equation the contributions due to the translational motion of the molecules are computed with $f_o$ or $f_o + f_1^{(0)}$, whereas in the case of the *generalized* Boltzmann equation $f_o$ or $f_o + f_1^{(0)} + n\, f_1^{(1)}$ respectively is used. Therefore, although in both cases the expressions (1.4a) or (1.4b) are used for the computation, the $f$ that is substituted into these formulae differs in the two cases.

The proportionality of the *off-diagonal elements of* $P_{ij}$: *the* $P_{ij,1}$ *and of the* $J_{i,1}$ with the gradients is again a direct consequence of the proportionality of $f_1$ itself with the gradients. The transport coefficients, however, contain now in addition to a density independent contribution, already computed in section 1, contributions *proportional to the density* due to a) triple collisions (from $f_1^{(1)}$) and b) collisional transfer ($\eta_1^{\varphi}$, $\lambda^{\varphi 1}$, $\lambda^{\varphi 2}$). The *bulk viscosity coefficient* $\eta_2$ that enters into the theory and that would contribute to the *diagonal elements of* $P_{ij}$, *vanishes* in the approximation considered i.e. including contributions up to terms proportional to $n$. In general, including the higher terms in the expansion of $A(x_1 | F_1)$ in powers of $1/v$, *a "virial" expansion in powers of n for the transport coefficients* would be obtained.

## Table 2

1. With $f_0$ the *Euler-equations*, which include terms up to the order of $\mu$, are obtained:

**Boltzmann equation**

$$\frac{Dn}{Dt} = - n \frac{\partial u_\alpha}{\partial q_\alpha} \quad (4.22a)$$

$$n m \frac{Du_i}{Dt} = - \frac{\partial p^K}{\partial q_i} \quad (4.22b)$$

$$n \frac{D(\frac{3}{2} kT)}{Dt} = - p^K \frac{\partial u_\alpha}{\partial q_\alpha} \quad (4.22c)$$

**General. Boltzmann equation**

$$\frac{Dn}{Dt} = - n \frac{\partial u_\alpha}{\partial q_\alpha} \quad (4.22a')$$

$$n m \frac{Du_i}{Dt} = - \frac{\partial p}{\partial q_i} \quad (4.22b')$$

$$n \frac{D(\frac{3}{2} kT)}{Dt} = - p \frac{\partial u_\alpha}{\partial q_\alpha} - R_1 \quad (4.22c')$$

$$n \frac{D\varepsilon_0}{Dt} = - p \frac{\partial u_\alpha}{\partial q_\alpha} \quad (4.22c'')$$

where we have used that the pressure tensor $P_{ij} = P_{ij,0}$ with

$$P_{ij,0} = P^K_{ij,0} = p^K \delta_{ij} = n kT \delta_{ij} \quad (4.23a)$$

$$P_{ij,0} = p \delta_{ij} = (p^K + p^\varphi) \delta_{ij} = = n kT [1 - \frac{n}{2} \beta_1(T) \dots] \delta_{ij} \quad (4.23a')$$

and that the heat flux vector $J_i = J_{i,0}$ with

$$J_{i,0} = J^K_{i,0} = 0 \quad (4.23b)$$

$$J_{i,0} = J^K_{i,0} = J^{\varphi 1}_{i,0} = J^{\varphi 2}_{i,0} = 0 \quad (4.23b')$$

2. With $f_0 + f_1$ the *Navier-Stokes equations*, which include terms up to the order of $\mu^2$, are obtained:

**Boltzmann equation**

$$\frac{Dn}{Dt} = - n \frac{\partial u_\alpha}{\partial q_\alpha} \quad (4.24a)$$

$$n m \frac{Du_i}{Dt} = - \frac{\partial p^K}{\partial q_i} - \frac{\partial P^K_{i\alpha,1}}{\partial q_\alpha} \quad (4.24b)$$

$$n \frac{D(\frac{3}{2} kT)}{Dt} = - (p^K \delta_{\alpha\beta} + P^K_{\alpha\beta,1}) D_{\alpha\beta} - \frac{\partial J^K_{\alpha,1}}{\partial q_\alpha} \quad (4.24c)$$

**General. Boltzmann equation**

$$\frac{Dn}{Dt} = - n \frac{\partial u_\alpha}{\partial q_\alpha} \quad (4.24a')$$

$$n m \frac{Du_i}{Dt} = - \frac{\partial p}{\partial q_i} - \frac{\partial P_{i\alpha,1}}{\partial q_i} \quad (4.24b')$$

$$n \frac{D(\frac{3}{2} kT)}{Dt} = - (p \delta_{\alpha\beta} + P_{\alpha\beta,1}) D_{\alpha\beta} - \frac{\partial J'_{\alpha,1}}{\partial q_\alpha} - R_1 - R_2 \quad (4.24c')$$

$$n \frac{D\varepsilon}{Dt} = - (p \delta_{\alpha\beta} + P_{\alpha\beta,1}) D_{\alpha\beta} - \frac{\partial J_{\alpha,1}}{\partial q_\alpha} \quad (4.24c'')$$

where we have used that the pressure tensor $P_{ij} = P_{ij,0} + P_{ij,1}$ with

$$P_{ij,1} = P^K_{ij,1} = - 2 \eta^K_1 (D_{ij} - \tfrac{1}{3} D_{\alpha\alpha} \delta_{ij}) \quad (4.25a)$$

$$P_{ij,1} = P^K_{ij,1} + P^\varphi_{ij,1} = \quad (4.25a')$$

$$= - 2 \eta_1 (D_{ij} - \tfrac{1}{3} D_{\alpha\alpha} \delta_{ij}) - \eta_2 D_{\alpha\alpha} \delta_{ij}$$

where $\eta_1 = \eta^K_1 + \eta^\varphi_1$ and $\eta_2 = 0 + O(n^2)$

and that the heat flux vector $J_i = J_{i,0} + J_{i,1}$ with

$$J_{i,1} = J^K_{i,1} = - \lambda^K \frac{\partial T}{\partial q_i} \quad (4.25b)$$

$$J'_{i,1} = J^{K'}_{i,1} + J^{\varphi 1}_{i,1} = - (\lambda^{K'} + \lambda^{\varphi 1}) \frac{\partial T}{\partial q_i} \quad (4.25b')$$

$$J_{i,1} = J^{K'}_{i,1} + J^{\varphi 1}_{i,1} + J^{\varphi 2}_{i,1} = - \lambda \frac{\partial T}{\partial q_i} \quad (4.25b'')$$

where $\lambda = \lambda^{K'} + \lambda^{\varphi 1} + \lambda^{\varphi 2}$.

As mentioned already before, the equation for the kinetic energy alone (equations (4.22c') and (4.24c')) cannot be given completely up to terms of the order $\mu$ or $\mu^2$ in the theory of the generalized Boltzmann equation. This is due to the dependence of $R_1$ on $f_1$ and of $R_2$ on $f_2$. There is, however, an equation for the total energy $\varepsilon$.

The *diagonal elements of* $P_{ij}$ (the local pressure $p$) and the local internal energy $\varepsilon$ have in the case of the ordinary Boltzmann equation their values for an ideal gas, which contain only the contributions due to the translational motion of the molecules. This is slightly inconsistent because the occurrence of binary collisions is essentially used in the computation of the transport coefficients. This inconsistency is eliminated in the case of the generalized Boltzmann equation where in the expression for the pressure the contributions due to binary collisions appear in the occurrence of the second virial coefficient $B(T)$ in the form $\beta_1(T) = -2\,B(T)/N$ in $p^{\varphi}$ and where similarly for $\varepsilon$, the $\varepsilon^{\varphi}$ contains the contribution $\beta_1'(T) = d\beta_1/dT$. These contributions arise from the term $J_1(ff)$ in the right hand side of the generalized Boltzmann equation (4.1), which stems from the functional $A_1^{(1)}(x_1|\,F_1)$, the term of the order $\mu$ in the expansion of $A^{(1)}(x_1\,|\,F_1)$ in powers of $\mu$.

In principle the complete *virial expansions* of $p$ and $\varepsilon$, known from equilibrium statistical mechanics, should be obtained if the complete expansion of $A(x_1|\,F_1)$ in powers of $1/v$ would be used. The $l$-th virial coefficient results then from the term of the order $\mu$ in the expansion of the functional $A^{(l-1)}(x_1|\,F_1)$ in powers of $\mu$. So would the third virial coefficient follow from a term $K_1(fff)$ in the right hand side of the equation (4.1), which is derived from the functional $A_1^{(2)}(x_1|\,F_1)$, that has not been given here. Thus the neglect of the difference in position of the two colliding molecules in the Stosszahlansatz, which is used in the ordinary Boltzmann equation, is the origin of the inconsistency mentioned above. Only the use of the proper boundary condition for $F_2(|\,F_1)$ for the spatially *in*homogeneous case and a subsequent expansion in powers of $\mu$ leads to consistent expressions for the local thermodynamic quantities and the local transport properties of the gas.

Formal expressions for the transport coefficients analogous to the equations (1.22) in section 1 have been derived on the basis of the generalized Boltzmann equation [9]. The viscosity $\eta_1$ is then determined by the functions $A^{(0)}(V)$ and $A^{(1)}(V)$, the heat conductivity $\lambda$ by the functions $B^{(0)}(V)$ and $B^{(1)}(V)$. No Sonine polynomial expansion leading to explicit expressions like (1.23) for $\eta_1$ and $\lambda$ as a function of temperature and density has yet been performed. Therefore no comparison with experiment of the density dependent correction terms to the transport coefficients has as yet been possible.

This must conclude our account of the theory of the hydrodynamical stage. For more details we refer to elsewhere [9].

*References*

1. D. Hilbert, Math. Ann. 72 (1912) 562.
D. Enskog, Kinetische Theorie der Vorgänge in mässig verdünnten Gasen, diss. Uppsala (1917).
2. S. Chapman, T. G. Cowling, The mathematical theory of non uniform gases, Cambridge Univ. Press (1953).
3. J. O. Hirschfelder, C. F. Curtiss, R. B. Bird, The molecular theory of gases and liquids, Wiley (1954).
4. L. Waldmann, Encycl. of Physics (ed. S. Flügge), Vol. XII.
5. G. E. Uhlenbeck, Higgins Lectures, Princeton University, 1954.
6. J. De Boer, J. Van Kranendonk, Physica 14 (1948) 442.
J. O. Hirschfelder, R. B. Bird, E. L. Spotz, J. chem. Phys. 12 (1948) 968; 17 (1950) 149.
J. De Boer, E. G. D. Cohen, Physica 17 (1951) 993.
E. G. D. Cohen, M. J. Offerhaus, J. De Boer, Physica 20 (1954) 501.
E. G. D. Cohen, M. J. Offerhaus, J. M. J. Van Leeuwen, B. W. Roos, J. De Boer, Physica 21 (1955) 737; 22 (1956) 791.
7. N. N. Bogolubov, Journ. Phys. U.S.S.R. 10 (1946) 265; Problems of a dynamical theory in statistical physics, transl. E. Gora, Providence College, Providence R.I. (1959).
8. G. E. Uhlenbeck, Boulder Lectures 1957.
9. S. T. Choh, G. E. Uhlenbeck, The kinetic theory of dense gases, Univ. of Michigan (1958).
10. M. Born, H. S. Green, Proc. roy. Soc. A 188 (1946) 10.
J. Yvon, La théorie statistique des fluides et l'équation d'état, Hermann et Cie, Paris (1935).
J. G. Kirkwood, J. chem. Phys. 14 (1946) 180.
11. H. Grad, Comm. Pure Appl. Math. 2 (1949) 331; Encycl. of Physics (ed. S. Flügge), vol. XII, sect. 28.
12. M. S. Green, J. chem. Phys. 25 (1956) 836.
13. M. S. Green, Physica 24 (1958) 393.
14. E. G. D. Cohen, Physica 27 (1961), to be published.
15. E. G. D. Cohen, T. H. Berlin, Physica 26 (1960) 717.
16. P. and T. Ehrenfest, The conceptual foundations of the statistical approach in mechanics, p. 54, Cornell Univ. Press (1959).

# MASTER EQUATION AND APPROACH TO EQUILIBRIUM FOR QUANTUM SYSTEMS

L. VAN HOVE
Instituut voor Theoretische Fysica
der Rijksuniversiteit, Utrecht

## 1. *Introduction*

In these lectures we restrict ourselves to the perturbation-theoretical approach to the derivation of the master equation. We assume that the Hamiltonian can be separated into two parts, the unperturbed one and a perturbation term being entirely responsible for the irreversible behaviour. The master equation was derived for the first time by Pauli [1]. In recent years there has been a revival of interest in the problem of deriving the master equation, by making less far going assumptions than Pauli but using more the special properties of the system at hand. Our main aim will be to review the work done in Utrecht since 1955 [2,3]. We restrict ourselves to the main features of the theory, leaving aside the detailed calculations. Other derivations, also using perturbation theory, have been given in recent years. Let us quote in this respect the work of Kohn and Luttinger [4] and of Prigogine and his collaborators, the latter making extensive use of diagram techniques.

There exists a close relation between the approach to equilibrium and the behaviour of correlation functions. These correlation functions play an important role in deriving expressions for transport coefficients, known as Kubo formulae [5]. We shall show the relation between these formulae and the master equation. This has been stressed first by Chester and Thellung [6] for the case of electrical conductivity.

## 2. *The Pauli master equation and the statistical assumptions involved*

As mentioned before Pauli derived the master equation for the occupation probability at time $t$,

$$\frac{\mathrm{d}}{\mathrm{d}t} p_t(\alpha) = \sum_{\alpha'} \{ W_{\alpha\alpha'}\, p_t(\alpha') - W_{\alpha'\alpha}\, p_t(\alpha) \} , \tag{2.1}$$

which is an equation of the Markoff type. $W_{\alpha\alpha'}$ is the transition probability per unit time between the unperturbed states $|\alpha'\rangle$ and $|\alpha\rangle$.

In the treatment of Pauli the Hamiltonian is separated into two parts

$$H' = H + \lambda V , \tag{2.2}$$

in which $H$ is the Hamiltonian of the unperturbed system with the eigenstates $|\alpha\rangle$ and eigenvalues $\varepsilon_\alpha$,

$$H|\alpha\rangle = \varepsilon_\alpha |\alpha\rangle ; \tag{2.3}$$

$\lambda V$ is the perturbation, $\lambda$ is a dimensionless expansion parameter measuring its strength. An essential point is that $H$ is simple and large in contrast to $V$, which has to be small for the applicability of perturbation theory but has complicated effects on the motion and mixes the unperturbed states so that it can give rise to irreversible behaviour. Examples can be found in solid state physics. Let us first take an insulating crystal. Then $H$ represents the harmonic part of the atomic motion. The anharmonic forces are described by $\lambda V$. Another example is that of a conducting crystal, where $H$ describes the free harmonic vibrations (phonons) and the conduction electrons in the periodic field of the ions at their equilibrium positions. In this case $\lambda V$ represents the electron phonon interaction. We stress again that the small interaction is very essential, because it is responsible for the approach to equilibrium and it gives rise to finite values of the transport coefficients.

The perturbation $V$ is characterized by its matrix elements $\langle\alpha|V|\alpha'\rangle$. The occupation probability at time $t$ is given by

$$p_t(\alpha) = |c_t(\alpha)^2| , \tag{2.4}$$

in which $c_t(\alpha)$ are the expansion coefficients of the state vector in terms of the unperturbed states,

$$\varphi_t = \sum_\alpha |\alpha\rangle\, c_t(\alpha) . \tag{2.5}$$

For the transition probability Pauli found

$$W_{\alpha\alpha'} = 2\pi\lambda^2\, \delta(\varepsilon_{\alpha'} - \varepsilon_\alpha)|\langle\alpha'|V|\alpha\rangle|^2 . \tag{2.6}$$

The $\delta$-function occurring in this expression has a meaning only, if the sum over the unperturbed states can be replaced by an integral, which is the case for a very large system. Let us take a cubic lattice with periodic boundary conditions. The wave vectors then have values of the form

$$k_{i,xyz} = \frac{2\pi}{\Omega^{1/3}} \times \text{integer}, \tag{2.7}$$

in which $\Omega$ represents the volume. If $\Omega$ is very large one has to sum over a very dense array of wave vectors, which makes it possible to replace the sum by an integral. In fact we have

$$\sum_{k} = \frac{\Omega}{8\pi^3} \int d_3k . \tag{2.8}$$

This transition from a sum to an integral is very significant for the procedure. In a crystal, the length of the cube $\Omega^{1/3}$ has to be large compared to the lattice distance; for a gas the significant length is the distance between particles.

The procedure of Pauli to derive (2.1) is very simple. Let us consider the state of the system at time $t$: $\varphi_t$. At time $t + \Delta t$ it is given by

$$\varphi_{t+\Delta t} = e^{-i(H+\lambda V)\Delta t} \varphi_t \tag{2.9}$$

(we always put $\hbar = 1$). The expansion coefficients at this later time $t + \Delta t$ can be easily found from

$$c_{t+\Delta t}(\alpha) = <\alpha | \varphi_{t+\Delta t}> . \tag{2.10}$$

The probability density is given by

$$p_{t+\Delta t}(\alpha) = |c_{t+\Delta t}(\alpha)|^2 . \tag{2.11}$$

Using the normalization condition $<\alpha|\alpha> = 1$ and the expression for the expansion coefficients of $\varphi_{t+\Delta t}$, we find

$$p_{t+\Delta t}(\alpha) = \sum_{\alpha'\alpha''} c_t^*(\alpha'')<\alpha''|e^{i(H+\lambda V)\Delta t}|\alpha><\alpha|e^{-i(H+\lambda V)\Delta t}|\alpha'>c_t(\alpha') . \tag{2.12}$$

We now assume random phases at time $t$ which amounts to dropping all terms with $\alpha' \neq \alpha''$. Indeed, random phases imply that the terms in the sum with $\alpha' \neq \alpha''$ oscillate very rapidly so that they will cancel out. By applying the random phase assumption we thus get from (2.12)

$$p_{t+\Delta t}(\alpha) = \sum_{\alpha'} |<\alpha|e^{-i(H+\lambda V)\Delta t}|\alpha'>|^2 \, p_t(\alpha') . \tag{2.13}$$

The expression $|<\alpha|e^{-i(H+\lambda V)\Delta t}|\alpha'>|^2$ can be calculated by first order time dependent perturbation theory, if $\Delta t$ is small enough. This yields the expression

$$p_{t+\Delta t}(\alpha) = p_t(\alpha')\,\delta_{\alpha\alpha'} + 2\pi\,\Delta t\,\lambda^2 \sum_{\alpha'} \delta(\varepsilon_\alpha - \varepsilon_{\alpha'})|<\alpha'|V|\alpha>|^2\,[p_t(\alpha) - p_t(\alpha')] \tag{2.14}$$

for $\Delta t > 0$. From this the Pauli equation follows immediately. For the derivation coarse graining is not required, but one has to assume the spectrum to be continuous, because otherwise the $\delta$-function in (2.14) will have no meaning. For applying perturbation theory to first order several conditions must be fulfilled. Before giving these conditions we define the following quantities

1. $\delta\varepsilon$: a characteristic energy of the unperturbed system (e.g. in a crystal the Debye-energy),
2. $\delta\varepsilon_\lambda = \lambda^2 \frac{|<|V|>|^2}{\delta\varepsilon}$, which is an energy related to the strength of the perturbation,
3. $\delta\varepsilon_\Omega$: the level separation of the individual excitations in the unperturbed system. The latter quantity is proportional to $\Omega^{-1/3}$, $\Omega$ being the volume of the crystal. For the validity of the derivation of the Pauli equation the following unequalities must hold

$$\delta\varepsilon >> \delta\varepsilon_\lambda >> \delta\varepsilon_\Omega ,$$

or in terms of the corresponding times

Debye period << relaxation time << traversal time

(the traversal time measures the time needed for an excitation to traverse the whole crystal).
$\delta\varepsilon_\Omega << \delta\varepsilon_\lambda$ has to be required for replacing the sum by an integral and the condition $\delta\varepsilon_\lambda << \delta\varepsilon$ permits to take only the lowest order in the perturbation expansion of $|<\alpha|e^{-i(H+\lambda V)\Delta t}|\alpha'>|^2$. One actually chooses $\Delta t$ such that

$$(\delta\varepsilon)^{-1} << \Delta t << (\delta\varepsilon_\lambda)^{-1} .$$

I want now to make an important observation: it is possible to derive from the random phase assumption more than Pauli did. In fact by calculating $p_{t-\Delta t}(\alpha)$ as we did for $p_{t+\Delta t}(\alpha)$ one finds that

$$\frac{d}{dt} p_t(\alpha) = 0 ,$$

in any point where $p_t(\alpha)$ has the same derivative in $t$ on the right as on the left, so that the random phase assumption can only hold at all times, if the system is in equilibrium. To get rid of this unsatisfactory fact one must require the random phase assumption only to hold in some limited fashion, namely as far as the calculation of $p_{t+\Delta t}$ for $\Delta t > 0$ is concerned, and not for the calculation of $p_{t-\Delta t}$.

### 3. *Special properties of the interaction*

It has been possible to derive the master equation assuming random phases at some initial time only by making additional assumptions about the perturbation, which appear to be fulfilled in the actual systems which we already mentioned as examples. As mentioned before, the transition to a large system is quite essential. In the limit of a large system we have

$$\sum_{k} = \frac{\Omega}{8\,\pi^3} \int d_3k \ , \tag{3.1}$$

$$\delta_{k,k'} = \frac{8\,\pi^3}{\Omega}\,\delta(k - k') \ . \tag{3.2}$$

The normalization of the states $|\alpha\rangle$ will now be taken such that

$$\langle\alpha|\alpha'\rangle = \delta(\alpha - \alpha') \ . \tag{3.3}$$

The expression in the right hand side of (3.3) means a product of $\delta$-functions, e.g. in the case of a conducting crystal the normalization condition is

$$\langle\alpha|\alpha'\rangle = \prod_i \delta(k_i - k_i') \prod_j \delta(q_j - q_j') \ , \tag{3.4}$$

if $|\alpha\rangle = |k_1\, k_2 \ldots q_1\, q_2 \ldots\rangle$ and $|\alpha'\rangle = |k_1'\, k_2' \ldots q_1'\, q_2' \ldots\rangle$, where the $k$ and $q$ represent the wave vectors of the phonons and electrons present in the states.

In second order perturbation theory the matrix elements to be calculated have the form $\langle\alpha|VAV|\alpha'\rangle$ where $A$ is a diagonal operator in the $|\alpha\rangle$-representation

$$A\,|\alpha\rangle = A(\alpha)\,|\alpha\rangle \ . \tag{3.5}$$

Let us consider such operators that $A_j(\alpha)$ is a smooth function of $\alpha$. For such operators the matrix element $\langle\alpha|VAV|\alpha'\rangle$ can be separated into two quite different parts

$$\langle\alpha|VAV|\alpha'\rangle = f_1(\alpha)\;\delta(\alpha - \alpha') + f_2(\alpha, \alpha') \tag{3.6}$$

in which the diagonal elements are larger by some power of $\Omega$ than the off-diagonal elements (compare with (3.2)). For the validity of (3.6) it is essential that $A(\alpha)$ is a smooth function of $\alpha$. In the following (3.6) will be assumed, which is in contrast to the derivation of Pauli, who did not use special properties of the system to derive the master equation. To indicate the diagonal part and the non-diagonal part of an operator we introduce the notation

$$\langle\alpha|VAV|\alpha'\rangle = \langle\alpha|\{VAV\}_d|\alpha'\rangle + \langle\alpha|\{VAV\}_{nd}\,|\alpha'\rangle \ . \tag{3.7}$$

Matrix elements of more complicated operators $VA_1V \ldots A_nV$, with all $A_j$ diagonal (they are of the type occurring in higher order perturbation theory) can also be divided in a diagonal (singular) and a non-diagonal part. This distinction between diagonal and non-diagonal parts can also be applied to the factors occurring in an operator product of the above form. Let us consider as an example the matrix element $\langle\alpha|VA_1VA_2VA_3V|\alpha'\rangle$. By separating diagonal parts one finds as some of the terms

$$\{VA_1V\}_d\, A_2\{VA_3V\}_d + \{VA_1V\}_d\, A_2\{VA_3V\}_{nd} + \\ + \{VA_1\, \{VA_2V\}_d\, A_3V\}_d + \{VA_1\, \{VA_2V\}_d\, A_3V\}_{nd} + \ldots \quad (3.8)$$

This will be clarified by diagrams, which enable us to select the several contributions. In fig. 1 the straight line represents the matrix element and the dots the initial, intermediate, and final states.

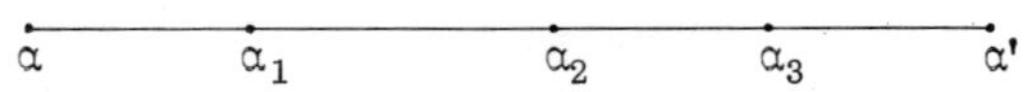


Fig. 1.

To classify the various contributions to à matrix element equal states (i.e. entering a δ-function) are connected by lines, which we call bridges; the connected states are called bound, the others free states. Fig. 2 gives the diagrams corresponding to the matrix elements in (3.8).

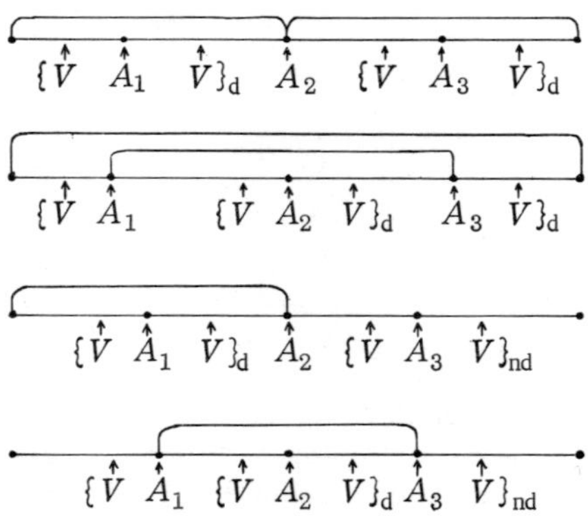


Fig. 2.

The general situation is examplified in fig. 3, where we have three groups of bound states, i.e. $\alpha$, $\alpha_2$, $\alpha_k$; $\alpha_3$, $\alpha_l$ and $\alpha_n$, $\alpha'$.

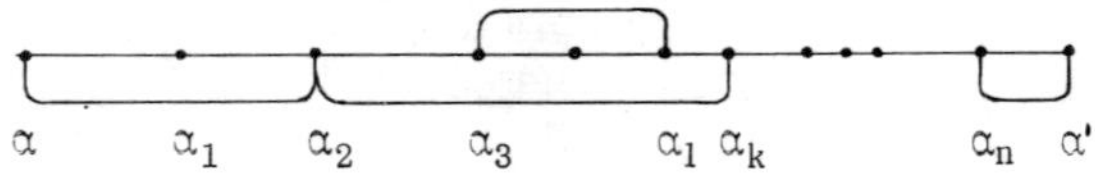


Fig. 3.

Moreover one verifies that terms with "crossing" bridges (as in fig. 4) do not contribute.

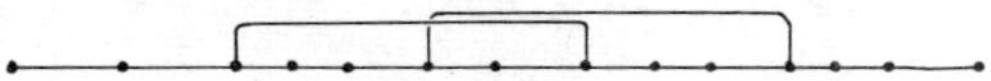

Fig. 4.

## 4. *Sketch of the improved derivation of the master equation*

In the following we make use of the special properties of the interaction, discussed in the foregoing section, in order to calculate $p_t(\alpha)$. This will be done without giving the mathematical details. We expand the state at time $t$ in eigenstates of $H$,

$$\varphi_t = \int_\alpha |\alpha> c_t(\alpha) . \tag{4.1}$$

At time $t = 0$ we have

$$\varphi_o = \int_\alpha |\alpha> c_o(\alpha) . \tag{4.2}$$

These two states are connected by the evolution operator

$$\varphi_t = e^{-i(H+\lambda V)t} \varphi_o . \tag{4.3}$$

For applying our techniques it is essential that we introduce a coarse grained probability $p_t(\alpha)$ which is defined by the identity

$$<\varphi_t|A|\varphi_t> = \int_\alpha A(\alpha)\; p_t(\alpha) , \tag{4.4}$$

where $A$ is any diagonal operator in the $|\alpha>$ representation with a smoothly varying eigenvalue $A(\alpha)$. Because of the smoothness of $A(\alpha)$ this definition of $p_t(\alpha)$ is equivalent to

$$p_t(\alpha) = \int_{\alpha'\varepsilon\Delta\alpha} |c_t(\alpha')|^2 \Bigg/ \int_{\alpha'\varepsilon\Delta\alpha} , \tag{4.5}$$

where $\Delta\alpha$ represents a small region around $\alpha$, which contains, however, many eigenstates. The averaging over this region $\Delta\alpha$ makes $p_t(\alpha)$ coarse grained. Starting from (4.4) one can calculate $p_t(\alpha)$. Using (4.2) and (4.3) one gets

$$<\varphi_t|A|\varphi_t> = <\varphi_o|e^{i(H+\lambda V)t} A\, e^{-i(H+\lambda V)t}|\varphi_o> =$$

$$= \int_{\alpha_o}\int_{\alpha_o'} c_o^*(\alpha_o') <\alpha_o'|e^{i(H+\lambda V)t} A\, e^{-i(H+\lambda V)t}|\alpha_o>.c_o(\alpha_o) . \tag{4.6}$$

Substituting in (4.6) the expansion

$$e^{-i(H+\lambda V)t} = e^{-itH} - i\lambda \int_0^t e^{-iH(t-t_1)}\, V e^{-iHt_1}\, dt_1 + \ldots \tag{4.7}$$

one gets for the general term of the series, thus found, an expression of the form

$$\langle \alpha_0' |\, V A_1 V A_2 V \ldots |\alpha_0\rangle\,, \tag{4.8}$$

where all $A_i$ except one ($A$ itself) are exponentials of the unperturbed Hamiltonian. Consequently we have

$$e^{i(H+\lambda V)t} A\, e^{-i(H+\lambda V)t} = \{\mathrm{I}\}_d + \{\mathrm{II}\}_{nd} \tag{4.9}$$

in which the first part corresponds to the contributions of the diagrams which have a bridge going from the initial to the final state, as illustrated in fig. 5.

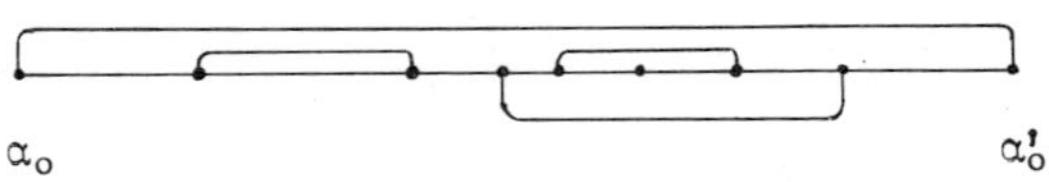


Fig. 5.

Using the property

$$\langle \alpha_0 |\{\quad\}_d| \alpha_0'\rangle = \delta(\alpha_0 - \alpha_0')\, \{\ldots\}\, (\alpha_0)\,, \tag{4.10}$$

where the bracket on the right hand side represents a function of $\alpha_0$, we get for (4.6) the following expression:

$$\langle \varphi_t | A | \varphi_t \rangle = \int_{\alpha_0} |c_0(\alpha_0)|^2 \,\{e^{i(H+\lambda V)t} A\, e^{-i(H+\lambda V)t}\}_d\, (\alpha_0)$$
$$+ \int_{\alpha_0, \alpha_0'} c_0^*(\alpha_0')\, c_0(\alpha_0) \langle \alpha_0' | \{e^{i(H+\lambda V)t} A\, e^{-i(H+\lambda V)t}\}_{nd} |\alpha_0\rangle\,. \tag{4.11}$$

The first term depends only on the initial probabilities. The second term can be supposed to vanish if the phases at the initial time are chosen sufficiently "random". There remains then to calculate the first term. This can be done explicitly for weak coupling, i.e. in the limit $\lambda \to 0$, $t \to \infty$, such that $\lambda^2 t$ is finite. Physically one can say, that $t$ has to be of the order of the relaxation time

$$t \propto \frac{\delta\varepsilon}{\lambda^2 |\langle |V|\rangle|^2}\,.$$

Under these special conditions only a special class of diagrams contributes. The most general diagram contributing to the function

$$\{e^{i(H+\lambda V)t} A\, e^{-i(H+\lambda V)t}\}_d\,(\alpha_o)$$

is illustrated in fig. 6.

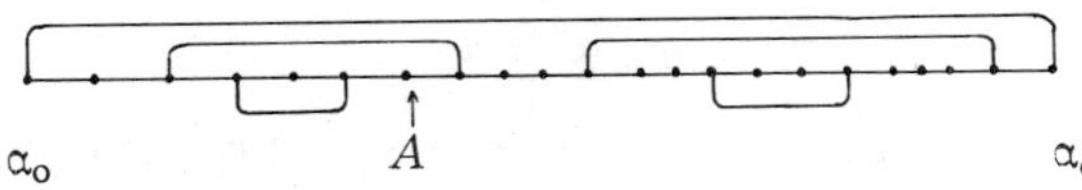

Fig. 6.

in which one intermediate state corresponds to the operator $A$ and the others to operators of the form $e^{-iH(t_{n-1}-t_n)}$. The structure of the particular diagrams which contribute for weak coupling ($\lambda \to 0$, $\lambda^2 t$ finite) is found to be

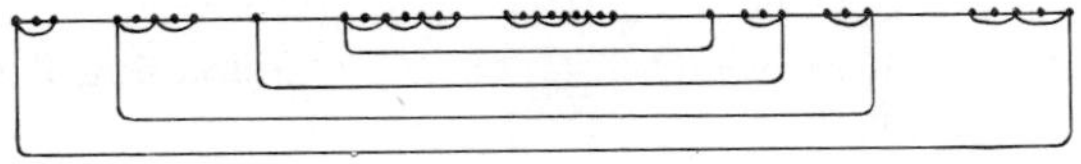

Fig. 7.

There are large and small bridges. Under the small bridges there is always exactly one point. If one calculates the contributions of the small and large bridges, one finds for $p_t(\alpha)$ an infinite series in which each term corresponds to a special diagram. Differentiating this result with respect to $t$ it is found that this quantity obeys the master equation

$$\frac{d}{dt} p_t(\alpha) = \int_{\alpha'} \{W_{\alpha\alpha'}\, p_t(\alpha') - W_{\alpha'\alpha}\, p_t(\alpha)\}$$

for $t > 0$, and for $t < 0$ the equation obtained by reversing the sign of the right hand side. $W_{\alpha\alpha'}$ is given by (2.6). It may be checked that other diagrams give rise to contributions of higher order in the coupling and can therefore be neglected.

## 5. *Relation to the Kubo formula*

Up till now we have studied the approach to equilibrium of systems left to themselves, making use of 1. the assumption that interference terms may be omitted, 2. special properties of the interaction. The Kubo formula on the other hand gives an expression for transport coefficients for systems in equilibrium, disturbed by a small external force. What is the relation between these two approaches? To be specific we shall study electrical conductivity. In a system containing charged particles disturbed by an external electric field a current is set up and one expects the following relation between the components $j_\mu$ of the current and the field components $E_\nu$ to hold

$$j_\mu = \sigma_{\mu\nu} E_\nu \,. \tag{5.1}$$

Kubo [5] and others have derived a formula expressing the conductivity tensor $\sigma_{\mu\nu}$ in terms of the equilibrium properties of the system,

$$\sigma_{\mu\nu} = \frac{\beta}{2}\int_{-\infty}^{\infty} dt \,\mathrm{Tr}\,[\rho\, J_\mu(t)\, J_\nu(0)] \,, \tag{5.2}$$

where $\rho$ is the equilibrium density matrix at temperature $T$,

$$\rho = \frac{e^{-\beta(H+\lambda V)}}{\mathrm{Tr}\,[e^{-\beta(H+\lambda V)}]} \,, \qquad \beta = \frac{1}{kT} \,, \tag{5.3}$$

$J_\mu$ is the current operator and $J_\mu(t)$ the corresponding Heisenberg operator

$$J_\mu(t) = e^{i(H+\lambda V)t}\, J_\mu\, e^{-i(H+\lambda V)t} \,. \tag{5.4}$$

If one tries to compute $\sigma_{\mu\nu}$, which contains a correlation operator calculated at equilibrium, one is led to a calculation, which is extremely close to the one of $p_t(\alpha)$ considered in the foregoing section. This was first remarked by Chester and Thellung and was used by them and by Verboven for an extensive calculation of the conductivity [6,7]. It can also be seen, that for the case of weak coupling this gives the relation between Kubo's formula and the usual calculation of the transport coefficient, employing the Boltzmann-Lorentz equation. Before demonstrating the close similarity of method in the derivation of the master equation and in the calculation of the conductivity by Kubo's formula, we introduce a quantity $P_t(\alpha\alpha_0)$, defined by

$$\{e^{i(H+\lambda V)t}\, A\, e^{-i(H+\lambda V)t}\}_d\,(\alpha_0) = \int_\alpha A(\alpha)\; P_t(\alpha\alpha_0) \tag{5.5}$$

for any $A$ diagonal in the $|\alpha\rangle$ representation with smooth eigenvalues $A(\alpha)$. Inserting this in (4.11) one finds

$$p_t(\alpha) = \int_{\alpha_0} P_t(\alpha\alpha_0)\,|\,C_0(\alpha_0)\,|^2 \,. \tag{5.6}$$

$P_t(\alpha\alpha_0)$ is a purely dynamic quantity depending only on the Hamiltonian and not on the initial state. For the case of weak coupling it is this quantity that enters in the expression for $\sigma_{\mu\nu}$.

In the example of electrons moving in a crystal lattice the current operator $J_\mu$ is diagonal in the unperturbed representation

$$J_\mu\,|\alpha\rangle = J_\mu(\alpha)\,|\alpha\rangle \,. \tag{5.7}$$

Substituting this in (5.2) and calculating the trace in the $|\alpha\rangle$-representation one obtains

$$\sigma_{\mu\nu} = \frac{\beta}{2\delta(o)} \int_{-\infty}^{\infty} dt \int_{\alpha\alpha_o} \langle\alpha_o|\rho|\alpha\rangle\langle\alpha|e^{i(H+\lambda V)t} J_\mu \, e^{-i(H+\lambda V)t}|\alpha_o\rangle \, J_\nu(\alpha_o) \; . \tag{5.8}$$

$\delta(o)$ enters into the denominator because we calculate the trace in a representation normalized to $\langle\alpha|\alpha'\rangle = \delta(\alpha - \alpha')$. Since $J_\mu$ is diagonal in the $|\alpha\rangle$-representation the expression

$$\langle\alpha|e^{i(H+\lambda V)t} J_\mu \, e^{-i(H+\lambda V)t}|\alpha_o\rangle \tag{5.9}$$

is just of the type (4.8) we have studied.

For weak coupling one has again to make a perturbation calculation for the case $\lambda$ small, $t$ large, $\lambda^2 t$ finite, since by the integration over time all $t$ values enter. However, in the case of weak coupling, the density matrix $\rho$ can be replaced by

$$\rho_o = \frac{e^{-\beta H}}{\mathrm{Tr}\, e^{-\beta H}} \, . \tag{5.10}$$

$\rho_o$ is of course diagonal in the $|\alpha\rangle$-representation

$$\langle\alpha_o|\rho_o|\alpha\rangle = \rho_o(\alpha_o)\, \delta(\alpha_o - \alpha) \; . \tag{5.11}$$

Eventually one finds

$$\begin{aligned} \sigma_{\mu\nu} &= \frac{\beta}{2} \int_{-\infty}^{\infty} dt \int_{\alpha\alpha_o} \rho_o(\alpha_o) \, \{e^{i(H+\lambda V)t} J_\mu \, e^{-i(H+\lambda V)t}\}_d (\alpha) \, J_\nu(\alpha_o) \\ &= \frac{\beta}{2} \int_{-\infty}^{\infty} dt \int_{\alpha\alpha_o} P_t(\alpha\alpha_o) \, J_\mu(\alpha) \, J_\nu(\alpha_o) \, \rho_o(\alpha_o) \; . \end{aligned} \tag{5.12}$$

The factor $\delta(o)$ present in (5.8) has now dropped out. $P_t(\alpha\alpha_o)$ is known from the derivation of the master equation and the integrations can be performed. But without calculating $P_t(\alpha\alpha_o)$ one can already show, that the result is equivalent to the result obtained by use of the Boltzmann-Lorentz equation. Even the mathematics is the same [2,6].

## 6. *Higher order effects*

So far we have restricted ourselves to calculations to lowest order of $\lambda$. In view of the fact, that the special properties of the perturbation made the analysis for the case of weak coupling quite easy, it is tempting to go further and use the same technique to look at higher order effects. This is particularly interesting, since the calculation of transport properties to higher order is a non-trivial matter which cannot be treated by the old techniques.

Since the conductivity tensor $\sigma_{\mu\nu}$ as discussed in section 5 is of the order $1/\lambda^2$ one expects that in general

$$\sigma_{\mu\nu} = \frac{\sigma^{(-2)}_{\mu\nu}}{\lambda^2} + \frac{\sigma^{(-1)}_{\mu\nu}}{\lambda} + \sigma^{(o)}_{\mu\nu} + \sigma^{(1)}_{\mu\nu}\,\lambda + \dots \tag{6.1}$$

(5.12) is the expression for $\lambda^{-2}\,\sigma^{(-2)}_{\mu\nu}$. In higher order the relation with the master equation is not as close as for weak coupling (eq. 5.12), but still the calculations are similar to a large extent. Again one has to deal with

$$\{e^{i(H+\lambda V)t}\, J_\mu\, e^{-i(H+\lambda V)t}\}_d(\alpha)\ , \tag{6.2}$$

that is, with expressions of the type

$$\int_{\alpha_o\alpha_o'} J_\mu(\alpha)\ Q_t(\alpha\alpha_o\alpha_o')\ . \tag{6.3}$$

Once more $Q_t(\alpha\alpha_o\alpha_o')$ is a purely dynamic quantity, which has to be calculated by continuing the systematic perturbational analysis we have given. It should be stressed, that the dynamics is completely separated from the statistics. The temperature operator only enters in taking the ensemble average over the dynamic quantity. It is a complicated affair to calculate the higher order terms in $\sigma_{\mu\nu}$, but closed expressions are obtained. Verboven has performed the calculations up to order $\sigma^{(o)}_{\mu\nu}$ [7].

If one is concerned with higher order effects, one can ask another question, relating to the time evolution of the system. For weak coupling we have obtained the master equation, but how do the higher order terms affect the time evolution of $p_t(\alpha)$? To find the answer one has to study the function $P_t(\alpha\alpha_o)$ to general order in the coupling. The simplifying assumption $\lambda \to 0$, $t \to \infty$, $\lambda^2 t$ finite, has to be abandoned and the function $P_t(\alpha\alpha_o)$ must be studied for both $\lambda$ and $t$ finite.

In all generality the question is much too difficult and we can not get such a detailed answer as for the case of weak coupling. We have proceeded in two steps. First, one can make an analysis to see whether the terms can be grouped together in a certain way. Indeed, it is possible to do this in a similar fashion as we already did for the weak coupling case. Second, one makes simplifying assumptions (not the weak coupling assumption, however) and studies a simple model to get a specific answer, which makes it possible to guess what the influence of the higher order terms on the time evolution will be in the general case.

First we give a classification of the higher order terms. The analysis is carried out by means of the resolvent operator, defined by

$$R_l = \frac{1}{H + \lambda V - l}, \tag{6.4}$$

where $l$ is a complex variable. The resolvent operator turns out to be an extremely convenient tool. The reason is that $R_l$ has a very simple expansion in powers of $\lambda$

$$R_l = R_l^o - \lambda R_l^o V R_l^o + \lambda^2 R_l^o V R_l^o V R_l^o + \ldots . \tag{6.5}$$

$R_l^o$ is the unperturbed resolvent operator

$$R_l^o = \frac{1}{H - l}. \tag{6.6}$$

The operator of motion is simply connected with the resolvent by means of a contour integral

$$e^{-i(H+\lambda V)t} = \frac{i}{2\pi} \oint R_l \, e^{-ilt} \, dl , \tag{6.7}$$

where the contour goes around all poles of $R_l$. The nature of the expansion for $R_l$, however, is much simpler than the one for the time evolution operator. The terms in the expansion for $R_l$ are of the type we have already studied. Each term contains the perturbation $V$ and the operator $R_l^o$, diagonal in the $|\alpha\rangle$-representation

$$R_l^o |\alpha\rangle = \frac{1}{\varepsilon_\alpha - l} |\alpha\rangle . \tag{6.8}$$

Hence the analysis in terms of diagonal parts can be carried out. From

$$\frac{-1}{(2\pi)^2} \{ \oint R_l \, e^{ilt} \, dl \, A \oint e^{-il't} R_{l'} \, dl' \} (\alpha_o) = \int_\alpha A(\alpha) \, P_t(\alpha\alpha_o) \tag{6.9}$$

one obtains $P_t(\alpha\alpha_o)$ in terms of a function $X_{ll'}(\alpha\alpha_o)$

$$P_t(\alpha\alpha_o) = \frac{-1}{(2\pi)^2} \oint \oint dl \, dl' \, e^{i(l-l')t} X_{ll'}(\alpha\alpha_o) . \tag{6.10}$$

A diagram analysis for the function $X_{ll'}(\alpha\alpha_o)$ yields

$$X_{ll'}(\alpha\alpha_o) = D_l(\alpha_o) D_{l'}(\alpha_o) \delta(\alpha - \alpha_o) + \lambda^2 D_l(\alpha) D_{l'}(\alpha) W_{ll'}(\alpha\alpha_o) D_l(\alpha_o) D_{l'}(\alpha_o)$$
$$+ \lambda^4 D_l(\alpha) D_{l'}(\alpha) \int_{\alpha_1} W_{ll'}(\alpha\alpha_1) D_l(\alpha_1) D_{l'}(\alpha_1) W_{ll'}(\alpha_1\alpha_o) D_l(\alpha_o) D_{l'}(\alpha_o) + \ldots . \tag{6.11}$$

The $W$'s, which are defined in terms of a first particular family of diagrams, are a natural generalization of the transition probabilities $w_{\alpha\alpha'}$, occurring in the master equation. $W_{ll'}(\alpha\alpha_o)$ contains the two states $\alpha$ and $\alpha_o$ and depends on $l$ and $l'$; this latter dependence reflects the fact that the transitions can no longer be considered instantaneous compared to the relaxation time. Here a time element enters in the transition rates. The function $D_l(\alpha)$ is a natural extension of

$$R_l^o(\alpha) = \frac{1}{\varepsilon_\alpha - l} \tag{6.12}$$

which describes the unperturbed motion. In the case of general coupling one has

$$D_l(\alpha) = \frac{1}{\varepsilon_\alpha - l - \lambda^2 G_l(\alpha)} \; . \tag{6.13}$$

$G_l(\alpha)$ plays the role of an energy shift and line width. The intuitive meaning of the general term of $X_{ll'}(\alpha\alpha_o)$ is now clear: Unperturbed motion in the state $\alpha_o$ changed to a motion involving self-energy and damping, generalized transition, etc. There are recipes to calculate $W_{ll'}(\alpha\alpha_o)$ and $G_l(\alpha)$. The functions are so constructed, that they have a bounded behaviour in the complex variable $l$. The functions $D$ and $W$ are not independent. In second order perturbation theory one has the familiar result that the unperturbed energies $\varepsilon_\alpha$ are replaced by

$$\varepsilon_\alpha \rightarrow \bar{\varepsilon}_\alpha = \varepsilon_\alpha + \delta_\alpha + i\,\gamma_\alpha \, , \tag{6.14}$$

where $\delta_\alpha$ is an energy shift and $\gamma_\alpha$ a line width. The magnitude of the shift and width are of course connected with the transition rates by well-known formulae. The relation between the functions $D$ and $W$ is a generalization of this well-known connection and is given by

$$G_l(\alpha) - G_{l'}(\alpha) = \int_{\alpha'} [D_l(\alpha') - D_{l'}(\alpha')]\, W_{ll'}(\alpha'\alpha) \; . \tag{6.15}$$

One can now try, using this information, to find the time evolution of $P_t(\alpha\alpha_o)$ in terms of an equation which would resemble the master equation and hence must be of the form

$$\frac{d}{dt} P_t(\alpha\alpha_o) = \;?\;?\; P_t(\alpha\alpha_o) \; .$$

I have not been able to find such an equation, but have found an equation for a partial $P$, a quantity $P_t^{(E)}(\alpha\alpha_o)$ of which $P_t(\alpha\alpha_o)$ is the integral over $E$ and for which an equation holds involving the times in the past,

$$\frac{d}{dt} P_t^{(E)}(\alpha\alpha_o) = \ldots + \int_o^t dt' \int_{\alpha'} K_{t-t'}^{(E)}(\alpha\alpha')\, P_t^{(E)}(\alpha'\alpha)$$
$$- \int_o^t dt' \int_{\alpha'} K_{t-t'}^{(E)}(\alpha'\alpha)\, P_t^{(E)}(\alpha\alpha_o) \, , \tag{6.16}$$

$$P_t(\alpha\alpha_o) = \int_{-\infty}^{\infty} P_t^{(E)}(\alpha\alpha_o)\, dE \; . \tag{6.17}$$

Hence we have found a non-markoffian equation for $P_t^{(E)}(\alpha\alpha_o)$. Prigogine and co-workers seem to have obtained a non-markoffian equation for $P_t(\alpha\alpha_o)$ itself. Now, however, one is in a rather ambiguous position, since an equation of non-markoffian type contains the past and can therefore be formulated in various ways. Furthermore, there is no example for which the equation has yet been solved. Actually, all we have done on higher order problems uses (6.11) rather than (6.16). For example one can study the behaviour of $P_t(\alpha\alpha_o)$ for long times and see whether equilibrium is reached. Indeed it can be shown that $P_t(\alpha\alpha_o)$ approaches a limit $P_\infty(\alpha\alpha_o)$ and that $P_\infty(\alpha\alpha_o)$ agrees with equilibrium theory. What one proves in fact is

$$\lim_{t\to\infty} \langle\varphi_t|A|\varphi_t\rangle = \langle A\rangle_{eq} \tag{6.18}$$

for smooth diagonal operators $A$. This long-time behaviour is not obtained from equation (6.16), but by studying the singularities of $X_{ll'}(\alpha\alpha_o)$. The rate or functional dependence upon time with which equilibrium is approached is not known in general, but has been studied by considering a very special example [3]. The essential simplification is that the transition rate $W_{ll'}(\alpha\alpha_o)$ is taken independent of the states $\alpha$, $\alpha_o$ for some region and zero elsewhere. This means that in the series expansion for $X_{ll'}(\alpha\alpha_o)$, eq. (6.11), $W_{ll'}$ can be taken in front of the integral signs. This leads to a geometric series and summation gives

$$X_{ll'}(\alpha\alpha_o) = D_l(\alpha_o)\, D_{l'}(\alpha_o)\, \delta(\alpha - \alpha_o) + \lambda^2 \frac{D_l(\alpha)\, D_{l'}(\alpha)\, D_l(\alpha_o)\, D_{l'}(\alpha)\, W_{ll'}}{1 - \lambda^2 W_{ll'} \int\limits_{\alpha_1} D_l(\alpha_1)\, D_{l'}(\alpha_1)} . \tag{6.19}$$

Further analytical approximations have been made in order to carry out the contour integrals of eq. (6.10), among these $W_{ll'}$ has been taken independent of $l$ and $l'$. The final result is of the form

$$P_t(\alpha\alpha_o) = P_\infty(\alpha\alpha_o) + e^{-2\gamma_\alpha t}\,\delta(\alpha - \alpha_o) - 2\,\lambda^2 W e^{-(\gamma_\alpha+\gamma_{\alpha_o})t}$$
$$\times \frac{[(\bar\varepsilon_\alpha - \bar\varepsilon_{\alpha_o})^2 + (\gamma_\alpha^2 - \gamma_{\alpha_o}^2)]\cos[(\bar\varepsilon_\alpha - \bar\varepsilon_{\alpha_o})t] + 2\gamma_{\alpha_o}(\bar\varepsilon_\alpha - \bar\varepsilon_{\alpha_o})\sin[(\bar\varepsilon_\alpha - \bar\varepsilon_{\alpha_o})t]}{[(\bar\varepsilon_\alpha - \bar\varepsilon_{\alpha_o})^2 + (\gamma_\alpha - \gamma_{\alpha_o})^2][(\bar\varepsilon_\alpha - \bar\varepsilon_{\alpha_o})^2 + (\gamma_\alpha + \gamma_{\alpha_o})^2]}$$
$$+ \lambda^2 W e^{-2\gamma_\alpha t} \frac{\gamma_\alpha - \gamma_{\alpha_o}}{\gamma_\alpha[(\bar\varepsilon_\alpha - \bar\varepsilon_{\alpha_o})^2 + (\gamma_\alpha - \gamma_{\alpha_o})^2]} . \tag{6.20}$$

The time rate with which equilibrium is approached is given by the $\gamma_\alpha$'s, certain linewidths, which, however, are no longer small compared to the differences $\bar\varepsilon_\alpha - \bar\varepsilon_{\alpha_o}$ between the shifted energies. The main result is that oscillating terms are found which may possibly be visible in certain cases. Numerical calculations which are being carried out, based on less stringent analytical approximations, indicate that in addition to the exponentially decreasing terms in (6.20) there are other terms which approach zero more slowly for $t \to \infty$. To conclude we write down, for comparison, the corresponding formula for $P_t(\alpha\alpha_o)$ in the weak coupling case

$$P_t(\alpha\alpha_o) = P_\infty(\alpha\alpha_o) + e^{-2\gamma_\alpha t}\,\delta(\alpha - \alpha_o) - \frac{\pi\lambda^2 W}{\gamma_\alpha} e^{-2\gamma_\alpha t}\,\delta(\varepsilon_\alpha - \varepsilon_{\alpha_o}). \quad (6.21)$$

## *References*

1. W. Pauli, Festschrift zum 60. Geburtstage A. Sommerfelds, Hirzel, Leipzig (1928) p. 30.
2. L. Van Hove, Physica 21 (1955) 517.
   Physica 23 (1957) 441.
   Physica 25 (1959) 268.
   L. Van Hove, La théorie des gaz neutres et ionisés, Ecole d'été de physique théorique, Les Houches, 1959, Hermann, Paris (1960) p. 149.
3. L. Van Hove and E. Verboven, Physica 27 (1961) 418.
4. W. Kohn and J. M. Luttinger, Phys. Rev. 108 (1957) 590.
5. R. Kubo, Can. J. of Phys. 34 (1956) 1274.
   J. Phys. Soc. Japan 12 (1957) 570.
6. G. V. Chester and A. Thellung, Proc. Phys. Soc. 73 (1959) 745.
7. E. Verboven, Physica 26 (1960) 1091.

# FUNDAMENTAL PROBLEMS IN STATISTICAL MECHANICS OF IRREVERSIBLE PROCESSES

N. G. VAN KAMPEN
Instituut voor theoretische fysica der Rijksuniversiteit, Utrecht, Netherlands

*Introduction*

1. Our purpose is to understand the following two observed facts.

(i) A collection of many similar particles (atoms, molecules etc.) can be described in a very rough, incomplete fashion by a small number of *macroscopic variables* ; yet this rough incomplete description is self-contained, in the sense that these macroscopic variables obey the phenomenological equations, which are differential equations of a *deterministic* type, i.e. they determine the future values from the initial ones.

(ii) In the microscopic complete description the motions of all individual particles are determined by the familiar differential equations of mechanics (Newton's equations or the Schrödinger equation), which are symmetrical with respect to past and future; yet the phenomenological equations for the macroscopic variables distinguish between past and future.

Fact (ii) makes clear that there cannot be a rigorous mathematical derivation of the macroscopic equations from the microscopic ones. Some additional information or assumption is indispensable. One cannot escape from this fact by any amount of mathematical funambulism. My policy will be to make these additional assumptions explicit rather than do disguise them.

2. Apart from some very special simple models, it is impossible to solve the microscopic equations. Statistical mechanics permits nevertheless to calculate macroscopic quantities from the properties of the individual particles. It should be regarded as a useful *approximation method*, applicable to systems with many degrees of freedom. The nature of this approximation, however, is of an unusual kind. It is not a numerical approximation to the true value of precisely defined quantities, like the approximations used in celestical mechanics. On the contrary, the macroscopic quantities one wants to compute are themselves only defined in the same approximate sense, i.e., in the limiting case of many degrees of freedom. This approx-

imate nature of the macroscopic quantities shows up in the existence of fluctuations.

*Summary*. The microscopic equations of motion completely determine the motion of all particles in full detail. Experience tells us, that a much less detailed, and yet self-contained, approximate description is possible. Statistical mechanics has the task to connect this approximate, macroscopic description with the precise, microscopic description; in particular, to answer the question: how does irreversibility enter?

3. An intermediate level between the microscopic equations and the macroscopic equations is the so-called *master equation*:

$$\frac{dP_J}{dt} = \sum_{J'} \{W_{JJ'} \ P_{J'} - W_{J'J} \ P_J\} \ . \tag{1}$$

Here $J$ denotes the various "states" (to be specified later) of the system; $P_J(t)$ the probability of the system to be in state $J$ at time $t$; $W_{JJ'}$ the probability per unit time to make a transition from $J'$ into $J$. The intuitive meaning of the several terms in (1) is clear.

Equation (1) is no longer invariant for time reversal and, in fact, it will be shown mathematically that its solutions tend to a fixed equilibrium distribution. Thus the main difficulty of statistical mechanics lies in the derivation of (1).

On the other hand, (1) is an equation for the probability distribution over the various states, rather than for the macroscopic state itself. The evolution of the system is here described as a *stochastic process*. In fact, the master equation (1) is nothing but the Chapman-Kolmogorov equation; hence the process is supposed to be a Markov process. In other words, equation (1) determines the probabilities at $t > 0$ once they are known at $t = 0$. Of course, that is not yet a deterministic description of the system in the usual sense.

4. Accordingly the passage from the microscopic (deterministic, reversible, exact) equations to the macroscopic (deterministic, irreversible, approximate) equations breaks up into two steps. The first step, from the microscopic equations to the master equation is the hard one, because it introduces irreversibility. The second step is the derivation of deterministic phenomenological equations from the random process described by (1).

In the first chapter we shall derive the master equation from the classical microscopic equations, and discuss at length the additional assumption to be made.

In the second chapter we shall derive the master equation for the quantum-mechanical case, which requires somewhat more mathematics. The derivation closely parallels the classical one.

In the third chapter we shall obtain some of the conclusions that follow from (1), in particular the approach to equilibrium, and the deterministic macroscopic equations.

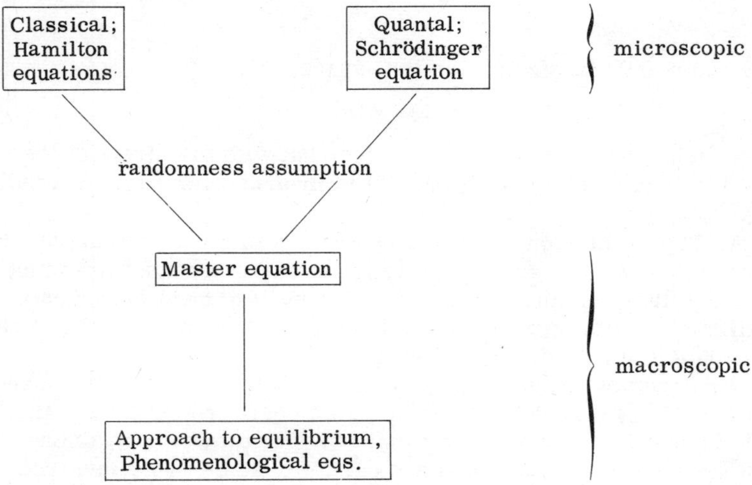


*Chapter I. Classical derivation of the master equation*

5. The instantaneous state of the system is described by $f$ coordinates $q_1, q_2, \ldots, q_f$ and $f$ canonically conjugate momenta $p_1, p_2, \ldots, p_f$. It can be represented by a dot in $2f$-dimensional phase space ("Γ-space"). The motion of the dot * is determined by the classical equations

$$\dot{q}_k = \frac{\partial E(q,p)}{\partial p_k}, \qquad \dot{p}_k = -\frac{\partial E(q,p)}{\partial q_k}. \tag{2}$$

This defines a set of non-intersecting curves in Γ-space; through each fixed point in Γ-space there passes one and only one such curve. The dots move along these trajectories; this constitutes a flow pattern, like a streaming fluid. Liouville's theorem tells that this is the flow pattern of an incompressible fluid.

Usually ** $E(q,p)$ is quadratic in the $p_k$, so that $E(q,-p) = E(q,p)$. It is then clear that from any solution

* I shall use the expression "point in Γ-space" to denote a given set of 2f values of the coordinates $q_1 \ldots p_f$, independent of time. I shall call a "dot in Γ-space" a set of values that change with time according to equation (2).

** This is not true when there is an external magnetic field, or when the whole system is rotating (e.g. centrifuge). In those cases time reversal involves reversal of the magnetic field and of the rotation. For the present discussion they may be excluded.

$$q_k = \varphi_k(t)\,, \qquad p_k = \psi_k(t)$$

of (1) it is possible to obtain a second solution by putting

$$q_k = \varphi_k(-t)\,, \qquad p_k = -\,\psi_k(-t)\,.$$

Thus the point transformation in Γ-space

$$(q,p) \longrightarrow (q,-p)$$

maps each trajectory into another one, but with the direction reversed. This is meant by saying that (2) is invariant for time reversal.

6. The conclusion from this invariance is, that one can never prove on the basis of (2) that entropy (or any other quantity) must tend to a maximum, in other words, that the system must approach an equilibrium state. For, if this is true for one solution of (2), it cannot be true for the reversed solution.

We have not excluded the possibility that, if one selects those solutions of (2) that at $t = 0$ are far from equilibrium, they will tend towards equilibrium. Yet this cannot be true either. For among these selected solutions there are certainly some that come from even further away. If one takes one of those and reverses all $p_k$ (i.e., all velocities) it will go back to where it came from. Thus we have constructed a solution that at $t = 0$ is far from equilibrium and yet does not move towards equilibrium for $t > 0$.

The saving feature is that solutions of this kind are rare. The overwhelming majority ("erdrückende Mehrzahl") of the solutions that at $t = 0$ are in some state far from equilibrium, will move to states nearer equilibrium. The others are exceptions, and describe large fluctuations, which are not impossible but highly improbable (in some sense).

These somewhat rough and preliminary considerations demonstrate that one cannot hope to derive the phenomenological laws (which describe the irreversible approach towards equilibrium) by studying individual solutions of (2). It will be necessary to speak about the average, or probable behavior. That is only possible if one studies collections ("ensembles") of solutions. By choosing such an ensemble one defines the a priori probability of the individual solutions to occur. Whether or not this agrees with the frequency with which they occur in nature can only be decided by experiment. That experiment has indeed an essential rôle in determining the correct a priori probability is shown by the existence of different kinds of statistics (Maxwell-Boltzmann, Bose-Einstein, Fermi-Dirac).

7. The intuitive picture of a flow pattern in Γ-space has to be qualified by an essential remark. The trajectories should not be visualized as smooth curves, but rather as very ragged and knotted. The flow in Γ-space has to be visualized as extremely turbulent.

Although the trajectories are continuous and differentiable (any number of times) on a sufficiently fine scale, the dots in Γ-space change their position and direction of motion so rapidly, that a somewhat coarser observation gives the impression that they jump back and forth discontinuously.

This is entirely analogous to the motion of a Brownian particle. A Brownian particle has, of course, at all times a finite velocity and acceleration. However, the experiments are much too coarse to observe this instantaneous velocity. Therefore one treats Brownian motion both experimentally and theoretically as if it consisted of a succession of instantaneous jumps.

Moreover, in the theory of Brownian motion an essential assumption is that the successive jumps are random, uncorrelated with one another. This is not unreasonable since the velocity varies so rapidly. We shall now describe an analogous property of our dots in Γ-space.

Take two dots in Γ-space very near to each other, representing two microscopically similar states. They will remain near each other for a short while; but, as they both jump about rapidly and irregularly, they will soon move apart. After a time of the order of the duration of a macroscopic observation they will represent two microscopically entirely different states. Hence, although it is true that according to (2) the initial values of $(q, p)$ determine the entire future, it is of little use to the macroscopic observer. The slightest uncertainty in the initial values magnifies so rapidly that soon one knows very little about the microscopic state; it is then only possible to indicate roughly a region in which the dot must be located. This is the reason why, in spite of the microscopic determinism exhibited by (2), we shall nevertheless be able to treat the process on a macroscopic scale as a random process, in complete analogy with Brownian motion.

To show that the state of affairs described here actually prevails, take the example of a dilute gas. As long as no collisions occur in the gas, the representing dot in Γ-space moves along a straight line with constant values of the $p_k$. A collision will have the effect that two $p$'s jump almost discontinuously to new values, after which the dot starts moving along a new straight line in a new direction. In ordinary air the mean time of free flight is of the order $10^{-10}$ sec., so that during any measurement (of temperature, density or composition) every single $p_k$ jumps many times. This shows that the path in Γ-space indeed varies extremely rapidly, as stated above.

Next consider two samples of a gas whose microscopic states are identical but for a slight difference in one $q$; that means that one molecule is slightly shifted. After the first collision there will be a slight difference between the two samples in the position and momentum of this molecule. This will have a magnified effect on the next collision, and after a few times of free flight collisions that take place in one gas are misses in the other and vice versa. This exhibits the *magnification of uncertainty* in the microscopic state.

*Summary*. The position of a dot in Γ-space, representing the state of a system at time $t$, is a function of $t$ and of the initial position at $t = 0$. Both these functional dependences are continuous and differentiable on a microscopic scale. However, the variation is so rapid, that on the coarse scale of the macroscopic observer the motion of the dot looks like a succession of random jumps, similar to Brownian motion.

8. A *physical quantity* is a function $A(q, p)$ in Γ-space. Each observation consists in determining the value of some $A(q,p)$ with a certain precision $\Delta a$. The scale of all values $a$ accessible to $A(q,p)$ may therefore be subdivided in intervals $\Delta a$,

$$\ldots,\ a_{-2},\ a_{-1},\ a_0,\ a_1,\ a_2,\ \ldots,\ a_\nu,\ a_{\nu+1},\ \ldots$$

Each interval $(a_\nu,\ a_{\nu+1})$ corresponds to a slice of Γ-space, determined by

$$a_\nu < A(q,p) < a_{\nu+1} \,.$$

A macroscopic observation of $A$ determines in which slice the dot (representing the microscopic state of our system) is located.

In most cases there will be a large number of macroscopic observables $A^{(r)}(q,p)$. If all their values have been observed, one knows that the dot lies in the region of Γ-space that is common to all the separate slices; that is the region determined by

$$a_\nu^{(r)} < A^{(r)}(q,p) < a_{\nu+1}^{(r)} \,. \qquad \text{(all } r) \tag{3}$$

These regions are called *phase cells*. We shall often label the phase cells with one subscript $J$. Each phase cell has a volume in Γ-space given by

$$G_J = \int_{(J)} dq_1 \ldots dq_f\, dp_1 \ldots dp_f \,,$$

where the integration extends over the $2f$-dimensional region determined by (3).

A complete measurement of all macroscopic quantities $A^{(r)}$ is equivalent with determining in which phase cell the representing dot is located. We shall say that this determines the *macroscopic state* of the system. Of course, there is an infinite number of microscopic states compatible with each macroscopic state, but they cannot be distinguished by an observation. This constitutes the well known *coarse graining*, which is essential in statistical mechanics, as pointed out by the Ehrenfests.

Now consider the motion of the system. From each point inside a particular phase cell a trajectory originates, but all these trajectories move out in different directions, jump back and forth and constitute an extremely complicated pattern. Hence, if the dot representing our system is somewhere inside the phase cell $J$ at $t = 0$, at a later time $t > 0$ it may be in any of the phase cells in the vicinity. The macroscopic observation of the system at $t = 0$ is not by far enough information to predict even only its macroscopic state in the future. Viewed from a macroscopic, or coarse-grained, point of view, the flow in phase space looks rather like a diffusion process.

9. It is also possible to construct a geometrical representation of the macroscopic states of the system by introducing "*a-space*". The coordinates $a_r$ of this new space correspond to the values of the respective macroscopic quantities $A^{(r)}$. Actually this $a$-space consists of discrete points, each of which is associated with one phase cell. Yet the $a$'s may be regarded as continuous variables, provided that only sufficiently smooth functions of the $A^{(r)}$ are taken into consideration. A small region $\Delta a_1 \Delta a_2 \ldots$ in $a$-space corresponds to a large volume in Γ-space, which may be denoted by

$$G(a)\ \Delta a_1\ \Delta a_2\ \ldots\ .$$

The *motion in a-space* has quite a different character than in Γ-space. Each Γ-point carries exactly one trajectory and therefore determines not only the microscopic state of the system, but also the whole future; but the different Γ-points in one phase cell carry different trajectories, which move out in all directions in Γ-space. Now a point in $a$-space represents an infinite number of Γ-points, each carrying its own trajectory. Hence, in contrast with the state of affairs in Γ-space, every $a$-point carries an infinity of trajectories in $a$-space, spreading out in all directions. There can be no equation of motion in $a$-space *. These considerations again lead to the idea of a diffusion process in $a$-space.

This description in terms of $a$-space is not needed for the present discussion, but will be used in section 30.

10. Let us boldly apply the methods of diffusion theory to the motion in Γ-space.

If at $t = 0$ the system is in cell $J'$, it has a *probability* $T_t(J|J')$ to be in any cell $J$ at time $t > 0$. Obviously one must have

$$T_0(J|J') = \delta_{JJ'}\ ; \qquad T_t(J|J') \geqslant 0\ ; \qquad \sum_J T_t(J|J') = 1\ . \tag{4}$$

* That is, of the usual type of an equation involving the quantities $a_r(t)$, $\dot{a}_r(t)$, $\ddot{a}_r(t)$, ..., up to some reasonable order of differentiation. It is easy to see that there does exist a differential equation if one includes derivatives of order $10^{23}$.

The precise definition of this probability is as follows. If we distribute at $t = 0$ a cloud of dots over cell $J$ *with constant density*, then at $t > 0$ a fraction $T_t(J|J')$ of these dots will have moved into cell $J$. In more familiar terms, if we construct an *ensemble* with constant density in $J$ and zero density outside, then a fraction $T_t(J|J')$ of the sample systems will be in $J$ at time $t$. In classical mechanics the concept of probability can only be introduced by means of an ensemble.

11. The definition of $T_t(J|J')$ does not involve any assumption. However, the success of the approach hinges on the validity of the identity

$$T_{t_1+t_2}(J|J') = \sum_{J''} T_{t_2}(J|J'')\ T_{t_1}(J''|J') \ . \tag{5}$$

Mathematical interpretation: The stochastic process used to describe the system macroscopically is a Markov process. Indeed, (5) is the Chapman-Kolmogorov equation.

Physical interpretation: Take at $t = 0$ an ensemble with constant density in $J'$, and zero outside. Let it move during $t_1$; Then in each cell $J''$ there is a fraction $T_{t_1}(J''|J')$ of the ensemble, distributed in some odd way (which could only be computed if the explicit solutions of the microscopic equations of motion were known). Now *redistribute* the dots inside each cell $J''$ so as to obtain a distribution with constant density. After this let the dots move again during $t_2$; finally count the fraction of the original ensemble that has arrived in any cell $J$ at the final time $t_1 + t_2$. Equation (5) asserts that this fraction should be the same as if the dots had moved freely, without being redistributed at some intermediate time. The intermediate redistributing should not affect the final fractions.

Before discussing the validity of (5) we show that it is equivalent with the master equation. For small $\Delta t$ one has *

$$T_{\Delta t}(J|J') = \delta_{JJ'} \{1 - \Delta t \sum_{J''} W_{J''J'}\} + \Delta t\, W_{JJ'} \ . \tag{6}$$

The coefficients $W_{JJ'}$ can be interpreted as the transition probabilities per unit time from $J'$ to $J$. By inserting (6) for $T_{t_2}(J|J'')$ into (5) one obtains the differential form of the Chapman-Kolmogorov equation

$$\frac{\partial}{\partial t} T_t(J|J') = \sum_{J''} \{W_{JJ''}\ T_t(J''|J') - W_{J''J}\ T_t(J|J')\} \ . \tag{7}$$

In order to write this in a more transparent fashion we introduce the occupation probability: $P_J(t)$ is the probability that at time $t$ the system is in cell $J$; hence

* $\Delta t$ should be so small that the macroscopic state of the system does not change very much, but so large that the microscopic variables have time to vary considerably.

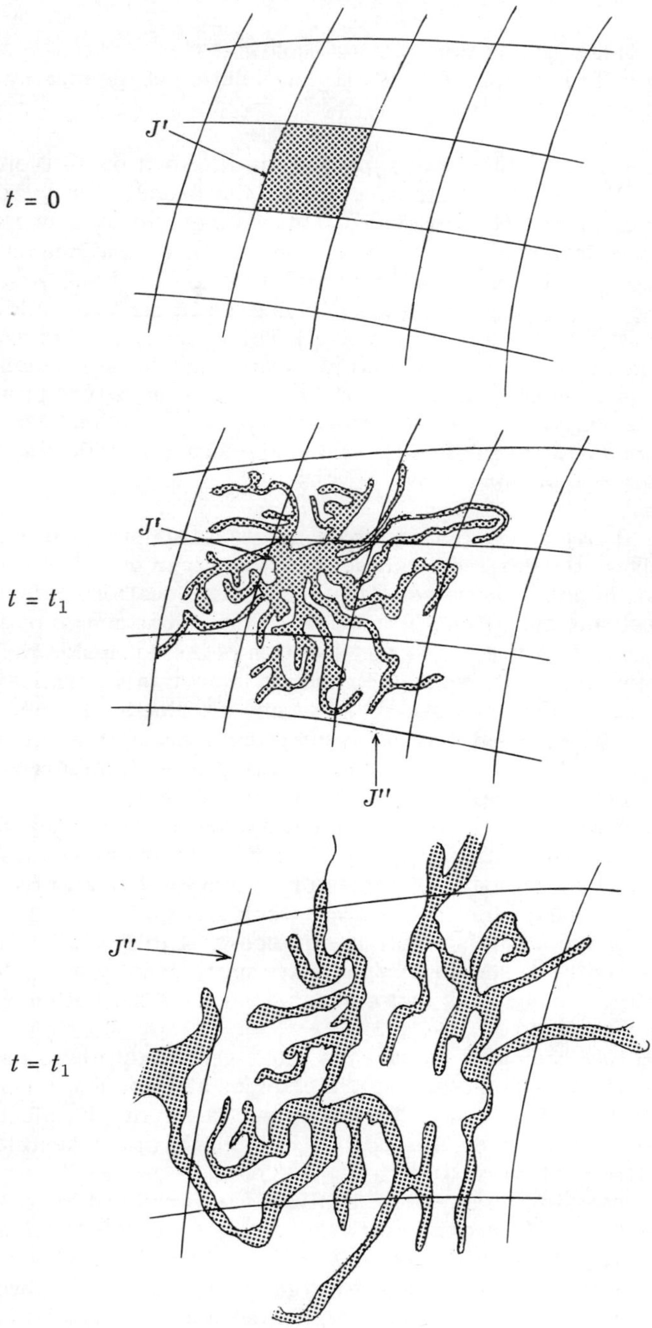
J'
t = 0
J'
t = t1
J''
J''
t = t1

$$P_J(t) = \sum_{J'} T_t(J\,|J')\; P_{J'}(0)\;.$$

Equation (7) now yields the "master equation" (1) for $P_J(t)$. Vice versa, it is easily seen that $T_t(J\,|J')$ is that solution of the master equation that for $t = 0$ reduces to $\delta_{JJ'}$.

*Summary*. The turbulent aspect of the flow pattern in Γ-space has led us to treat the motion as a diffusion process, thus introducing the theory of stochastic processes. This is only fruitful if one also assumes the Chapman-Kolmogorov equation (5), so that one deals with Markov processes. The redistributing erases all information concerning the past history which is contained in the exact microscopic location of the dots in Γ-space.) This assumption is the additional element by which statistical mechanics has to be supplemented in order to obtain irreversible equations. It has the nature of a randomness assumption. It is the general form of the various assumptions known as: Stosszahlansatz, molecular chaos hypothesis, or random phase approximation.

12. a. It was shown that the randomness assumption (5) is sufficient to obtain the desired results. It is also clear that it is necessary. For, in order that a set of deterministic equations exist for the macroscopic quantities, it must be so that the exact microscopic state of the system (i.e., the exact location of the dot inside the phase cell) is immaterial. If that is the case, it is certainly permissible to move around the dots in a phase cell at any intermediate time.

b. It may be argued that the assumption is not unreasonable. It is true that $T_{t_2}(J\,|J'')$ was defined by means of an ensemble which at $t_2 = 0$ had constant density inside $J''$. But it can be seen that actually the fractions arriving in the various cells $J$ are largely independent of the precise distribution at $t_2 = 0$. Indeed, from every small (but not too small) subregion of $J''$ trajectories move out in all directions, so that even an ensemble that at $t_2 = 0$ would be confined to one such subregion, would still distribute itself according to $T_{t_2}(J\,|J'')$. Hence it does not matter if some subregions are more densily populated than others. This suggests that smearing out the distribution will never affect the future.

c. Yet this cannot be universally true. One might for example pick out in $J''$ those points whose trajectories at $t = t_1 + t_2$ arrive in one given phase cell, $J_1$ say. Then construct a density distribution in $J''$ which overemphasizes these points. Obviously this ensemble will not distribute itself according to $T_{t_2}(J\,|J'')$.

The answer to this objection is, that it is very hard to construct such an ensemble. In order to pick those points one has to know all the individual trajectories between $t_1$ and $t_1 + t_2$ and trace them backwards. Moreover these points are not gathered in certain subregions of $J''$, but are arranged on very complicated thin threads. To find them is as hard as to arrange a deck of cards (before shuffling) in such order, that after the shuffling they will appear according to suit and value.

The assumption (5) must now be read as follows. If one starts at $t = 0$ from a constant density in $J'$, then at $t_1 > 0$ each cell $J''$ will contain some of the dots, arranged in some odd, complicated pattern. It is assumed that this odd, complicated pattern is not one of the exceptional distributions mentioned above. The distribution of the dots that arrive in $J''$ from $J'$ is assumed to be uncorrelated with the distribution of those dots in $J''$ that will be in any given $J_1$ after any time $t_2$.

d. This can only be true if the set of macroscopic observables $A^{(r)}$ are chosen just right. In case there are too many $A^{(r)}$ the phase cells are too small, so that the randomness assumptions cannot be correct. If there are too few $A^{(r)}$, they do not fully describe the macroscopic aspect of the system, so that one cannot, of course, expect to have a Markov process. For, some information concerning the past would be contained in the macroscopic quantities that one has neglected to include in the macroscopic equations. That would show up by the fact that the dots are not sufficiently randomly distributed in the phase cells, so that the randomness assumption cannot be right.

e. Thus we have arrived at the crucial question of how to choose the set of macroscopic variables $A^{(r)}$. This seems to me the main problem in statistical mechanics of irreversible processes. In practice one usually knows for given systems from experience which quantities are to be taken. For instance, for a gas: the density, the flow velocity, and the temperature at each point in space. Sometimes one finds additional quantities; for instance in hydrogen gas the ratio of para- and orthohydrogen *. This is not just a matter of asking: which quantities can be observed? Some of the quantities are impossible to observe, and yet show their existence by the presence of a relaxation time.

f. The general problem is: given a Hamiltonian of a many-body system, determine the complete set of macroscopic variables $A^{(r)}$. This problem has not been solved. It may be stated that the $A^{(r)}$ must be slowly varying and that they usually involve the coordinates of many particles. (These same features are also responsible for the fact that they are often just those variables that are accessible to macroscopic observations.) It is usually not so hard to test whether a given quantity is macroscopic; but it is not known how to investi-

* Another example is Hahn's spin-echo experiment. The possibility of re-aligning spins, after they have been de-aligned by their different precession rates, demonstrates that the total magnetization alone does not provide a complete set of macroscopic variables. This case has caused some undue misgivings about the foundations of statistical mechanics [J. M. Blatt, Prog. Theor. Phys. 22, 745 (1959); J. E. Mayer, J. Chem. Phys. 34, 1207 (1961)].

gate whether or not a set of $A^{(r)}$ provides a complete macroscopic description (except experimentally).

g. Many systems possess several "levels of macroscopicality". For slow variations a smaller number of macroscopic variables suffices than for rapid variations. This is what is observed in relaxation experiments (e.g. paramagnetic relaxation). In such systems the spectrum of relaxation times breaks up in several groups, separated by large gaps. If one is interested in variations of a certain periodicity (often induced by external forces), all relaxation times that are much smaller than the period may be neglected. The more rapid the variations are, the more variables have to be taken into account and the smaller the phase cells are.

Another example of this phenomenon is the Brownian particle. As far as actual observations are concerned, a master equation for the position alone suffices. If one also wants to describe the rapid fluctuations in the velocity (which are responsible for the observed displacements), a master equation for both the velocity and the position is necessary (Rayleigh particle).

13. Example: one-component monatomic gas. Subdivide the gas volume in elements, large enough to contain many atoms, but so small that the macroscopic state of the gas is practically homogeneous in each volume element. The elements will be labelled by their position vectors $\vec{R}$. The macroscopic variables $A^{(r)}$ are the mean density $n(\vec{R})$, the mean velocity $\vec{u}(\vec{R})$, and the mean energy per particle $\varepsilon(\vec{R})$ in each volume element.

Why only these? Why not, for instance, the mean cubed velocity? Because $n(\vec{R})$, $\vec{u}(\vec{R})$, $\varepsilon(\vec{R})$ are not affected by the collisions, and hence vary only through interaction with neighboring volume elements, which is a much slower process. The higher moments of the velocity distribution in the volume element, however, are affected by the collisions inside the volume element itself; they therefore vary rapidly and may be regarded as microscopic variables.

Accordingly, a phase cell corresponds to given values of all the $n(\vec{R})$, $\vec{u}(\vec{R})$, $\varepsilon(\vec{R})$. When these quantities are given, there are still many microscopic states possible, corresponding to the many different $\Gamma$-points inside the phase cell. Our general assumption is that for calculating the evolution of the system one need not know the actual microscopic state. It is sufficient to calculate the average behavior for an ensemble with constant density inside the phase cell. Such an ensemble has given values for $n(\vec{R})$, $\vec{u}(\vec{R})$, $\varepsilon(\vec{R})$, but has a Maxwellian velocity distribution inside each volume element. Thus one is led to the idea that the gas may be described by the macroscopic quantities $n(\vec{R})$, $\vec{u}(\vec{R})$, $\varepsilon(\vec{R})$; the missing knowledge concerning the microscopic state is replaced by assuming a local Maxwell distribution. From this picture one may calculate the slow change in

$n(\vec{R})$, $\vec{u}(\vec{R})$, $\varepsilon(\vec{R})$, which yields the ordinary macroscopic hydrodynamical equations of motion (Euler equations).

14. If one also wants to find how the approach to local equilibrium takes place, one has to study the effect of the collisions on the particles in one volume element. In line with the general method of statistical mechanics, one would like to select additional "macroscopic" variables in order to describe these rapid variations of the system. Bogolyubov has suggested that one should take for this the one-particle distribution function $f_1(\vec{r}, \vec{v})$. The remaining microscopic variables are the $f_2$, $f_3$, ..., which describe correlations between positions and velocities of more particles. Our previous phase cells are now subdivided in smaller phase cells, each of them corresponding to one out of all possible forms of $f_1(\vec{r},\vec{v})$. The process of smearing out the distribution inside each of these phase cells then amounts to replacing $f_2$, $f_3$, ... with certain definite functions, which no longer depend on the previous history of the system, but only on the instantaneous form of $f_1(\vec{r},\vec{v})$. That is, one replaces them with functionals of $f_1$, for example

$$f_2(r_1, v_1, r_2, v_2; t) \longrightarrow F_2(r_1, v_1, r_2, v_2 | f_1) .$$

This idea can only be correct if the approach of $f_1$ to the Maxwell distribution is actually much slower than the rate at which $f_2$, $f_3$, ... approach these functionals. The former process requires a time of the order of the time between collisions. Bogolyubov claims that the latter process requires only a time of the order of a duration of a collision. I have not been able to convince myself that this is true.

## *Chapter II. Quantummechanical derivation of the master equation*

15. Our treatment will closely parallel the classical treatment. There is again a microscopic state, obeying a microscopic equation of motion (Schrödinger equation), which is invariant for time reversal. We shall define macroscopic variables and phase cells. At the expense of an unproven but reasonable randomness assumption we shall be able to obtain the master equation.

A difference with the classical theory is that we shall not need to introduce ensembles, since in quantum mechanics probabilities are already built in. More precisely: a single pure wave function (as distinct from a density matrix) for the whole system corresponds to a classical ensemble. However, the microscopic equation involves probability amplitudes, rather than probabilities. Our passage to a macroscopic description will therefore simultaneously perform the transition from complex probability amplitudes to real, non-negative probabilities.

*Warning*. In a naive approach to quantum statistics one sometimes writes down the master equation (1) and identifies the states $J$ with the individual eigenstates of some unperturbed Hamiltonian. The transition probabilities $W_{JJ'}$ are identified with squares of matrix elements of the perturbation. This interpretation of the master equation is incorrect (although it often leads to correct results): it is essential that $J$ refers to macroscopic states, which are collections of numerous eigenstates, as will be shown.

16. The microscopic state of the system is described by a wave function $\psi(q_1, q_2, q_3, \ldots, q_f; t)$ or, in more general terminology, by a vector $\psi(t)$ in Hilbert space. It varies with time according to the quantummechanical equation of motion

$$-\frac{\hbar}{i}\frac{\partial\psi}{\partial t} = E\psi\,, \tag{8}$$

where $E$ is the Hamiltonian operator.

Invariance for time reversal follows from the remark that in most cases $E$, when written in the $q$-representation, is a real (differential) operator *. Hence one has

$$\frac{\hbar}{i}\,\frac{\partial\psi^*}{\partial t} = E\psi^*\,.$$

Consequently, if $\psi(t)$ is a solution of (8), another solution is $\psi^*(-t)$. It follows again, that it is impossible to show from pure quantum mechanics alone, that entropy increases.

17. In classical theory one could talk about the solutions in terms of trajectories in $\Gamma$-space, even though it was entirely impossible to compute them. Similarly one can now study the solutions of (8) by introducing the eigenfunctions $\varphi_n$ and eigenvalues $E_n$ of the operator $E$. All solutions of (8) are obtained by writing

$$\psi(t) = \sum_n a_n\,\varphi_n\,e^{-itE_n/\hbar}\,,$$

where the $a_n$ are arbitrary complex constants.

It is again essential to remark that the $\varphi_n$ are extremely complicated functions of the $f$ coordinates $q_1, \ldots, q_f$. To compute them is even more impossible than to compute the classical solutions.

Moreover it might look natural to list the $\varphi_n$ according to increasing** $E_n$. However, in such a list two successive $\varphi_n$ do not at all look

* For a more general definition of "real" see E. P. Wigner, Gött. Nachr. 31, 546 (1932).

** Since E includes the interaction, there is no degeneracy, except by accident. We shall assume that there is no degeneracy at all, but it is easy to see that accidental degeneracy does not affect the results.

alike. $\varphi_n$ and $\varphi_{n+1}$ describe two physical states in which all physical quantities have entirely different values, except the energy. It will be necessary to order the wave functions according to their macroscopic physical aspect rather than their microscopic energy value.

18. In order to demonstrate this fact, consider one single particle in a rectangular box with periodic boundary conditions. The energy eigenvalues are (in obvious notation)

$$E_{n_x n_y n_z} = \frac{2\pi^2 \hbar^2}{m} \left\{ \frac{n_x^2}{L_x^2} + \frac{n_y^2}{L_y^2} + \frac{n_z^2}{L_z^2} \right\} . \tag{9}$$

The eigenfunctions can be conveniently represented by the lattice points of a rectangular lattice in $k$-space with lattice distances $2\pi/L_x$, $2\pi/L_y$, $2\pi/L_z$. If two lattice points are near to each other the corresponding states are similar, because the three components of the momentum differ little between them.

The energy of each state is, according to (9) proportional to the square of the distance of the corresponding lattice point to the origin. The energies of similar eigenstates are not very different. However, if one lists the lattice points according to their exact distances from the origin, it is geometrically evident that two successive points will, in general, not at all be similar *.

Clearly it is more natural to group similar eigenstates together, even if this entails a slight reshuffling in the energy scale. This will be done for our system with many degrees of freedom. First, however, we have to select the quantities that, like the three components of the momentum in this example, will be used to define "similarity" of states.

19. A physical quantity is an operator $A$. The matrix elements (in the representation in which the total energy $E$ is diagonal) are

$$A_{nm} = \langle \varphi_n | A | \varphi_m \rangle .$$

The time-dependent operator $A(t)$ is the operator with matrix elements

$$A(t)_{nm} = A_{nm}\, e^{\frac{i}{\hbar}(E_n - E_m)}$$

This shows that $A(t)$ is a constant of the motion if $A_{nm}$ is a diagonal matrix, that is, if $A$ commutes with $E$. It also shows that elements $A_{nm}$ that lie far from the diagonal are associated with rapid fluctuations. In fact, a matrix element with $|E_n - E_m|$ of the order of a macroscopically observable energy difference, is associated with

* In this example each energy eigenvalue is eight-fold degenerate, but there is no other degeneracy unless $L_x$, $L_y$, $L_z$ happen to be rationally dependent.

extremely rapid fluctuations. The period of such fluctuations is $\hbar/|E_n-E_m|$, which is much too small to be observed directly as the variation of an observable quantity (owing to the smallness of $\hbar$).

This remark opens up a possibility of disentangling the connection between energy measurements and time measurements, which is typical for quantum mechanics but does not show up in macroscopic observations. *Macroscopic energy differences are connected with time variations that are too rapid to be observed; macroscopic time variations are connected with energy differences that are too small to be measured.*

Accordingly we shall suppose that the macroscopic variables whose rate of change we are interested in, are operators that only have appreciable matrix elements close to the diagonal, corresponding to energy differences much smaller than observable energy differences. This means that these variables are *slowly varying* compared to the other, microscopic ones. Although I am again unable to give a definite characterization of macroscopic variables, the above property seems to be the essential one.

This property may also be stated as follows. The matrix elements of the commutator $[E,A]$ are

$$[E,A]_{nm} = (E_n - E_m)\, A_{nm} \,.$$

This is small since $A_{nm}$ is negligible unless $|E_n - E_m|$ is small. Thus macroscopic quantities are those that almost commute with $E$, that is their commutator with $E$ is much smaller than for microscopic quantities.

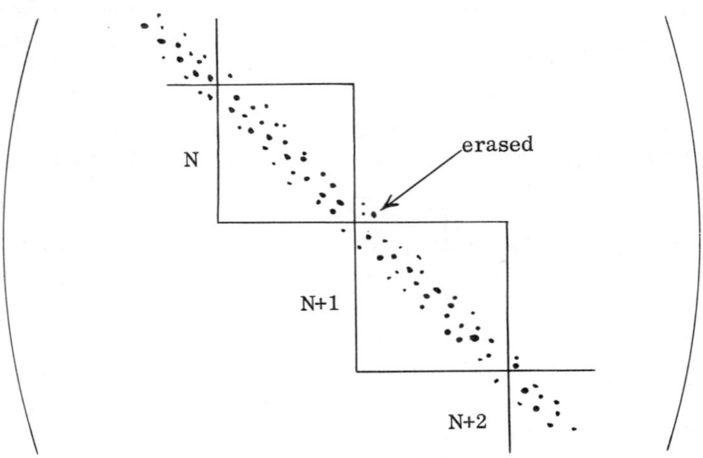

The operator A in the E-representation

20. We shall now construct the analog of classical phase cells. Only the case of one macroscopic quantity $A$ will be considered; the extension to more than one is straightforward.

a. Subdivide the energy spectrum in intervals $\Delta E$, where $\Delta E$ is smaller than observable energy differences, but still so large that it contains many individual levels. These intervals will be called *energy shells*, and distinguished by an index $N$. Inside each shell we take an intermediate energy value $E_N$. A "mutilated energy operator" $\{E\}$ may then be defined by replacing all eigenvalues $E_n$ inside shell $N$ with $E_N$,

$$E = \sum_n |\varphi_n\rangle E_n \langle\varphi_n| = \sum_N \Big[ \sum_{n\in N} E_n |\varphi_n\rangle\langle\varphi_n| \Big] ,$$

$$\{E\} = \sum_N E_N \Big[ \sum_{n\in N} |\varphi_n\rangle\langle\varphi_n| \Big] .$$

Clearly the act of measuring the energy may be described by the operator $\{E\}$ as well as by $E$. However, in the Schrödinger equation describing the evolution of the system, $E$ can not be replaced with $\{E\}$.

b. Erase all matrix elements $A_{nm}$ that link two different energy shells. The new operator $A'$ is given by

$$A = \sum_n \sum_m |\varphi_n\rangle A_{nm} \langle\varphi_m| ,$$

$$A' = \sum_N \Big[ \sum_{n\in N} \sum_{m\in N} |\varphi_n\rangle A_{nm} \langle\varphi_m| \Big] .$$

As $A_{nm}$ is only appreciable if $|E_n - E_m| \ll \Delta E$, there are only a few elements to be erased; hence $A'$ may be used as well as $A$ to describe a macroscopic measurement.

c. The new operator $A'$ commutes with $\{E\}$, so that they may be diagonalized simultaneously. This requires a unitary transformation inside each separate energy shell. The resulting complete set of Hilbert vectors $\chi_\lambda^{(N)}$ has the properties

$$\{E\}\, \chi_\lambda^{(N)} = E_N\, \chi_\lambda^{(N)} ,$$

$$A'\, \chi_\lambda^{(N)} = A_\lambda^{(N)}\, \chi_\lambda^{(N)} .$$

d. Subdivide the eigenvalue spectrum $A_\lambda^{(N)}$ in intervals $\Delta A$, where $\Delta A$ is smaller than observable differences in $A$, but so large that (for each fixed $N$) there are still many $\bar{A}_\lambda^{(N)}$ in every interval. These intervals will be distinguished by the index $\Lambda$. The mutilated, or coarse-grained, operator $\{A\}$ is defined by replacing all eigenvalues $A_\lambda^{(N)}$ with $\lambda\in\Lambda$ by some intermediate value $A_\Lambda^{(N)}$

$$\{A\} = \sum_N \sum_\Lambda A_\Lambda^{(N)} \Big[ \sum_{\lambda\in\Lambda} |\chi_\lambda^{(N)}\rangle\langle\chi_\lambda^{(N)}| \Big] .$$

The act of measuring the physical quantity $A$ may be described by the operator $\{A\}$ as well as by $A$.

e. The $\chi_\lambda^{(N)}$ with $\lambda\in\Lambda$ span a linear subspace; this is the quantum-mechanical phase cell. For convenience and generality we relabel the phase cells with $J$ instead of $N,\Lambda$ and write $\xi_{Ji}$ for an orthogonal set spanning the cell $J$. (One may, of course, take the $\chi_\Lambda^{(N)}$ for this set.) The subscript $i$ runs from $i = 1$ to the number of dimensions, $G_J$ say, of the cell. They constitute a common set of eigenvectors of $\{E\}$ and $\{A\}$,

$$\{E\}\ \xi_{Ji} = E_J\ \xi_{Ji}\ , \qquad \{A\}\ \xi_{Ji} = A_J\ \xi_{Ji}\ .$$

Here $E_J$ and $A_J$ are new notations for $E_N$ and $A_\Lambda^{(N)}$.

*Summary*. We have supposed that $A$ is slowly varying, that is, the commutator $[E,A]$ is small. It was then possible to construct mutilated operators $\{E\}$ and $\{A\}$ such that:

(i) for macroscopic observations $\{E\}$ and $\{A\}$ are indistinguishable from $E$ and $A$;

(ii) $\{E\}$ and $\{A\}$ commute;

(iii) the common eigenvectors belonging to given eigenvalues $E_J, A_J$ constitute a linear subspace, the phase cell. The phase cell is spanned by $G_J$ orthonormal vectors $\xi_{Ji}$, where $G_J$ is still very large.

21. The wave function $\psi$ describing the instantaneous state of the total system may be expanded in the complete set of $\xi_{Ji}$,

$$\psi = \sum_J \sum_i b_{Ji}\, \xi_{Ji}\ .$$

According to the standard interpretation of quantum mechanics, $|b_{Ji}|^2$ is the probability of finding the system in the state $\xi_{Ji}$; hence

$$P_J = \sum_{i=1}^{G_J} |b_{Ji}|^2$$

is the probability of finding the system in phase cell $J$. The quantum-mechanical expectation value of the quantity $A$ is

$$\langle\psi|A|\psi\rangle \approx \langle\psi|\{A\}|\psi\rangle = \sum_J A_J\ P_J\ .$$

Thus we have obtained a macroscopic description of the system, in terms of the probabilities $P_J$ of finding the system in the various phase cells, just like in the classical case. This description is much coarser than the microscopic description, provided by the total set of $b_{Ji}$. Nevertheless it is sufficient to compute expectation values of macroscopic quantities. Note that there is no need to introduce an ensemble of systems: one single wave function $\psi$ already leads to the concept of a probability distribution over phase cells.

22. The main problem remains: to find the time dependence of the $P_J$ from that of $\psi$. One may write

$$\psi(t) = \sum_{\mathrm{Ji}} b_{\mathrm{Ji}}(t)\, \xi_{\mathrm{Ji}} \;.$$

It is easy to see that the time dependence of the coefficients $b$ is of the general form

$$b_{\mathrm{Ji}}(t) = \sum_{\mathrm{J'i'}} \langle Ji\,|U(t)|\,J'i'\rangle\; b_{\mathrm{J'i'}}(0) \;,$$

where $U(t)$ is a unitary matrix, which could only be found explicitly by solving the Schrödinger equation. It follows that

$$P_{\mathrm{J}}(t) = \sum_{\mathrm{J'i'}} \sum_{\mathrm{J''i''}} [\sum_{\mathrm{i}} \langle Ji\,|U(t)|\,J'i'\rangle\langle Ji\,|U(t)|\,J''i''\rangle^{*}]\; b_{\mathrm{J'i'}}(0)\; b^{*}_{\mathrm{J''i''}}(0) \;. \quad (10)$$

This equation shows that $P_{\mathrm{J}}(t)$ is *not* determined uniquely by the $P_{\mathrm{J}}(0)$; rather one needs the full microscopic information about the system at $t = 0$. The reason is that so far we have been doing pure quantum mechanics, without injecting additional assumptions.

23. At this stage one often introduces an ensemble. It is supposed that the equation for $P_{\mathrm{J}}(t)$ should be averaged over a large number of systems with different microscopic states at $t = 0$, but with one and the same macroscopic state. The ensemble is further specified by postulating

$$\overline{b_{\mathrm{J'i'}}(0)\; b^{*}_{\mathrm{J''i''}}(0)}^{\,\mathrm{ens.}} = \delta_{\mathrm{J'J''}}\;\delta_{\mathrm{i'i''}}\;\frac{P_{\mathrm{J}}(0)}{G_{\mathrm{J}}} \;. \quad (11)$$

This amounts to postulating random phases, and equal probabilities for the several states $\xi_{\mathrm{Ji}}$ of a phase cell (in analogy with the constant distribution in the classical case). Clearly this has the desired effect:

$$P_{\mathrm{J}}(t) = \sum_{\mathrm{J'}} [\sum_{\mathrm{ii'}} \langle Ji\,|U(t)|\,J'i'\rangle^{2}\,]\; P_{\mathrm{J'}}(0)/G_{\mathrm{J'}} \;. \quad (12)$$

This equation has the same form that was found in the classical case.

$$P_{\mathrm{J}}(t) = \sum_{\mathrm{J'}} T_{\mathrm{t}}(J\,|\,J')\; P_{\mathrm{J'}}(0) \;. \quad (13)$$

It is easy to check that the $T_{\mathrm{t}}(J\,|J')$ defined here obey the same conditions (4).

24. However, I want to argue that the introduction of an ensemble is neither necessary nor sufficient. As mentioned above (section 21) one does not need an ensemble in quantum mechanics to define the probability distribution over phase cells. It was only used to reduce the double sum over pairs of $b$'s in (10) to a single sum over $P_{\mathrm{J}}(0)$. This can be done in a more economical way.

First note that the right-hand side of (10) consists of an extremely large number of terms, with wildly varying phases. Unless the

$b_{Ji}(0)$ are chosen in a very special way, these terms will cancel each other. The only surviving contribution stems from the terms with $J' = J''$, $i' = i''$, because they are necessarily non-negative and cannot therefore cancel. This remark reduces (10) to

$$P_J(t) = \sum_{J'} \sum_{i'} [\sum_{i} |\langle Ji\,|U(t)|J'i'\rangle|^2]\,|b_{J'i'}(0)|^2 . \tag{14}$$

Secondly we note that for each fixed $J'$ the right-hand side is a sum of the type

$$\sum_{i=1} \alpha_i \, \beta_i \, ,$$

with a large number $G$ of positive quantities $\alpha_i$ and $\beta_i$. The magnitudes of the successive $\alpha$'s vary rapidly, and the same is true for the successive $\beta$'s. Hence one has with good approximation

$$\sum_{i=1} \alpha_i \, \beta_i = \frac{1}{G} (\sum_{i=1} \alpha_i)\, (\sum_{i=1} \beta_i) .$$

Applying this device to (14) one obtains again (12).

Admittedly two rather bold assumptions had to be made to replace the introduction of an ensemble *. However, I feel that these assumptions correspond more accurately with the actual mechanism that is responsible for the validity of (12). For instance, they exhibit clearly the necessity of phase cells containing many microscopic states (coarse graining); whereas the ensemble assumption (11) would justify (12) even if one identifies $J$ with individual states - which is certainly incorrect. Rather *one should regard the use of ensembles as a trick for obtaining the correct results, which is justified by the properties we assumed concerning the terms in (10).*

25. We have now arrived at the same equation (13) of the classical theory. Again this equation is of little use unless one also has the Markov property (5). In the classical case it was argued that this property holds, because the probability for a transition from $J''$ to $J$ is largely independent of the special way in which the dots are distributed in $J''$. A similar argument applies to quantum mechanics. It has been shown that (12) holds for practically all sets of values for the $b_{Ji}(0)$. Hence one may apply the same equation again to $b_{Ji}(t_1)$, with the result

$$P_J(t_1 + t_2) = \sum_{J''} T_{t_2}(J\,|J'')\, P_{J''}(t_1) . \tag{15}$$

* Actually the two assumptions are inseparable, because the result is only independent of the special choice of the $\xi_{Ji}$ in each cell when both assumptions are made. If one would only assume the phases of the $b_{Ji}(0)$ to be random, but not their magnitudes, the result would not be invariant for a unitary transformation of the $\xi_{Ji}$ inside each phase cell.

This equation, together with (13) yields the desired property (5).

It is possible to select values for the $b_{\mathrm{Ji}}$ such that our assumptions are not correct. These are the exceptional states, which are just in the process of building up a large fluctuation (see section 6). We had to assume that (for all times in the period that we observe the system) the $b_{\mathrm{Ji}}(t)$ never assume these exceptional values. This may be called a *"repeated" randomness assumption*, as it assumes something about the $b_{\mathrm{Ji}}(t)$ at a series of successive moments.

It should be emphasized that, owing to the coarse graining, this repeated randomness assumption is considerably weaker than the one used in Pauli's derivation. There it had to be assumed that, at least at a series of successive moments, the phase correlations between *individual* (unperturbed) states are zero. This is manifestly at variance with the Schrödinger equation. Some authors try to justify this strong (too strong) form of the repeated randomness assumption by remarking that any measurement will destroy the phase correlations. However, heat is conducted, diffusion takes place, even if nobody is measuring it. Moreover, in order to destroy the phase correlations between individual states, one would have to measure the unperturbed energy with an accuracy corresponding to the level distance!

Of course, since $b_{\mathrm{Ji}}(t)$ is uniquely determined by the $b_{\mathrm{Ji}}(0)$ through the Schrödinger equation, one may claim that our repeated randomness assumption is actually only an assumption concerning the initial state. However, this is only verbal, not material progress. The fact remains that nobody is able to tell from given initial values $b_{\mathrm{Ji}}(0)$, whether or not the system will obey the macroscopic equation during the next hour, say (unless one could solve the Schrödinger equation explicitly). *In every statistical-mechanical derivation of macroscopic laws the repeated randomness assumption (or some equivalent assumption) is implied.*

It can now also be seen why the introduction of an ensemble is not sufficient. If one has invoked an ensemble, specified by (11), to arrive at (13), it is not possible to use the same device again for deriving (15). For the $b_{\mathrm{J''i''}}(t_1)$ are uniquely determined by the $b_{\mathrm{J'i''}}(0)$ and there is no reason why they should obey the same condition (11). Thus one is forced either to resort to the same randomness assumption we made; or else to suppose that at an arbitrary intermediate time $t_1$ one may replace the ensemble with a new one. As mentioned above, the latter hypothesis is implicit in most of the older work. However, such a drastic intervention in quantum mechanics can only be justified by showing that it does not affect the results. To show this was just the purpose of our work.

*Chapter III. Consequences of the master equation*

26. The result of the preceding work is the master equation

$$\dot{P}_J = \sum_{J'} (W_{JJ'} P_{J'} - W_{J'J} P_J) . \tag{16}$$

The $W_{JJ'}$ refer to the *structure* of the system, while the $P_J(t)$ describe the situation at time $t$. All these quantities are defined in such a way as to be non-negative. Moreover one has at all times

$$\sum_J P_J(t) = 1 , \tag{17}$$

which is clearly compatible with (16). The master equation is on a macroscopic, or coarse grained, level, and is true both classically and quantum-mechanically. (Of course, the *values* of the $W_{JJ'}$ do depend on the underlying theory.) Once this master equation is established, the real difficulty of statistical mechanics - irreversibility - has been mastered. It will be our first task to-day to show that, indeed, the master equation implies approach to equilibrium.

However, the master equation describes the evolution of a probability distribution, and not yet, like true phenomenological equations, the deterministic time dependence of macroscopic quantities themselves. It will be our second task to show that under suitable conditions such deterministic equations can be derived from the master equation.

27. The properties of the solutions of (16) have been studied extensively in the mathematical litterature. In particular, it has been shown that for $t \to \infty$ they (usually) tend to stable asymptotic distributions $P_J(\infty)$. However, we can simplify the proof by invoking an important symmetry property of the transition probabilities $W_{JJ'}$,

$$W_{JJ'} G_{J'} = W_{J'J} G_J . \tag{18}$$

This symmetry relation is also the foundation of the Onsager relations. It can be derived both classically and quantum-mechanically from the invariance for time reversal of the microscopic equations. It is necessary * that both $E$ and $A$ (and any other macroscopic quantities used in the construction of the phase cells) are even functions of the $p$'s. We here give only the quantum-mechanical proof.

In the $q$-representation, both $E$ and $A$ are real. Hence the $\varphi_n$ and the $\chi_\lambda^{(N)}$ are real, so that it is possible to choose real functions for

* Otherwise (18) has to be replaced with a more general relation, see ref. 12. The fact that (18) is not universally true shows that the usual argument, based on the hermiticity of the perturbation term in the Hamiltonian, is incorrect.

the $\xi_{Ji}$. Then the scalar product $\langle\varphi_n|\xi_{Ji}\rangle$ is real, and

$$\langle Ji|U(t)|J'i'\rangle = \sum_n \langle\xi_{Ji}|\varphi_n\rangle\, e^{-iE_n t/\hbar} \langle\varphi_n|\xi_{J'i'}\rangle$$

is a symmetrical matrix. According to its definition (equations (12) and (13)) one therefore has for $T_t(J|J')$

$$T_t(J|J')\, G_{J'} = T_t(J'|J)\, G_J\,.$$

From this follows immediately (18).

28. A second remark concerning the properties of the $W_{JJ'}$ is that they must vanish if $J$ and $J'$ belong to different energy shells. Classically this follows directly from the fact that no trajectories in Γ-space cross from one energy shell to another. Quantum-mechanically it can be shown by studying the operator $U(t)$. An easier proof, however, is the following.

For any function $f$ one has

$$\langle f(E)\rangle \approx \langle f(\{E\})\rangle = \sum_J f(E_J)\, P_J\,.$$

This must be constant during the motion:

$$\sum_{JJ'} f(E_J)\, W_{JJ'}\, P_{J'} = \sum_J f(E_J)\, P_J \sum_{J'} W_{J'J}\,.$$

Take $P_J = \delta_{JJ_0}$, and take for $f(E_J)$ any function that vanishes in the shell $E_{J_0}$. Then the right-hand side is zero, so that

$$\sum_J f(E_J)\, W_{JJ_0}$$

must be zero for all $f$ that vanish on $E_{J_0}$. This can only be true provided that $W_{JJ_0}$ vanishes as soon as $J$ and $J_0$ belong to different energy shells.

Consequently, the master equation breaks up in separate equations for the separate energy shells, unconnected with each other. Each separate equation has the same form as the total master equation, but refers only to phase cells $J$ that belong to a single shell.

If there are additional macroscopic variables that are constants of the motion, the master equation breaks up in even smaller parts. Vice versa, if the master equation breaks up in smaller disconnected parts, it is easy to see that there must be at least one other macroscopic constant of the motion. However, for simplicity we shall suppose that this is not the case.

Thus we have to study the solutions of equations of type (16) where the matrix $W_{JJ'}$ satisfies (18) and cannot be reduced; that is, the cells $J$ do not decompose in two sets between which all transition probabilities vanish.

29. It follows from (18) that a special solution is $P_J(t) = \text{const.}\ G_J$. This solution does not depend on $t$ and therefore describes an equilibrium state *,

$$P_J^{eq} = G_J/G \qquad \text{where} \qquad G = \sum_J G_J \ .$$

We shall now show that any other solution tends to this equilibrium solution for $t \longrightarrow \infty$.

Define an entropy of the probability distribution by

$$S(P_J) = -\sum_J P_J \log (P_J/G_J) = \log G - \sum_J P_J \log (P_J/P_J^{eq}) \ .$$

First note that this is never greater than $\log G$; indeed the second term is equal to

$$\sum_J (P_J \log P_J - P_J \log P_J^{eq} - P_J + P_J^{eq}) \ ,$$

and it can easily be seen that each term of this sum is non-negative (Klein's lemma). Only for $P_J = P_J^{eq}$ does $S$ attain its maximum value $\log G$.

Next one finds with some algebra

$$\begin{aligned} \frac{dS}{dt} &= -\sum_J \dot{P}_J \log \frac{P_J}{G_J} \\ &= -\sum_{JJ'} W_{JJ'}\, G_{J'} \left(\frac{P_{J'}}{G_{J'}} - \frac{P_J}{G_J}\right) \log \frac{P_J}{G_J} \\ &= \tfrac{1}{2} \sum_{JJ'} W_{JJ'}\, G_{J'} \left[\left(\frac{P_{J'}}{G_{J'}} - \frac{P_J}{G_J}\right) \left(\log \frac{P_{J'}}{G_{J'}} - \log \frac{P_J}{G_J}\right)\right] . \end{aligned}$$

There appear two factors, $W_{JJ'} G_{J'}$ and the factor [ ], which are both non-negative. Hence $dS/dt \geqslant 0$. Moreover, $dS/dt = 0$ can only occur if for any $J, J'$ either $W_{JJ'}$ or [ ] vanishes. Thus one finds $P_J/G_J = P_{J'}/G_{J'}$ for any two phase cells that are connected by non-zero transition probabilities. Since we supposed that there are no other macroscopic constants of the motion than the energy, $P_J/G_J$ must be equal for all phase cells in a given energy shell.

Thus we have found that in equilibrium each phase cell in an energy shell is occupied with a probability proportional to its size **. The distribution over the various energy shells is, of course, not determined by the equation, but only by the initial distribution. This constitutes the foundation of equilibrium statistical mechanics.

*Note* that equilibrium has been defined as a special form of the

* Hence equation (18) is nothing but detailed balancing.

** Remember, in classical mechanics the size $G_J$ is identical with the volume in $\Gamma$-space; in quantum-mechanics with the number of dimensions in Hilbert space.

*macroscopic* probability distribution. One usually defines equilibrium in terms of a density matrix diagonal in some "unperturbed" Hamiltonian, which means that all phase correlations between individual microscopic states vanish. This postulates much more detailed information than can be derived from the macroscopic observation that the system is in equilibrium. Actually one should define the familiar ensembles of statistical mechanics in terms of the $P_J$,

$$\frac{P_J^{\text{can.}}}{G_J} = \text{const.}\ e^{-\beta E_J}\ , \qquad \frac{P_J^{\text{microcan.}}}{G_J} = \text{const.}\ \delta_{E_J,E}\ .$$

30. Our second task is to show that this approach to the equilibrium distribution takes place in such a way that it can be described by phenomenological equations, i.e., deterministic differential equations for reasonably well-defined values of the macroscopic variables.

For this purpose we introduce the continuous description (section 9) by putting

$$P_J \longrightarrow P(a_1, a_2, \ldots)\ \Delta a_1\, \Delta a_2 \ldots = P(a)\ \Delta a\ ,$$

$$G_J \longrightarrow G(a)\ \Delta a\ ,$$

$$W_{JJ'} \longrightarrow W(a|a')\ \Delta a\ .$$

This is not just for convenience, but it has the essential effect of ordering the phase cells according to their macroscopic significance. That is, if two phase cells belong to values of $(a_1, a_2, \ldots)$ that are not very different, they describe macroscopic states that are not very different. Hence one may expect $P(a)$, $G(a)$ and $W(a|a')$ to be smooth functions of the $a$'s.

For simplicity we shall again suppose that there is only one macroscopic variable $a$. The master equation now takes the form

$$\frac{\partial P(a,t)}{\partial t} = \int \{W(a|a')\, P(a',t) - W(a'|a)\, P(a,t)\}\ da'\ . \qquad (19)$$

The limits of integration need not be specified. For convenience we assume them to be $-\infty$ and $+\infty$, but this is not essential. The symmetry relation becomes

$$W(a|a')\, G(a') = W(a'|a)\, G(a)\ .$$

31. Let me first describe in an intuitive way how the stochastic process (19) will give rise to a deterministic phenomenological equation for $a$. The equilibrium distribution $P(a,\infty) = G(a)$ is some Gaussian-like * distribution about the equilibrium value of $a$ (for which

* In most cases it is exactly, or almost exactly, a Gaussian.

may be taken zero). The width of $G(a)$ determines the magnitude of the equilibrium fluctuations.

Let us suppose that the initial state is also described by some bell-shaped distribution $P(a, 0)$. If the initial state is *macroscopically different from equilibrium*, the distance between these two distributions must be much larger than the width of each of them. It will be shown that at any intermediate time, the $P(a,t)$ will be some bell-shaped distribution at some intermediate position, with a shape intermediate between those of $P(a,0)$ and $P(a,\infty)$. Thus the whole distribution slides bodily along the $a$-axis, while its width does not vary materially.

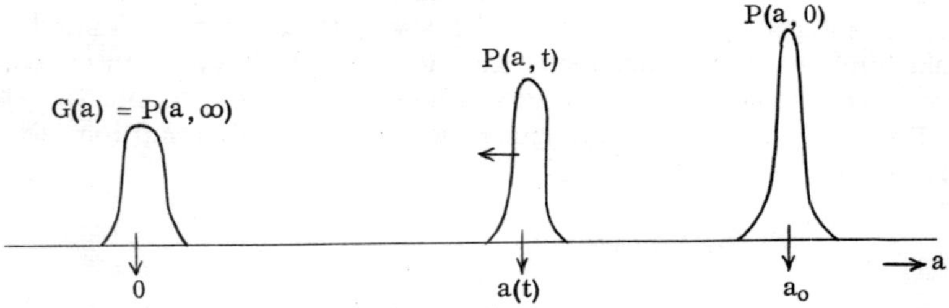


From this picture it becomes clear that there is a pretty well-defined value of $a$, which varies from its initial value $a_0$ to the final value $a = 0$. One may take, for instance, the average

$$\langle a \rangle_t = \int a\, P(a,t)\, da \,.$$

We shall have to show that the fluctuations about this value, determined by the variance

$$\sigma^2 \equiv \langle (a - \langle a \rangle_t)^2 \rangle_t = \langle a^2 \rangle_t - \langle a \rangle_t^2 \,,$$

are relatively small for all $t > 0$.

32. To show this it is convenient to use the abbreviations

$$\alpha_1(a') = \int (a - a')\, W(a\,|a')\, da \,,$$

$$\alpha_2(a') = \int (a - a')^2\, W(a\,|a')\, da \,,$$

which are sometimes called the first and second "derivate moments". One then finds with some algebra

$$\frac{d}{dt} \langle a \rangle = \langle \alpha_1(a) \rangle \,,$$

$$\frac{d}{dt} \langle a \rangle^2 = \langle \alpha_2(a) \rangle + 2\, \langle a\, \alpha_1(a) \rangle \,.$$

These equations are rigorous, but do not constitute a self-contained set, because the averages occurring on the right cannot be expressed in $\langle a \rangle$ and $\langle a^2 \rangle$ alone.

Yet, if $\alpha_1(a)$ and $\alpha_2(a)$ do not vary much *over the range of the fluctuations*, one may expand

$$\alpha_1(a) = \alpha_1(\langle a \rangle) + (a - \langle a \rangle)\, \alpha_1'(\langle a \rangle) + \tfrac{1}{2}\, (a - \langle a \rangle)^2\, \alpha_1''(\langle a \rangle) + \dots .$$

This gives, if terms with second and higher derivatives are omitted,

$$\frac{d}{dt} \langle a \rangle = \alpha_1(\langle a \rangle) . \tag{20}$$

Similarly,

$$\frac{d}{dt} \sigma^2 = \alpha_2(\langle a \rangle) + 2\, \sigma^2\, \alpha_1'(\langle a \rangle) . \tag{21}$$

Equation (20) is the desired phenomenological equation for $\langle a \rangle$ as a function of $t$. Equation (21) serves to verify that, indeed, $\sigma^2$ remains small during the process. The term $\alpha_2(\langle a \rangle)$ on the right exhibits the tendency of $\sigma^2$ to increase linearly with $t$, due to the random jumps of $a$. This tendency, however, is kept in check by the second term ($\alpha_1'$ is always negative). Hence $\sigma^2$ will tend to become equal to

$$\frac{\alpha_2(\langle a \rangle)}{-\, 2\, \alpha_1'(\langle a \rangle)} ,$$

rather than increase indefinitely. The condition for the approximate validity of (20) is therefore

$$\left| \frac{\alpha_2\, \alpha_1''}{4\, \alpha_1'} \right| \ll \alpha_1 .$$

This condition will be satisfied if $\alpha_1''$ is small, that is, if the phenomenological equation (20) does not deviate too strongly from a linear law.

The above somewhat qualitative considerations can be made more rigorous by using an actual power series expansion in a parameter, see reference 18.

33. The most important case is that of *linear phenomenological equations*. Then $\alpha_1(a)$ is a linear function; since it must vanish for the equilibrium value $a = 0$, one has

$$\alpha_1(a) = -\, Ca ,$$

Alternatively one may say that $\alpha_1(a)$ is now expanded *in the whole range* of $a$-values under consideration, and that only the first term is retained. Experimentally it is known that this is correct in a very wide range (e.g., Ohm's law). Similarly we shall suppose that $\alpha_2(a)$ may also be replaced with the first term in its expansion, viz.,

$$\alpha_2(a) \longrightarrow \alpha_2(0) \equiv D \ .$$

In this linear case equations (20) and (21) can be solved explicitly. One finds

$$\langle a \rangle_t = \langle a \rangle_o \, e^{-Ct} \ , \tag{22}$$

$$\sigma_t^2 = \sigma_o^2 \, e^{-2Ct} + \frac{D}{2C} (1 - e^{-2Ct}) \ . \tag{23}$$

The first term of this expression for $\sigma_t^2$ exhibits the fading away of the initial variance; the second term exhibits the building up of the equilibrium value $\sigma_\infty^2 = D/2C$. Clearly if $\sigma_o^2$ is of the same order as $\sigma_\infty^2$ then $\sigma_t^2$ will also be of that order at all times *.

34. In this connection one remark may be added. Suppose one includes in equation (20) the next term,

$$\frac{d}{dt} \langle a \rangle = \alpha_1(\langle a \rangle) + \tfrac{1}{2} \, \sigma^2 \, \alpha_1''(\langle a \rangle) \ .$$

This is no longer an equation for $\langle a \rangle$ alone; rather the rate at which $\langle a \rangle$ decreases also depends on the width of the distribution. Since according to (23) $\sigma^2$ involves a term with $\sigma_o^2$, the way in which $\langle a \rangle_t$ varies with time depends not only on the initial $\langle a \rangle_o$, but also on the width of the initial distribution. Moreover, this term with $\sigma_o^2$ does not die out more rapidly than $\langle a \rangle_t^2$ itself, so that it is *not* even true that after a short initial transient time $\langle a \rangle_t$ obeys a unique law. This fact exhibits the limit of validity of the description in terms of phenomenological equations.

As a consequence one has to be very careful when treating the non-linear case, in particular non-linear fluctuations. For example, the auto-correlation function of $a$ is defined by

$$\langle a(0) \, a(t) \rangle \equiv \int a_o \, P^{eq}(a_o) \, da_o \int a \, P(a_o | a, t) \, da \ .$$

Here $P(a_o | a, t)$ denotes that solution of the master equation that is determined by the initial condition

$$P(a_o | a, 0) = \delta(a - a_o) \ .$$

It is not permissible, as is sometimes done, to use instead of $P(a_o | a, t)$ another solution, namely the one that for $t = 0$ reduces to a

* In fact $\sigma_t^2$ always lies between $\sigma_o^2$ and $\sigma_\infty^2$.

Gaussian distribution of $a$ with mean value $a_0$. Only in the linear case do both procedures amount to the same.

*Literature*

The literature about these problems is enormous, but should be handled with a good deal of critical sense. A comprehensive survey is:

1. D. ter Haar, Revs. Mod. Phys. 27, 289 (1955).

Foundations of kinetic theory with discussions of irreversibility are found in the classic

2. L. Boltzmann, Vorlesungen über Gastheorie (Leipzig 1896 and 1898).

A profound critical discussion of the classical theory is given in

3. P. und T. Ehrenfest, Encyklopädie der mathematischen Wissenschaften 4, no. 32 (1911); English translation: The conceptual foundations of the statistical approach in mechanics (Cornell Univ. Press, Ithaca, N.Y. 1959).

A later discussion, also quantum mechanics, in

4. R. C. Tolman, The principles of statistical mechanics (Oxford 1938).

Some textbooks contain relevant information:

5. D. ter Haar, Elements of statistical mechanics (Wiley, New York 1954).
6. R. Becker, Theorie der Wärme (Berlin 1955).
7. G. Ludwig, Die Grundlagen der Quantenmechanik (Berlin 1954), ch. 5.

Articles that are in particular connected with the present developments:

8. W. Pauli, Sommerfeld Festschrift (Leipzig 1928) p. 30.
9. J. von Neumann, Zeits. Phys. 57, 30 (1929).
10. W. Pauli and M. Fierz, Zeits. Phys. 106, 572 (1937).
11. G. Ludwig, Zeits. Phys. 135, 483 (1953).
12. N. G. van Kampen, Physica 20, 603 (1954).
13. Idem, Fortschritte der Physik 4, 405 (1956).

Mathemetical proofs that the solutions of the master equation tend to an equilibrium distribution can be found in

14. R. von Mises, Wahrscheinlichkeitsrechnung und ihre Anwendung in der Statistik und theoretischen Physik (Leipzig 1931), § 16; M. Fréchet, Méthode des fonctions arbitraires, théorie des événements en chaîne dans le cas d'un nombre fini d'états possibles (Paris 1938); W. Feller, An introduction to probability theory and its applications, Vol. 1 (Wiley, New York 1950), ch. 15; J. L. Doob, Stochastic processes (Wiley, New York 1953).

The symmetry of the transition probabilities for the quantum-mechanical case has been shown in ref. 12, for the classical case in

15. E. P. Wigner, J. Chem. Phys. 22, 1912 (1954); P. Mazur, these proceedings.

See also the discussions in

16. E. C. G. Stueckelberg, Helv. Phys. Acta 25, 577 (1952); J. S. Thomsen, Phys. Rev. 91, 1263 (1953).

The derivation of phenomenological equations from the master equation is given in

17. N. G. van Kampen, Physica 23, 707 and 816 (1957); Physica 25, 1294 (1959).
18. Idem, Can. J. Phys. 39, 551 (1961).

For the theory of Brownian motion, which is the oldest and most instructive illustration, see

19. S. Chandrasekhar, Revs. Mod. Phys. 15, 1 (1953); M. C. Wang and G. E. Uhlenbeck, Revs. Mod. Phys. 17, 323 (1945).

These review articles also contain further references; they have been reprinted in

20. Selected papers on noise and stochastic processes (Edited by N. Wax, Dover, New York 1945).

# STATISTICAL CONSIDERATIONS ON THE BASIS OF NON-EQUILIBRIUM THERMODYNAMICS

P. MAZUR
Instituut Lorentz, University of Leiden, Holland

## 1. *Introduction*

In these lectures we shall discuss the foundation of a number of postulates used in non-equilibrium thermodynamics. We shall not consider the problem of establishing irreversible equations from first principles, but rather take the irreversible behaviour for granted. On this basis we shall derive by statistical mechanical methods the Onsager reciprocal relations, and discuss the so-called fluctuation-dissipation theorem. We shall also discuss the use of thermodynamic functions outside equilibrium.

## 2. *Time reversal invariance*

We consider an adiabatically insulated system, consisting of $N$ point particles *. We shall apply to this system the concepts of classical statistical mechanics. The microscopic state of the system is described by a point

$$(\vec{r}^{\mathrm{N}}, \vec{p}^{\mathrm{N}}) = (\vec{r}_1, \vec{r}_2, \ldots, \vec{r}_{\mathrm{N}}; \vec{p}_1, \vec{p}_2, \ldots, \vec{p}_{\mathrm{N}}) \tag{1}$$

in phase space. Here $\vec{r}_1, \vec{r}_2, \ldots, \vec{r}_{\mathrm{N}}$ are the 3 dimensional position vectors of the particles and $\vec{p}_1, \vec{p}_2, \ldots, \vec{p}_{\mathrm{N}}$ their momenta.

The behaviour of the system is studied by considering a representative ensemble of systems with a probability density $\rho(\vec{r}^{\mathrm{N}}, \vec{p}^{\mathrm{N}}, t)$. The rate of change with time of this probability is given by the Liouville equation

$$\frac{\partial \rho}{\partial t} = [H, \rho] \tag{2}$$

where the right hand side represents the Poisson bracket of the Hamiltonian $H(\vec{r}^{\mathrm{N}}, \vec{p}^{\mathrm{N}})$ of the system and the probability density.

If one introduces the so-called Liouville operator

* The assumption of point particles is not a necessary one. It is made here only for the sake of simplicity in the presentation.

$$-iL(\vec{r}^N,\vec{p}^N) \equiv [H(\vec{r}^N,\vec{p}^N),\ \ ] = \sum_{i=1}^{N} \left(\frac{\partial H}{\partial \vec{r}_i}\frac{\partial}{\partial \vec{p}_i} - \frac{\partial H}{\partial \vec{p}_i}\frac{\partial}{\partial \vec{r}_i}\right) \tag{3}$$

one may rewrite equation (2) in the form

$$\frac{\partial \rho}{\partial t} = -iL\rho \, . \tag{4}$$

This equation has the formal solution

$$\rho(\vec{r}^N,\vec{p}^N,t) = e^{-iLt}\rho(\vec{r}^N,\vec{p}^N,0) \tag{5}$$

where the operator exp $(-iLt)$ is defined through its power series expansion.

On the other hand the rate of change with time of any dynamical function $\alpha(\vec{r}^N,\vec{p}^N)$ is given by

$$\frac{d\alpha}{dt} = iL\alpha \, . \tag{6}$$

This equation has the solution

$$\alpha(\vec{r}^N,\vec{p}^N,t) = e^{iLt}\alpha(\vec{r}^N,\vec{p}^N) \, , \tag{7}$$

where $\vec{r}^N$, and $\vec{p}^N$ are the initial coordinates and momenta of the system.

Since the Hamiltonian is invariant for the transformation $\vec{p} \to -\vec{p}$, the Liouville operator has the property

$$L(\vec{r}^N,\vec{p}^N) = -L(\vec{r}^N,-\vec{p}^N) \, . \tag{8}$$

It follows therefore that

$$e^{iL(\vec{r}^N,\vec{p}^N)t} = e^{-iL(\vec{r}^N,-\vec{p}^N)t} \, . \tag{9}$$

This equality expresses the time reversal invariance of the mechanical equations of motion. It is the basic relation for a derivation of the Onsager reciprocal relations. If an external magnetic field $\vec{B}$ acts on the system the Hamiltonian will be invariant for the transformation $\vec{p} \to -\vec{p}$, only if the magnetic field is also reversed. One then has instead of (8)

$$L(\vec{r}^N,\vec{p}^N;\vec{B}) = -L(\vec{r}^N,\vec{p}^N,-\vec{B}) \, . \tag{10}$$

A similar situation arises for Corioli forces where the angular velocity vector $\vec{\omega}$ must be reversed.

Finally let us consider the time dependent average value $\bar{\alpha}(t)$ of $\alpha(\vec{r}^N,\vec{p}^N)$ in a non-stationary ensemble with probability density $\rho(\vec{r}^N,\vec{p}^N,t)$.

This average value is given by

$$\bar{\alpha}(t) = \langle \rho(\vec{r}^N,\vec{p}^N,t)\,\alpha(\vec{r}^N,\vec{p}^N)\rangle \, , \tag{11}$$

where the brackets denote the integration over phase space, and where we have assumed that $\rho$ is normalized to unity:

$$<\rho> = 1 \ . \tag{12}$$

In view of (5) we may rewrite (11) as

$$\bar{\alpha}(t) = <[e^{-iL(\vec{r}^N, \vec{p}^N)} \ \rho(\vec{r}^N, \vec{p}^N, 0)] \ \alpha(\vec{r}^N, \vec{p}^N)> \ . \tag{13}$$

By using the "unitary" character of exp $(-iLt)$ we may convert this expression into

$$\bar{\alpha}(t) = <\rho(\vec{r}^N, \vec{p}^N, 0) \ e^{iLt} \ \alpha(\vec{r}^N, \vec{p}^N)> = <\rho(\vec{r}^N, \vec{p}^N, 0) \ \alpha(\vec{r}^N, \vec{p}^N, t)> \ . \tag{14}$$

We may therefore also average the values of the dynamical functions at time $t$ over the initial distribution (at time $t = 0$).

The "unitary" property of exp $(-iLt)$ is obtained by partial integration.

### 3. *Extensive state variables and their fluctuations*

An adiabatically insulated system with energy $E$ in statistical equilibrium is represented by the microcanonical ensemble with the stationary probability density

$$\rho^m(\vec{r}^N, \vec{p}^N) = \frac{1}{\Omega(E)} \delta\{H(\vec{r}^N, \vec{p}^N) - E\} \tag{15}$$

where the normalizing factor $\Omega(E)$, the structure function of the system, is given by

$$\Omega(E) = <\delta(H - E)> \ . \tag{16}$$

For the macroscopic description of the system one is not interested in the complete set of mechanical variables describing its microscopic state, but only in a much more restricted number of variables. We may choose for these variables the extensive properties (such as the energies, masses, electric charges, etc.) of macroscopic infinitesimal sub-regions within the system. These regions should on a microscopic scale contain a large enough number of particles so that the principles of statistical mechanics may be applied to them. Let us then denote this restricted set of variables by $\alpha_1, \alpha_2, \ldots, \alpha_n$, where $n$ is much smaller than $N$.

For convenience, a matrix or tensor notation will be introduced. We shall consider the quantities $\alpha_i$ with $i = 1, 2, \ldots, n$ as the components of a vector $\vec{\alpha}$. The variables $\vec{\alpha}$ are functions of the dynamical variables $\vec{r}^N$ and $\vec{p}^N$ of the system:

$$\vec{\alpha} = \vec{\alpha}(\vec{r}^N, \vec{p}^N) \ . \tag{17}$$

Moreover we normalize these variables in such a way that their mean values are zero

$$\bar{\vec{\alpha}} = \langle \vec{\alpha}(\vec{r}^N, \vec{p}^N)\ \rho^m(\vec{r}^N, \vec{p}^N) \rangle = 0 \ . \tag{18}$$

We now introduce the probability $f(\vec{a})\, d\vec{a}$ that the system is in a state for which $\vec{a} < \vec{\alpha} < \vec{a} + d\vec{a}$. For the probability density $f(\vec{a})$ we have the formal expression

$$f(\vec{a}) = \langle \rho^m\ \delta\{\vec{\alpha}(\vec{r}^N, \vec{p}^N) - \vec{a}\} \rangle \ . \tag{19}$$

We shall consider such variables $\vec{\alpha}$ that are sums of large numbers of microscopic (molecular) variables. For the type of extensive properties discussed here, this is always the case. This is clear at once for the number of particles (or the mass) contained in a small region. For the energy of a small subsystem this is also the case since one may neglect the interaction energy between groups of say 100 ÷ 1000 particles (this interaction energy might be considered as a surface effect) and since a large number of such groups occur inside the regions for which we defined the variable $\vec{\alpha}$ (such regions might contain $10^{12} \div 10^{16}$ particles). It can then be shown that the variables $\vec{\alpha}$ behave like sums of large numbers of independent random variables so that the central limit theorem of probability theory may be applied, which leads to the following Gaussian distribution:

$$f(a_1, a_2, \ldots, a_n) = \sqrt{\frac{\mathrm{Det}\,(g_{ij})}{(2\pi k)^n}}\ \exp\left[-\left(\frac{1}{2k}\right) \sum_{i,j=1}^{n} g_{ij}\,(a_i\, a_j)\right] , \tag{20}$$

or in matrix notation:

$$f(\vec{a}) = \sqrt{\frac{|\vec{\vec{g}}|}{(2\pi k)^n}}\ \exp\left[-\left(\frac{1}{2k}\right) \vec{\vec{g}} : (\vec{a}\,\vec{a})\right] . \tag{21}$$

In eqs. (20) and (21) $\mathrm{Det}\,(g_{ij}) = |\vec{\vec{g}}|$ is the determinant of the symmetric positive definite matrix $\vec{\vec{g}}$ with elements $g_{ij}$, $\vec{a}\,\vec{a}$ a dyadic product with elements $a_i\, a_j$, $i, j = 1, \ldots, n$ and $k$ is Boltzmann's constant.

The distribution (21) has variances

$$\overline{\vec{a}\,\vec{a}} = k\, \vec{\vec{g}}^{-1} \ , \tag{22}$$

where $\vec{\vec{g}}^{-1}$ is the reciprocal matrix of $\vec{\vec{g}}$ (i.e. $\vec{\vec{g}}^{-1} \cdot \vec{\vec{g}} = \vec{\vec{g}} \cdot \vec{\vec{g}}^{-1} = \vec{\vec{U}}$, $\vec{\vec{U}}$ being the unit matrix).

In order to introduce a connection between the probability of a state and its entropy $S(\vec{a})$, we make use of Boltzmann's entropy postulate

$$S(\vec{a}) = k \ln f(\vec{a}) + \text{constant} \ . \tag{23}$$

Hence

$$f(\vec{a}) = c \exp\left[\frac{S(\vec{a})}{k}\right] , \tag{24}$$

where $c$ is a constant, or

$$f(\vec{a}) = f(0) \exp \left[\frac{\Delta S}{k}\right] , \tag{25}$$

$$f(0) = c \exp \left[\frac{S(0)}{k}\right] , \tag{26}$$

and

$$\Delta S = S(\vec{a}) - S(0) . \tag{27}$$

By comparing eq. (25) with eq. (21), we find:

$$\Delta S = - \tfrac{1}{2} \overleftrightarrow{g} : \vec{a}\, \vec{a} . \tag{28}$$

Eq. (28) may be regarded as a Taylor series expansion up to order $\vec{a}^2$ of $S(\vec{a})$ around its maximum value $S(0)$. We have the following relation between the matrix $\overleftrightarrow{g}$ and the thermodynamic quantity $S(\vec{a})$:

$$g_{ij} = - \left(\frac{\partial^2 S(\vec{a})}{\partial a_i\, \partial a_j}\right)_{\vec{a}=0} . \tag{29}$$

The state of maximum probability corresponding to the entropy $S(0)$ is often referred to as the "equilibrium state" of the system. On the other hand the word equilibrium may be used in connection with a distribution over possible states: one may call (21) the equilibrium distribution of states. In thermodynamics we do not distinguish between these two concepts of equilibrium although we know that a system in equilibrium does fluctuate around an average state. That the two concepts of equilibrium are almost equivalent may be seen by calculating the average value of $S(\vec{a})$ with (28) and (22):

$$\overline{S(\vec{a})} = S(0) - \tfrac{1}{2} \overleftrightarrow{g} : \overline{\vec{a}\,\vec{a}} = S(0) - \tfrac{1}{2} k\, \overleftrightarrow{g} : \overleftrightarrow{g}^{-1} = S(0) - \tfrac{1}{2} n\, k . \tag{30}$$

Since $S(0)$ is of order $Nk$ and since $N >> n$, the difference between $\overline{S(\vec{a})}$ and $S(0)$ is negligible. In other words, the distribution of states (21) is so sharply peaked that virtually only the "equilibrium state" contributes to the average of $S(\vec{a})$.

For later use we also introduce intensive variables $\vec{X}$ conjugate to the extensive variables $\vec{a}$:

$$\vec{X} = \frac{\partial \Delta S}{\partial \vec{a}} = - \overleftrightarrow{g} \cdot \vec{a} \tag{31}$$

where eq. (28) has been used. From eqs. (22) and (31) we obtain:

$$\overline{\vec{a}\,\vec{X}} = - \overline{\vec{a}\,\vec{a}} \cdot \overleftrightarrow{g} = - k\, \overleftrightarrow{g}^{-1} \cdot \overleftrightarrow{g} = - k\, \overleftrightarrow{U} ,$$

for the correlation between extensive variables and their thermodynamically conjugate intensive variables.

## 4. *Microscopic reversibility; detailed balance*

In section 3 we have discussed the "first" distribution function $f(\vec{a})$. In order to discuss the time behaviour of the macroscopic variables $\vec{\alpha}$, let us now consider the second distribution function $f(\vec{a},\vec{a}';\tau)$. By definition $f(\vec{a},\vec{a}';\tau)\,d\vec{a}\,d\vec{a}'$ is the joint probability to find the system at some initial time in a state for which $\vec{a} \leq \vec{\alpha}(\vec{r}^N,\vec{p}^N) \leq \vec{a} + d\vec{a}$ and at a time $\tau$ later in a state for which $\vec{a}' \leq \vec{\alpha}(\vec{r}^N,\vec{p}^N) \leq \vec{a}' + d\vec{a}'$.

We have for this joint probability density in the microcanonical ensemble the formal expression

$$f(\vec{a},\vec{a}';\tau) = <\rho^m\, \delta(\vec{\alpha} - \vec{a}')\, e^{-iL\tau}\, \delta(\vec{\alpha} - \vec{a})> . \tag{32}$$

This function has the following properties:

1. $$\int f(\vec{a},\vec{a}';\tau)\, d\vec{a}' = f(\vec{a}) \tag{33}$$

2. $$\int f(\vec{a},\vec{a}';\tau)\, d\vec{a} = f(\vec{a}') \tag{34}$$

3. $$f(\vec{a},\vec{a}';0) = f(\vec{a})\, \delta(\vec{a} - \vec{a}') \tag{35}$$

4. $$f(\vec{a},\vec{a}';\tau) = f(\vec{a}',\vec{a};-\tau) . \tag{36}$$

Properties (33) - 35) are trivial consequences of the definition of a joint probability density. They follow immediately from expression (32). The relation (36) expresses the stationarity of the system and is obtained using the unitary character of exp $(-iLt)$ and the fact that $\rho^m$ is stationary. It follows indeed from (32), that

$$f(\vec{a},\vec{a}';\tau) = <[e^{iL\tau}\, \rho^m\, \delta(\vec{\alpha} - \vec{a}')]\, \delta(\vec{\alpha} - \vec{a})> =$$

$$\doteq <\rho^m\, [e^{iL\tau}\, \delta(\vec{\alpha} - \vec{a}')]\, \delta(\vec{\alpha} - \vec{a})> = f(\vec{a}',\vec{a};-\tau) . \tag{37}$$

Furthermore $f(\vec{a},\vec{a}';\tau)$ possesses the property

$$f(\vec{a},\vec{a}';\tau) = f(\vec{a},\vec{a}';-\tau) \tag{38}$$

provided that the $\alpha$'s are even functions of the particle velocities, i.e. provided that

$$\vec{\alpha}(\vec{r}^N,\vec{p}^N) = \vec{\alpha}(\vec{r}^N,-\vec{p}^N) . \tag{39}$$

This is the important property of microscopic reversibility which can be established in the following way. Applying the property of time reversal invariance eq. (9) we have from (32), and using also (39), and the fact that $\rho^m$ is even in the momenta,

$$f(\vec{a},\vec{a}';\tau) = <\rho^m(\vec{r}^N,-\vec{p}^N)\, \delta\{\vec{\alpha}(\vec{r}^N,-\vec{p}^N) - \vec{a}'\}\, e^{iL(\vec{r}^N,-\vec{p}^N)\tau}\, \delta\{\vec{\alpha}(\vec{r}^N,-\vec{p}^N) - \vec{a}\}> . \tag{40}$$

With the transformation of variables $\vec{p}^N \rightarrow -\vec{p}^N$ we then get

$$f(\vec{a},\vec{a}';\tau) = <\rho^m(\vec{r}^N,\vec{p}^N)\ \delta\{\vec{\alpha}(\vec{r}^N,\vec{p}^N) - \vec{a}'\}\ e^{iL(\vec{r}^N,\vec{p}^N)\tau}\ \delta\{\vec{\alpha}(\vec{r}^N,\vec{p}^N) - \vec{a}\}>. \tag{41}$$

Now the last member of (41) is according to (32) equal to $f(\vec{a},\vec{a}';-\tau)$. This establishes the property (38).

As a consequence of both (36) and (38) one also has (for $\vec{\alpha}$'s of the type 39)

$$f(\vec{a},\vec{a}';\tau) = f(\vec{a}',\vec{a};\tau) \tag{42}$$

which expresses the property of detailed balance in $\vec{\alpha}$-space.

It is useful for our further discussion to introduce also the conditional probability density

$$P(\vec{a}\,|\vec{a}';\tau) = \frac{f(\vec{a},\vec{a}';\tau)}{f(\vec{a})}, \qquad (\tau \geq 0). \tag{43}$$

It follows from (33) - (35) and from relation (42) that this function has the properties

$$\int P(\vec{a}\,|\vec{a}';\tau)\,.\,d\vec{a}' = 1 \tag{44}$$

$$\int f(\vec{a})\,P(\vec{a}\,|\vec{a}';\tau)\,d\vec{a} = f(\vec{a}') \tag{45}$$

$$P(\vec{a}\,|\vec{a}';0) = \delta(\vec{a} - \vec{a}') \tag{46}$$

$$f(\vec{a})\,P(\vec{a}\,|\vec{a}';\tau) = f(\vec{a}')\,P(\vec{a}'\,|\vec{a};\tau). \tag{47}$$

The last relation expresses the property of detailed balance in a more usual form. In the derivation of this property we used the time reversal invariance and the even character of $\vec{\alpha}(\vec{r}^N,\vec{p}^N)$. It may occur that one needs, for the macroscopic description of the system, also variables which are odd functions of the particle velocities (e.g. momentum densities). Denoting such variables, in our vector notation, by $\vec{\beta}(\vec{r}^N,\vec{p}^N) = -\vec{\beta}(\vec{r}^N,-\vec{p}^N)$ and their values by $\vec{b}$, one can show from time reversal invariance, that one then has instead of (47)

$$f(\vec{a},\vec{b})\,P(\vec{a},\vec{b}\,|\vec{a}',\vec{b}';\tau) = f(\vec{a}',\vec{b}')\,P(\vec{a}',-\vec{b}'\,|\vec{a},-\vec{b};\tau). \tag{48}$$

The generalization to the case that an external magnetic field acts on the system is straightforward. We shall not consider this case in these lectures.

Up to this point we only considered a stationary ensemble. In our discussion of the mean regression of fluctuations we shall have to consider also non-stationary ensembles. For this reason we now introduce a joint probability density $f(\vec{a},t;\vec{a}',t+\tau)$ for the $\alpha$'s in a non-stationary ensemble, with probability density $\rho(t)$. This joint distribution function is given by the formal expression

$$f(\vec{a},t;\vec{a}',t+\tau) = <\delta(\vec{\alpha} - \vec{a}')\,e^{-iL\tau}\,\rho(t)\,\delta(\vec{\alpha} - \vec{a})> \tag{49}$$

or using the unitary property of exp $(-iL\tau)$, as

$$f(\vec{a},t;\vec{a}';t+\tau) = \langle\rho(t)\ \delta(\vec{\alpha} - \vec{a})\ e^{iL\tau}\ \delta(\vec{\alpha} - \vec{a}')\rangle\ . \tag{50}$$

Corresponding to this non-stationary joint distribution function we define a non-stationary conditional distribution function

$$P(\vec{a},t\,|\vec{a}';t+\tau) \equiv \frac{f(\vec{a},t;\vec{a}',t+\tau)}{f(\vec{a},t)}\ , \qquad \tau \geq 0\ . \tag{51}$$

where $f(\vec{a},t) = \langle\rho(t)\ \delta(\vec{\alpha} - \vec{a})\rangle$ is thw time dependent probability density for the $\vec{\alpha}$ variables.

The conditional probability density (51) will at some specific time $t = t_o$ reduce to the stationary conditional probability density (43), provided that at $t = t_o\ \rho(\vec{r}^N, \vec{p}^N, t)$ is of the form

$$\rho(\vec{r}^N, \vec{p}^N, t_o) = \varphi\{\vec{\alpha}(\vec{r}^N, \vec{p}^N)\}\ \delta\{H(\vec{r}^N, \vec{p}^N) - E\}\ , \tag{52}$$

where $\varphi\{\vec{\alpha}(\vec{r}^N, \vec{p}^N)\}$ is some function which depends on $\vec{r}^N$ and $\vec{p}^N$ only through its dependence on $\vec{\alpha}(\vec{r}^N, \vec{p}^N)$. We have indeed in that case, using (50) and (41)

$$P(\vec{a},t_o\,|\vec{a}';t_o+\tau) = \frac{\langle\rho(t_o)\ \delta(\vec{\alpha} - \vec{a})\ e^{iL\tau}\ \delta(\vec{\alpha} - \vec{a}')\rangle}{\langle\rho(t_o)\ \delta(\vec{\alpha} - \vec{a})\rangle}$$

$$= \frac{\langle\rho^m\ \delta(\vec{\alpha} - \vec{a})\ e^{iL\tau}\ \delta(\vec{\alpha} - \vec{a}')\rangle}{\langle\rho^m\ \delta(\vec{\alpha} - \vec{a})\rangle} = P(\vec{a}\,|\,\vec{a}';\tau) \tag{53}$$

or

$$P(\vec{a}, t_o|\,\vec{a}';\, t_o+\tau) = P(\vec{a}\,|\,\vec{a}';\,\tau)\ , \tag{54}$$

a relation (true at $t = t_o$, but not at all times), between two quantities of which one pertains to a non-equilibrium ensemble and the other to a non-equilibrium ensemble. In particular eq. (54) will be valid if a measurement had indicated that the system is at a given initial time $t_o$ in a specified state $\vec{a}_o$ and is therefore represented in phase space, according to the postulate of equal a priori probability, by an ensemble with uniform non-zero density for those points in phase space for which $\vec{a}_o \leq \vec{a}(\vec{r}^N, \vec{p}^N) \leq \vec{a}_o + d\vec{a}_o$ and with zero density elsewhere.

## 5. *Time correlation functions*

In the microcanonical ensemble all simultaneous distribution functions for values of the $\alpha$'s, at any finite number of distinct times are stationary, i.e. invariant under a translation in time. (We have discussed this property for the joint distribution function at two time points). We may therefore say that the microcanonical ensemble generates a stationary stochastic (vector) process $\vec{\alpha}(t)$. A central quantity for the discussion of such a stationary vector process is the time correlation function matrix defined as

$$\vec{\vec{R}}(\tau) = E\ \{\vec{\alpha}(t)\ \vec{\alpha}(t+\tau)\} = \langle\rho^m\ \vec{\alpha}\ e^{iL\tau}\ \vec{\alpha}\rangle \tag{55}$$

where $E$ denotes the expectation value. Since we have already defined the stationary joint distribution function $f(\vec{a},\vec{a}';\tau)$ (cf. eq. (32)), we may also write for $\vec{\vec{R}}(\tau)$ instead of (55)

$$\vec{\vec{R}}(\tau) = \int\int \vec{a}\,\vec{a}'\, f(\vec{a},\vec{a}';\tau)\, d\vec{a}\, d\vec{a}' \tag{56}$$

or, for $\tau \geq 0$ in terms of the conditional distribution function

$$\vec{\vec{R}}(\tau) = \int\int \vec{a}\,\vec{a}'\, f(\vec{a})\, P(\vec{a}\,|\vec{a}',\tau)\, d\vec{a}\, d\vec{a}' . \tag{57}$$

Now from the property of stationarity (36), we find that

$$\begin{aligned}\vec{\vec{R}}(\tau) &= \int\int \vec{a}\,\vec{a}'\, f(\vec{a},\vec{a}';\tau)\, d\vec{a}\, d\vec{a}' = \\ &= \int\int \vec{a}\,\vec{a}'\, f(\vec{a}',\vec{a};-\tau)\, d\vec{a}\, d\vec{a}' = \int\int \widetilde{\vec{a}\,\vec{a}'}\, f(\vec{a},\vec{a}';-\tau)\, d\vec{a}\, d\vec{a}' \\ &= \widetilde{\vec{\vec{R}}}(-\tau) ,\end{aligned} \tag{58}$$

where $\widetilde{\vec{\vec{R}}}$ and $\widetilde{\vec{a}\,\vec{a}}$ are the transposed matrices of $\vec{\vec{R}}$ and $\vec{a}\,\vec{a}$ respectively ($\widetilde{R}_{ij} = R_{ji}$).

On the other hand we have for $\vec{\alpha}$ type variables (i.e. even in the velocities), using the property of detailed balance in $\vec{\alpha}$ space).

$$\begin{aligned}\vec{\vec{R}}(\tau) &= \int\int \vec{a}\,\vec{a}'\, f(\vec{a},\vec{a}';\tau)\, d\vec{a}\, d\vec{a}' = \int\int \vec{a}\,\vec{a}'\, f(\vec{a}',\vec{a};\tau)\, d\vec{a}\, d\vec{a}' \\ &= \int\int \widetilde{\vec{a}\,\vec{a}'}\, f(\vec{a},\vec{a}';\tau)\, d\vec{a}\, d\vec{a}' = \widetilde{\vec{\vec{R}}}(\tau) .\end{aligned} \tag{59}$$

In the more general case, when both $\vec{\alpha}$ and $\vec{\beta}$ type variables are taken into account, one finds in a similar way the relations

$$\vec{\vec{R}}_{\alpha\alpha}(\tau) = \widetilde{\vec{\vec{R}}}_{\alpha\alpha}(\tau) \tag{60}$$

$$\vec{\vec{R}}_{\alpha\beta}(\tau) = -\widetilde{\vec{\vec{R}}}_{\beta\alpha}(\tau) \tag{61}$$

$$\vec{\vec{R}}_{\beta\beta}(\tau) = \widetilde{\vec{\vec{R}}}_{\beta\beta}(\tau) , \tag{62}$$

where

$$\vec{\vec{R}}_{\alpha\beta}(\tau) = E\,\{\vec{\alpha}(t)\;\vec{\beta}(t+\tau)\} \tag{63}$$

$$\vec{\vec{R}}_{\beta\beta}(\tau) = E\,\{\vec{\beta}(t)\;\vec{\beta}(t+\tau)\} . \tag{64}$$

It is by means of the relations (60) - (62) which are a consequence of time reversal invariance that we shall establish in the next section the reciprocal relations. Let us first however discuss another important property of the correlation function matrix. According to the

Wiener-Khinchin theorem the matrix $\vec{\vec{R}}$ has a Fourier integral representation

$$\vec{\vec{R}}(\tau) = \tfrac{1}{2}\int_{-\infty}^{+\infty} \vec{\vec{S}}(\omega)\, e^{-i\omega\tau}\, d\omega \tag{65}$$

where the matrix

$$\vec{\vec{S}}(\omega) = \frac{1}{\pi}\int_{-\infty}^{+\infty} \vec{\vec{R}}(\tau)\, e^{i\omega\tau}\, d\tau \tag{66}$$

is called the spectral density matrix, and has the property that

$$\vec{\eta}\,.\,\vec{\vec{S}}(\omega)\,.\,\vec{\eta}^{*} \geq 0 \tag{67}$$

where $\vec{\eta}$ is an arbitrary $n$ dimensional complex vector and $\vec{\eta}^{*}$ the complex conjugate vector. (For the case of a single variable $\alpha$ (67) simply implies that the spectral density function $S(\omega)$ is positive.) From (66) it also follows with the stationarity condition (58) that

$$\vec{\vec{S}}(\omega) = \tilde{\vec{\vec{S}}}{}^{*}(\omega) \tag{68}$$

where $\vec{\vec{S}}^{*}(\omega)$ is the complex conjugate of $\vec{\vec{S}}(\omega)$.

Furthermore it follows from (66) and the fact that $\vec{\vec{R}}(\tau)$ is real, that

$$\vec{\vec{S}}(\omega) = \vec{\vec{S}}^{*}(-\omega)\ . \tag{69}$$

Let us now write $\vec{\vec{S}}(\omega)$ in the form

$$\vec{\vec{S}}(\omega) = \vec{\vec{G}}(\omega) + i\,\vec{\vec{H}}(\omega)\ . \tag{70}$$

We then have from (68) and (69),

$$\vec{\vec{G}}(\omega) = \tilde{\vec{\vec{G}}}(\omega) = \vec{\vec{G}}(-\omega) \tag{71}$$

and

$$\vec{\vec{H}}(\omega) = -\,\tilde{\vec{\vec{H}}}(\omega) = -\,\vec{\vec{H}}(-\omega)\ . \tag{72}$$

Hence the real part of $\vec{\vec{S}}$ is a symmetric matrix and an even function of $\omega$, while the imaginary part is antisymmetric and an odd function of $\omega$. From (67) one finds that $\vec{\vec{G}}(\omega)$ satisfies the inequality

$$\vec{\zeta}\,.\,\vec{\vec{G}}\,.\,\vec{\zeta} \geq 0 \tag{73}$$

where $\vec{\zeta}$ is a real $n$ dimensional vector.

We may now rewrite (65) with (70), (71) and (72) in the form

$$\vec{\vec{R}}(\tau) = \int_{0}^{\infty} \vec{\vec{G}}(\omega)\cos\omega\tau\; d\omega + \int_{0}^{\infty} \vec{\vec{H}}(\omega)\sin\omega\tau\; d\omega\ . \tag{74}$$

Up to this point we have not yet investigated the influence of microscopic reversibility on the spectral density matrix. For $\vec{\alpha}$ type variables microscopic reversibility leads to the symmetric character of $\vec{\vec{R}}(\tau)$ (cf. eq. (59)). It follows from (66) that $\vec{\vec{S}}(\omega)$ must then also be symmetric so that one has for this case in view of (70) - (72)

$$\vec{\vec{H}}(\omega) = 0 \, . \tag{75}$$

The generalization to the case of both even and odd ($\vec{\alpha}$- and $\vec{\beta}$-type) variables and the case, that an external magnetic field is present, is straight forward. We wish at this point to make a few remarks concerning correlation functions in general. The spectrum of a correlation function may be either continuous or discrete. If it is discrete, it follows from (65) that the correlation function is an almost periodic function. If it is (absolutely) continuous it follows from the Lemma of Riemann Lebesgue that the correlation function $\vec{\vec{R}}(\tau)$ has the property

$$\lim_{\tau\to\infty} \vec{\vec{R}}(\tau) = 0 \, . \tag{76}$$

In order for a process to be irreversible its correlation function must surely satisfy the requirement (76). Now for a finite conservative mechanical system any of the correlation functions for the macroscopic variables $\vec{\alpha}$ must be an almost periodic function as a consequence of Poincaré's theorem (the existence of Poincaré cycles). The corresponding spectrum must therefore be discrete. One may, however, expect that for a proper choice of $\vec{\alpha}$-variables the spectrum becomes continuous in the limit of an infinite system so that the property (76) then holds. For our further discussion we shall proceed on this assumption.

## 6. *Reciprocal relations*

It is empirically known that for macroscopic values $\vec{a}_0$ of the variables $\vec{a}$ (i.e. for $\vec{a}_0\,\vec{a}_0 >> k\,\vec{\vec{g}}^{-1}$), the conditional averages of the $\vec{a}$ obey linear differential equations of the first order.

$$\frac{\partial \bar{\vec{a}}^{\vec{a}_0}(t)}{\partial t} = - \vec{\vec{M}} \cdot \bar{\vec{a}}^{\vec{a}_0}(t) \, . \tag{77}$$

Here the conditional averages are defined as

$$\bar{\vec{a}}^{\vec{a}_0}(t) = \int \vec{a}\, P(\vec{a}_0|\vec{a};t)\, \mathrm{d}\vec{a} \, , \tag{78}$$

and $\vec{\vec{M}}$ is a matrix of phenomenological coefficients. The conditional probability in eq. (78) refers to a non-stationary ensemble with initially (at time $t = 0$) uniform non-zero density in the range $(\vec{a}_0, \vec{a}_0 + \mathrm{d}\vec{a}_0)$ and with zero density elsewhere in phase space. According to our previous discussion (cf. eq. (54)) this function is identical with the stationary conditional probability.

For the derivation of the Onsager reciprocal relations, it will be necessary to make the assumption that eq. (77) is also valid in the fluctuation region ($\vec{a}_0\,\vec{a}_0 \simeq k\,\vec{\vec{g}}^{-1}$). This assumption is in agreement, for instance, with Svedberg's and Westgren's experiments on colloid statistics. These experiments show that the average behaviour of "small" density fluctuations is in perfect agreement with the macroscopic law of diffusion.

The solution of eq. (77) has the form

$$\vec{\vec{a}}^{\vec{a}_0}(t) = \exp[-\vec{\vec{M}}t] \cdot \vec{a}_0 , \tag{79}$$

where the matrix

$$\exp[-\vec{\vec{M}}t] \equiv \sum_{n=0}^{\infty} \frac{(-\vec{\vec{M}}t)^n}{n!} , \tag{80}$$

operates on $\vec{a}_0$.

It follows from (78) and (79) that

$$\int \vec{a}\, P(\vec{a}_0|\vec{a};t)\, d\vec{a} = e^{-\vec{\vec{M}}t} \cdot \vec{a}_0 . \tag{81}$$

We now multiply both members of (81) by $\vec{a}_0 f(\vec{a}_0)$ and integrate over $\vec{a}_0$. We then get, with (57) and (22)

$$\vec{\vec{R}}(t) = k\, \vec{\vec{g}}^{-1} \cdot e^{-\tilde{\vec{\vec{M}}}t} , \qquad (t \geq 0) . \tag{82}$$

The linear laws (77) therefore imply that the correlation function matrix for the stationary processes $\vec{a}(t)$, is of the special form (82). For the case of variables that are even in the particle velocities, it follows from microscopic reversibility (60) that

$$\vec{\vec{g}}^{-1} \cdot e^{-\tilde{\vec{\vec{M}}}t} = e^{-\vec{\vec{M}}t} \cdot \vec{\vec{g}}^{-1} \tag{83}$$

or

$$\vec{\vec{g}}^{-1} \cdot \tilde{\vec{\vec{M}}} = \vec{\vec{M}} \cdot \vec{\vec{g}}^{-1} . \tag{84}$$

These symmetry relations satisfied by the matrix $\vec{\vec{M}}$ represent the Onsager reciprocity theorem and are therefore a consequence of microscopic reversibility, or more fundamentally of time reversal invariance. It is seen that the relations (84) involve the matrix $\vec{\vec{g}}$, whose elements are thermodynamic quantities. If one introduces the matrix

$$\vec{\vec{L}} = \vec{\vec{M}} \cdot \vec{\vec{g}}^{-1} \tag{85}$$

the reciprocal relations (84) take the simple form

$$\vec{\vec{L}} = \tilde{\vec{\vec{L}}} \tag{86}$$

while the regression laws (77) may be written in the form

$$\frac{\partial \vec{\vec{a}}^{\vec{a}_0}(t)}{\partial t} = -\vec{\vec{M}} \cdot \vec{\vec{g}}^{-1} \cdot \vec{\vec{g}}\, \vec{\vec{a}}^{\vec{a}_0}(t) = +\vec{\vec{L}} \cdot \vec{\vec{X}}^{\vec{a}_0}(t) \tag{87}$$

where we have used the definition (31) of the variables $\vec{X}$, thermodynamically conjugate to the variables $\vec{a}$. We have then established a connection between the form (87) of the regression law for which the symmetry (86) is valid, and thermodynamics. One of the problems of non-equilibrium thermodynamics is precisely to find the proper "thermodynamic forces" $\vec{X}$ conjugated to the $\vec{\alpha}$ variables.

The generalization of (86) to the case of both $\vec{\alpha}$- and $\vec{\beta}$-type variables is straightforward and follows with (60) (61) (62). One then gets the full scheme of the so-called Onsager-Casimir reciprocal relations.

The content of the derivation given above is the following: if linear regression laws are accepted to exist, then necessarily, the Onsager reciprocal relations hold; the problem of establishing these laws from first principles lies outside the scope of this discussion. It is a problem which one has to face already for the case of one single variable $\vec{\alpha}$, when the problem of the reciprocal relations does not yet arise.

We may also remark that linear laws of the type (77) can only hold on a "microscopic time scale". Indeed it follows from (82) and (85), and considering for simplicity, the case of a single variable $\vec{\alpha}$, that

$$L = -k^{-1} \lim_{t \downarrow 0} \frac{\partial R(t)}{\partial t} . \tag{88}$$

On the other hand, we find from the definition (55) of $R(t)$

$$\lim_{t \downarrow 0} \frac{\partial R(t)}{\partial t} = \langle \rho^m \, \alpha \, \dot{\alpha} \rangle = 0 \tag{89}$$

due to the stationarity of the microcanonical ensemble. Hence the assumption of a linear equation of the type (77) or equivalently, of the existence of the finite limit (88) can only hold on a "macroscopic type scale" (i.e. for times larger than some characteristic small time $\tau_0$ but small compared to the relaxation time $M^{-1}$). One could also say that the regression laws can only hold for "sufficiently" long times.

## 7. *The fluctuation dissipation theorem*

In the previous section we have discussed the influence of microscopic reversibility on the coefficients of the "regression laws" (77). On the other hand for systems in which some properties (described by our variables $\vec{\alpha}$) "respond" to an external driving force, we can study the influence of microscopic reversibility without making any specific assumption concerning the mean regression of fluctuations. Let us indeed assume that the Hamiltonian of the system subjected to external forces is of the form

$$H(\vec{r}^N, \vec{p}^N, t) = H^o(\vec{r}^N, \vec{p}^N) - \vec{\alpha}(\vec{r}^N, \vec{p}^N) \cdot \vec{F}(t) \tag{90}$$

where $H^o$ is the Hamiltonian of the system in the absence of external forces. The function $\vec{\alpha}(\vec{r}^N, \vec{p}^N)$ forms a set of dynamical functions of the type discussed previously, while $\vec{F}(t)$ represents a set of time dependent forces, such that

$$\vec{F}(-\infty) = 0 \; . \tag{91}$$

The Liouville equation (2) may now be written as

$$\frac{\partial\rho}{\partial t} = [H^o, \rho] - [\vec{\alpha}, \rho] \, . \, \vec{F}(t) = - i\, L^o \rho - [\vec{\alpha}, \rho] \, . \, \vec{F}(t) \; . \tag{92}$$

Let us consider the case that the system under consideration was in statistical equilibrium and adiabatically insulated before the forces were switched on, and therefore represented by a microcanonical ensemble with density $\rho^m$ at time $t \to -\infty$. With such an initial value for $\rho$ (92) has the solution

$$\begin{aligned} \rho(t) &= \rho^m - \int_{-\infty}^{t} e^{-iL^o(t-t')} [\vec{\alpha}, \rho(t')] \, . \, \vec{F}(t') \, dt \\ &= \rho^m - \int_0^{\infty} e^{-iL^o\tau} [\vec{\alpha}, \rho(t-\tau)] \, . \, \vec{F}(t-\tau) \, d\tau \end{aligned} \tag{93}$$

where the new variable $\tau = t - t'$ has been introduced.

Let us now compute the average value of $\vec{\alpha}$ in the non-stationary ensemble (93). One obtains for this average

$$\bar{\vec{\alpha}}(t) = \langle\rho(t)\, \alpha\rangle = - \int_0^{\infty} \langle\vec{\alpha}\, e^{-iL^o\tau} [\vec{\alpha}, \rho(t-\tau)]\rangle \, . \, \vec{F}(t-\tau) \, d\tau \tag{94}$$

where we have made use of the fact that the average of $\vec{\alpha}$ vanishes in the microcanonical ensemble. Since we shall only be interested in a "response" of $\vec{\alpha}$ linear in $\vec{F}$ we may replace in the right hand side of (94) $\rho(t - \tau)$ by its initial value $\rho^m$, so that we get

$$\bar{\vec{\alpha}}(t) = \int_{-\infty}^{\infty} \vec{\vec{\varkappa}}(\tau) \, . \, \vec{F}(t - \tau) \, d\tau \tag{95}$$

with

$$\vec{\vec{\varkappa}}(\tau) \; \begin{aligned} &= 0 && \text{for} \quad \tau < 0 \\ &= - \langle\vec{\alpha}\, e^{iL^o\tau} [\vec{\alpha}, \rho^m]\rangle && \text{for} \quad \tau \geq 0 \; . \end{aligned} \tag{96}$$

Note that the matrix $\vec{\vec{\varkappa}}(\tau)$ fulfills a so-called causality condition.

Due to the form (15) of $\rho^m$ we have (for $\tau \geq 0$)

$$\begin{aligned} \vec{\vec{\varkappa}}(\tau) &= - \langle\vec{\alpha}\, e^{iL^o\tau} [\vec{\alpha}, H^o] \frac{\partial\rho^m}{\partial H^o}\rangle \\ &= - \langle\vec{\alpha}\, e^{-iL^o\tau}\, iL^o\, \vec{\alpha} \frac{\partial\rho^m}{\partial H^o}\rangle = + \langle\frac{\partial\rho^m}{\partial H^o} (iL^o\, e^{iL^o\tau}\, \vec{\alpha})\, \vec{\alpha}\rangle \\ &= \frac{\partial}{\partial\tau} \langle\frac{\partial\rho}{\partial H^o} (e^{iL^o\tau}\, \vec{\alpha})\, \vec{\alpha}\rangle \end{aligned} \tag{97}$$

where use has been made of the definition (3) of the Liouville operator.

On the other hand one has

$$\frac{\partial \rho^m}{\partial H^o} = \frac{1}{\Omega(E)} \frac{\partial}{\partial H^o} \{\delta(H^o - E)\} = - \frac{\partial \ln \Omega(E)}{\partial E} \rho^m - \frac{\partial \rho^m}{\partial E} , \qquad (98)$$

Since according to statistical thermodynamics

$$\frac{\partial \ln \Omega(E)}{\partial E} = \frac{1}{kT} , \qquad (99)$$

where $T$ is the absolute temperature of the system in the absence of driving forces, we may rewrite eq. (98) in the form

$$\frac{\partial \rho^m}{\partial H^o} = - \frac{1}{kT} \rho^m - \frac{\partial \rho^m}{\partial E} . \qquad (100)$$

Introducing this expression for $\partial \rho^m / \partial H^o$ into (97) we get with (55)

$$\vec{\vec{\varkappa}}(\tau) = - \frac{\partial}{\partial \tau} \{\frac{1}{kT} \tilde{\vec{\vec{R}}}(\tau) + \frac{\partial}{\partial E} \tilde{\vec{\vec{R}}}(\tau)\} . \qquad (101)$$

In the limit as $N \rightarrow \infty$, $E \rightarrow \infty$. $T$ remaining finite, the second term on the right hand side of (101) becomes negligeable compared to the first term. One therefore finally gets for macroscopic systems

$$\vec{\vec{\varkappa}}(\tau) \begin{cases} = 0 & \text{if} \quad \tau < 0 \\ = - \dfrac{1}{kT} \dfrac{\partial \tilde{\vec{\vec{R}}}}{\partial \tau} & \text{if} \quad \tau \geq 0 . \end{cases} \qquad (102)$$

Let us now define a generalized susceptibility matrix as the Fourier transform $\vec{\vec{\varkappa}}(\omega)$ of $\vec{\vec{\varkappa}}(\tau)$

$$\vec{\vec{\varkappa}}(\omega) = \int_0^\infty \varkappa(\tau) \, e^{i\omega\tau} \, d\tau = - (kT)^{-1} \int_0^\infty \frac{\partial \tilde{\vec{\vec{R}}}}{\partial \tau} e^{i\omega\tau} \, d\tau . \qquad (103)$$

This generalized susceptibility determines the response of $\vec{\alpha}$ to a driving force of frequency $\omega$ (cf. eq. (95)). The expression (101) demonstrates that the susceptibility matrix is completely determined by the time correlation function matrix $\vec{\vec{R}}(t)$. By introducing the Wiener Khinchin theorem (65) into (191) we obtain the alternative expression

$$\begin{aligned} \vec{\vec{\varkappa}}(\omega) &= \tfrac{1}{2} (kT)^{-1} \int_0^\infty \int_{-\infty}^{+\infty} i\omega' \, \tilde{\vec{\vec{S}}}(\omega') \, e^{i(\omega - \omega')} \, d\tau \, d\omega' \\ &= \tfrac{1}{2} (kT)^{-1} \int_{-\infty}^{+\infty} i\omega' \, \tilde{\vec{\vec{S}}}(\omega') \, \{\pi \, \delta(\omega - \omega') + \frac{1}{i(\omega - \omega')}\} \, d\omega' \\ &= \frac{\pi i\omega}{2kT} \tilde{\vec{\vec{S}}}(\omega) + \frac{1}{2kT} \int_{-\infty}^{+\infty} \frac{\omega' \, \tilde{\vec{\vec{S}}}(\omega)}{\omega - \omega'} \, d\omega' . \end{aligned} \qquad (102)$$

In the integral at the right hand side of this equation, which connects the susceptibility matrix to the spectral density matrix, one has to take the principal value. The consequences of microscopic reversibility for $\vec{\vec{\kappa}}(\omega)$ can now easily be established. In fact the symmetry properties of $\vec{\vec{\kappa}}(\omega)$ will be the same as those of the spectral density matrix $\vec{\vec{S}}(\omega)$. For $\vec{\alpha}$ type variables the matrix $\vec{\vec{S}}$ reduces to its symmetric real part $\vec{\vec{G}}$ (cf. eqs. (70) and (71)) so that one has for this case

$$\vec{\vec{\kappa}}(\omega) = \tilde{\vec{\vec{\kappa}}}(\omega) . \tag{103}$$

If we now split $\vec{\vec{\kappa}}(\omega)$ into its real and imaginary parts

$$\vec{\vec{\kappa}}(\omega) = \vec{\vec{\kappa}}'(\omega) + i\, \vec{\vec{\kappa}}''(\omega) \tag{104}$$

we obtain from (104) (with $S(\omega) = G(\omega)$, a real symmetric matrix)

$$\vec{\vec{\kappa}}'(\omega) = (2\,kT)^{-1} \int_{-\infty}^{+\infty} \frac{\omega'\, G(\omega')}{\omega - \omega'}\, d\omega' \tag{105}$$

and

$$\vec{\vec{\kappa}}''(\omega) = \pi (2\,kT)^{-1}\, \omega\, \vec{\vec{G}}(\omega) . \tag{106}$$

Note that by combining these two equations we get

$$\vec{\vec{\kappa}}'(\omega) = \frac{1}{\pi} \int_{-\infty}^{+\infty} \frac{\vec{\vec{\kappa}}''(\omega')}{\omega - \omega'}\, d\omega' . \tag{107}$$

This equation which connects the real and the imaginary parts of the generalized susceptibility matrix, is in fact one of the Kramers-Kronig (or dispersion) relations for $\vec{\vec{\kappa}}(\omega)$, and a consequence of the fact that $\vec{\vec{\kappa}}(\tau)$ satisfies a causality condition.

With the result (106) we may now rewrite the Wiener-Khinchin theorem eq. (74) for the case of $\vec{\alpha}$ type variables, in the form

$$\vec{\vec{R}}(\tau) = \frac{kT}{\pi} \int_{-\infty} \cos \omega t \frac{\vec{\vec{\kappa}}''(\omega)}{\omega}\, d\omega \tag{108}$$

and

$$\frac{\vec{\vec{\kappa}}''(\omega)}{\omega} = \frac{1}{2\,kT} \int_{-\infty}^{+\infty} \vec{\vec{R}}(\tau) \cos \omega\tau\, d\omega . \tag{109}$$

These formulae represent the fluctuation dissipation theorem due to Callen and Greene: it connects the correlation function matrix, which characterizes the time behaviour of fluctuations in an equilibrium system to the imaginary part of the susceptibility matrix which is a measure for the dissipation of energy in the system. (Indeed the absorption of energy per period is for a single variable, proportional to $\frac{1}{2}\, \omega\kappa''(\omega)$). Note that a reversible, almost periodic, character of a correlation function corresponds to a discrete spectrum of $\kappa''(\omega)$ (i.e. an absorption spectrum with sharp lines). Irreversible behaviour is characterized by the property (76) of a correlation function and therefore by a continuous absorption spectrum (true dissipation). The property of microscopic reversibility leads as we have seen to a number of relations amongst the elements of the susceptibility matrix which are determined from absorption experiments.

Let us finally consider the case of external forces such that

$$\vec{F}(t)\begin{cases} = \vec{F} & \text{if} \quad t \leq 0 \\ = 0 & \text{if} \quad t > 0 \end{cases} \tag{110}$$

when $\vec{F}$ is a vector with constant components.

We then find from (95) and (102)

$$\begin{aligned} \bar{\vec{\alpha}}(t) &= \int_t^{\infty} \vec{\vec{\varkappa}}(\tau)\, d\tau \;.\; \vec{F} = -\frac{1}{kT}\int_t^{\infty} \frac{\partial \tilde{\vec{\vec{R}}}(\tau)}{\partial t}\, d\tau \;.\; \vec{F} \\ &= \frac{1}{kT}\tilde{\vec{\vec{R}}}(t) \;.\; \vec{F} \end{aligned} \tag{111}$$

where we have also assumed that $\vec{\vec{R}}(t)$ satisfies the requirement (76) i.e. that the process $\vec{\alpha}(t)$ is irreversible. Thus the correlation function matrix describes the relaxation of the mean value $\bar{\vec{\alpha}}(t)$ after the forces are switched off.

Equation (111) may alternatively be written as

$$\bar{\vec{\alpha}}(t) = \tilde{\vec{\vec{R}}}(t) \;.\; \tilde{\vec{\vec{R}}}^{-1}(0) \;.\; \bar{\vec{\alpha}}(0) \;.$$

These linear relations are equivalent to the regression laws (77) only if $\vec{\vec{R}}(t)$ has the exponential form (82). This demonstrates the more general character of the analysis of the present section.

## 8. *The gaussian Markoff process*

In the preceding sections only some aspects of the processes $\vec{a}(t)$ have been considered. In particular, besides some results derived from the principles of statistical mechanics, it was postulated in section 6 that the empirical laws, valid for the mean regression of large fluctuations, also hold for the mean regression of small fluctuations.

On the other hand, one may be interested in the explicit form of the conditional probability density $P(\vec{a}_0\,|\,\vec{a};t)$ itself which describes the processes in more detail. One could then, for instance, calculate the variance $\overline{(a - \bar{\vec{a}}^{\vec{a}_0})(a - \bar{\vec{a}}^{\vec{a}_0})}^{\vec{a}_0}$ in order to discuss the time behaviour of entropy. In view of a more detailed description of this kind it is necessary to make further assumptions about the nature of the process $\vec{a}(t)$.

We shall first establish an integral relation obeyed by the conditional probability in $\vec{\alpha}$-space. It follows from eqs. (50) that:

$$\int f(\vec{a}', t; \vec{a}, t+\tau)\, d\vec{a}' = f(\vec{a}, t+\tau) \;, \tag{112}$$

and therefore, by introducing the conditional probability density $P(\vec{a}', t\,|\vec{a}\,; t+\tau)$, that:

$$f(\vec{a}, t+\tau) = \int f(\vec{a}', t)\; P(\vec{a}'; t\,|\,\vec{a}; t+\tau)\, d\vec{a}' \;. \tag{113}$$

In particular,

$$f(\vec{a}, t+\tau) = \int f(\vec{a}'', 0)\; P(\vec{a}''; 0\,|\,\vec{a}; t+\tau)\, d\vec{a}'' \;, \tag{114}$$

and

$$f(\vec{a}, t) = \int f(\vec{a}'', 0)\; P(\vec{a}''; 0\,|\,\vec{a}; t)\, d\vec{a}'' \;. \tag{115}$$

Let us now suppose that

$$f(\vec{a}, 0) = \delta(\vec{a} - \vec{a}_\circ) \tag{116}$$

and that the density in phase space $\rho(\vec{r}^N, \vec{p}^N; 0)$ is uniform in the region $(\vec{a}_o, \vec{a}_o + d\vec{a}_o)$. The eq. (114) becomes:

$$f(\vec{a}, t+\tau) = P(\vec{a}_o; 0\,|\,\vec{a}; t+\tau) = P(\vec{a}_o\,|\,\vec{a}; t+\tau) \tag{117}$$

and eq. (115):

$$f(\vec{a}, t) = P(\vec{a}_o; 0\,|\,\vec{a}; t) = P(\vec{a}_o\,|\,\vec{a}; t) \;, \tag{118}$$

where eq. (54) has been used. Substituting these results into eq. (113) we obtain:

$$P(\vec{a}_o\,|\,\vec{a}; t+\tau) = \int P(\vec{a}_o\,|\,\vec{a}'; t)\; P(\vec{a}'; t\,|\,\vec{a}; t+\tau)\, d\vec{a}' \;. \tag{119}$$

This equation is an exact relation between the stationary and the non-stationary probability densities. The conditional probability density $(P(\vec{a}'; t\,|\vec{a}; t+\tau)$ is connected with a distribution $\rho(\vec{r}^N, \vec{p}^N, t)$ in phase space which is not uniform over the region $(\vec{a}', \vec{a}' + d\vec{a}')$ and which has developed from the initial distribution $\rho(\vec{r}^N, \vec{p}^N, 0)$. We shall now assume that the non-stationary conditional probability density $P(\vec{a}'; t\,|\vec{a}; t+\tau)$ in the relation (136) may be replaced by the stationary conditional probability density $P(\vec{a}'\,|\,\vec{a}; \tau)$ so that this equation becomes the so-called Smoluchowski equation:

$$P(\vec{a}_o\,|\,\vec{a}; t+\tau) = \int P(\vec{a}_o\,|\,\vec{a}'; t)\; P(\vec{a}'\,|\,\vec{a}; \tau)\, d\vec{a}' \;. \tag{120}$$

This means that we suppose that the density $\rho(\vec{r}^N, \vec{p}^N, t)$ remains "sufficiently" uniform in the course of time over regions $(\vec{a}, \vec{a} + d\vec{a})$ in phase space. This repeated assumption of "randomness" can certainly not rigorously be justified for all times. Its validity is assumed here for time intervals which are macroscopically of interest. It should be noted that the variables a refer to subsystems which are in some sense "weakly coupled". (The interaction energies between the subsystems are surface effects), so that certainly some of the conditions under which Van Hove has actually derived an equation of this type (for sufficiently long time intervals) are satisfied. Processes for which eq. (120) is valid are called Markoff processes.

We shall furthermore assume that in short time intervals the variables change only by small amounts; more specifically we shall assume that:

$$\lim_{t\to 0} \frac{\overline{\Delta\vec{a}}^{\mathrm{a}}}{t} = \lim_{t\to 0} t^{-1} \int \Delta\vec{a}\; P(\vec{a}\,|\vec{a}';t)\; d\vec{a}' = -\vec{\vec{M}} \cdot \vec{a}\;, \tag{121}$$

$$\lim_{t\to 0} \frac{\overline{\Delta\vec{a}\;\Delta\vec{a}}^{\mathrm{a}}}{t} = \lim_{t\to 0} t^{-1} \int \Delta\vec{a}\;\Delta\vec{a}\; P(\vec{a}\,|\vec{a}';t)\; d\vec{a}' = 2\,\vec{\vec{Q}}\;, \tag{122}$$

$$\lim_{t\to 0} \frac{\overline{\Delta\vec{a}\;\Delta\vec{a}\;\Delta\vec{a}\;\ldots\;\Delta\vec{a}}}{t} = 0\;, \tag{123}$$

where $\Delta\vec{a} = \vec{a}' - \vec{a}$ and where in the last formula $\Delta\vec{a}\;\Delta\vec{a}\;\ldots\;\Delta\vec{a}$ stands for an ordered product of more than two factors. The left-hand sides of (121) to (123) contain the "moments" of various orders $\overline{\Delta\vec{a}}^{\mathrm{a}}$, $\overline{\Delta\vec{a}\;\Delta\vec{a}}^{\mathrm{a}}$, etc., of the changes $\Delta\vec{a}$. Of course eq. (121) is identical with the previously introduced regression hypothesis (77). The factor 2 in front of the constant symmetrical positive definite matrix $\vec{\vec{Q}}$ of (122) has been introduced for convenience.

It can now be shown from the Smoluchowski equation (120) together with the conditions (121) - (123) that $P(\vec{a}_0|\,\vec{a};t)$ must satisfy the following partial differential equation:

$$\frac{\partial P}{\partial t} = \tilde{\vec{\vec{M}}} : \frac{\partial}{\partial\vec{a}}(P\vec{a}) + \vec{\vec{Q}} : \frac{\partial^2 P}{\partial\vec{a}\;\partial\vec{a}}\;. \tag{124}$$

Eq. (124) is a generalized Fokker-Planck equation and must be solved with the initial condition, (cf. eq. (46))

$$P(\vec{a}_0|\vec{a};0) = \delta(\vec{a} - \vec{a}_0)\;. \tag{125}$$

Since, according to eq. (45) we have

$$\int f(\vec{a}_0)\; P(\vec{a}_0|\vec{a};t)\; d\vec{a} = f(\vec{a})\;, \tag{126}$$

It follows by multiplying (124) with $f(\vec{a}_0)$ and integrating over $\vec{a}_0$ that $f(\vec{a})$ is a stationary solution of the Fokker-Planck equation:

$$\tilde{\vec{\vec{M}}} : \frac{\partial}{\partial\vec{a}}(f\vec{a}) + \vec{\vec{Q}} : \frac{\partial^2 f}{\partial\vec{a}\;\partial\vec{a}} = 0\;. \tag{127}$$

Inserting the explicit form (21) of the distribution function $f(\vec{a})$, this equation becomes:

$$(\vec{\vec{Q}} - k\,\vec{\vec{M}} \cdot \vec{\vec{g}}^{-1}) : (k^{-2}\,\vec{\vec{g}} \cdot \vec{a}\vec{a} \cdot \vec{\vec{g}} - k^{-1}\,\vec{\vec{g}}) = 0\;. \tag{128}$$

This is a relation which the second moment $\vec{\vec{Q}}$ must satisfy. Since (128) must hold for all values of $\vec{a}$ and since the second factor is a symmetric matrix, it follows that the symmetric part of the first factor must vanish. This gives, since $\vec{\vec{Q}}$ is symmetric:

$$\vec{\vec{Q}} = \tfrac{1}{2}\, k(\vec{\vec{M}} \cdot \vec{\vec{g}}^{-1} + \vec{\vec{g}}^{-1} \cdot \vec{\vec{M}}) \tag{129}$$

or, with the matrix $\vec{\vec{L}}$ of eq. (85):

$$\vec{\vec{Q}} = \tfrac{1}{2}\, k(\vec{\vec{L}} + \tilde{\vec{\vec{L}}}) \,. \tag{130}$$

This relation determines the matrix $\vec{\vec{Q}}$ uniquely.

The fundamental solution of eq. (124) can be shown to be the Gaussian distribution

$$P(\vec{a}_0 | \vec{a}; t) = \sqrt{\frac{|\vec{\vec{v}}|}{(2\pi k)^n}} \exp\left[-\frac{1}{2k}\, \vec{\vec{v}} : (\vec{a} - \overline{\vec{a}}^{\vec{a}_0})(\vec{a} - \overline{\vec{a}}^{\vec{a}_0})\right], \tag{131}$$

with first moments,

$$\overline{\vec{a}}^{\vec{a}_0} = \exp\left[-\vec{\vec{M}}t\right] \cdot \vec{a}_0 \tag{132}$$

and with variances

$$\overline{(\vec{a} - \overline{\vec{a}}^{\vec{a}_0})(\vec{a} - \overline{\vec{a}}^{\vec{a}_0})}^{\vec{a}_0} = k\, \vec{\vec{v}}^{-1} = k(\vec{\vec{g}}^{-1} - \exp\left[-\vec{\vec{M}}t\right] \cdot \vec{\vec{g}}^{-1} \cdot \exp\left[-\tilde{\vec{\vec{M}}}t\right]) \,. \tag{133}$$

The foregoing treatment shows that the assumption on the Markoffian character of the process with the conditions (121) - (123) on the moments leads to a Gaussian conditional probability density (131). A process of this kind is called a Gaussian Markoff process.

It should, at this point, be stressed that real irreversible processes can very well have non-Gaussian and (or) non-Markoffian character. Such processes do not satisfy all the conditions (120 - (123). On the other hand the Gaussian Markoff processes, based on these conditions, may not only serve as a useful illustration of irreversible behaviour, but could also represent a good approximation to a class of irreversible processes.

The irreversible character of the Gaussian Markoff processes can best be demonstrated in the following way. Consider the function

$$H(t) = \int P(\vec{a}_0 | \vec{a}; t) \ln \left\{\frac{P(\vec{a}_0 | \vec{a}; t)}{f(\vec{a})}\right\} d\vec{a} \,. \tag{134}$$

This function may alternatively be represented by the integral

$$H(t) = \int \left[P(\vec{a}_0 | \vec{a}; t) \ln \left\{\frac{P(\vec{a}_0 | \vec{a}; t)}{f(\vec{a})}\right\} - P(\vec{a}_0 | \vec{a}; t) + f(\vec{a})\right] d\vec{a} \,, \tag{135}$$

since it follows from normalization that

$$\int \{f(\vec{a}) - P(\vec{a}_0|\vec{a};t)\}\, d\vec{a} = 0\ . \tag{136}$$

We thus have

$$H(t) \geq 0\ , \tag{137}$$

as the integrand in (135) is of the form $x \log x - x \log y - x + y$ and satisfies the inequality

$$x \log x - x \log y - x + y \geq 0\ . \tag{138}$$

This inequality holds when $x$ and $y$ are themselves quantities which cannot assume negative values (as is true for $f(\vec{a})$ and $P(\vec{a}_0|\vec{a};t)$ which are probability densities).

It follows now from (134) that the time derivative of $H$ may be written as

$$\frac{\partial H}{\partial t} = \int \frac{\partial P}{\partial t} \ln (P/f)\, d\vec{a}\ , \tag{139}$$

where use has been made of the normalization condition for $P(\vec{a}_0|\vec{a};t)$. Introducing into this expression the Fokker-Planck equation (124) one obtains

$$\frac{\partial H}{\partial t} = \int \{\tilde{\vec{\vec{M}}} : \frac{\partial}{\partial \vec{a}}(P\vec{a}) + \vec{\vec{Q}} : \frac{\partial^2 P}{\partial \vec{a}\, \partial \vec{a}}\} \ln (P/f)\, d\vec{a} =$$

$$= \int \{\tilde{\vec{\vec{L}}} : \frac{\partial}{\partial \vec{a}}(P\vec{a}) \cdot \vec{\vec{g}} + k\, \tilde{\vec{\vec{L}}} : \frac{\partial^2 P}{\partial \vec{a}\, \partial \vec{a}}\} \ln (P/f)\, d\vec{a}\ . \tag{140}$$

The last member follows by means of (85), expression (130) for $\vec{\vec{Q}}$, and by observating that the matrix $\partial^2 P/\partial\vec{a}\, \partial\vec{a}$ is symmetric. Integrating by parts one gets

$$\frac{\partial H}{\partial t} = -\, \tilde{\vec{\vec{L}}} : \int P(\vec{a} \cdot \vec{\vec{g}} + k \frac{\partial \ln P}{\partial \vec{a}}) \frac{\partial \ln (P/f)}{\partial \vec{a}}\, d\vec{a}\ . \tag{141}$$

On the other hand we have the relation

$$\vec{a} \cdot \vec{\vec{g}} = -k \frac{\partial \ln f}{\partial \vec{a}}\ , \tag{142}$$

as can easily be seen from the explicit form (21) for $f(\vec{a})$. With this relation equation (141) becomes

$$\frac{\partial H}{\partial t} = -k \int P\, \tilde{\vec{\vec{L}}} : \frac{\partial \ln (P/f)}{\partial \vec{a}} \frac{\partial \ln (P/f)}{\partial \vec{a}}\, d\vec{a} \leq 0\ . \tag{143}$$

The inequality follows from the fact that the quadratic form in the integrand is positive definite (cf. (122) and (130)), and that the function $kP(\vec{a}_0|\vec{a};t)$ is itself positive for all values of $\vec{a}$. We have here a so-called $H$-theorem for the case of a multidimensional Markoff-process. The usual conclusions may be drawn from this theorem.

The positive function $H(t)$ must decrease monotonously in the course of time. Since this function cannot become negative it follows for all values of $\vec{a}$, that

$$\lim_{t\to 0} \frac{\partial(P/f)}{\partial \vec{a}} = 0 \ , \tag{144}$$

or

$$\lim_{t\to 0} P = af \ , \tag{145}$$

where $a$ is a constant. From normalization it follows that the constant must be unity, so that finally

$$\lim_{t\to\infty} P(\vec{a}_0|\vec{a};t) = f(\vec{a}) \ . \tag{146}$$

This result makes apparent the irreversible nature of the processes described. It implies also that the variances $\vec{\vec{v}}^{-1}$, which are zero initially, tend to their equilibrium values $\vec{\vec{g}}^{-1}$ as $t \to \infty$. This means that the distribution function $P(\vec{a}_0|\vec{a};t)$ remains sharply peaked in the course of time. Therefore, from the macroscopic point of view, that is if one does not measure any deviation of the order of the equilibrium fluctuations, there is an overwhelming probability for any macroscopic deviation $\vec{a}_0$ to decay on a single occasion already according to the average linear laws. This is in agreement with the phenomenological laws of macroscopic physics.

The theory discussed in the present section applies to both $\vec{\alpha}$ and $\vec{\beta}$ variables.

## 9. *Entropy and random fluctuations*

According to the second law of thermodynamics the entropy of an adiabatically insulated system must increase monotonically until thermodynamic equilibrium is established within the system. In the present section we shall investigate, whether this behaviour of entropy may be obtained on the basis of the definition of entropy in terms of random variables $\vec{a}$ and of the properties of the Gaussian Markoff process $\vec{a}(t)$.

For a state $\vec{a}$ the deviation $\Delta S$ of the entropy from its maximum value is according to Boltzmann's entropy postulate, (cf. eq. (28)):

$$\Delta S = -\tfrac{1}{2}\vec{\vec{g}} : \vec{a}\,\vec{a} \ . \tag{147}$$

The conditional average which describes the mean time behaviour of this quantity when the initial specified state was $\vec{a}_0$ is given by

$$\overline{\Delta S}^{\vec{a}_0} = -\tfrac{1}{2}\vec{\vec{g}} : \int \vec{a}\,\vec{a}\; P(\vec{a}_0|\vec{a};t)\,\mathrm{d}\vec{a} \ . \tag{148}$$

Substituting the expression (131) for $P(\vec{a}_0|\vec{a};t)$ into this equation we obtain:

$$\overline{\Delta S}^{\vec{a}_0} = -\tfrac{1}{2}\,\vec{\vec{g}} : \int (\vec{a} - \bar{\vec{a}}^{\vec{a}_0})(\vec{a} - \bar{\vec{a}}^{\vec{a}_0})\, P(\vec{a}_0|\vec{a};t)\,\mathrm{d}\vec{a} - \tfrac{1}{2}\,\vec{\vec{g}} : \bar{\vec{a}}^{\vec{a}_0}\,\bar{\vec{a}}^{\vec{a}_0} =$$

$$= -\tfrac{1}{2}\,\vec{\vec{g}} : (k\,\vec{\vec{v}}^{-1} + \bar{\vec{a}}^{\vec{a}_0}\,\bar{\vec{a}}^{\vec{a}_0}), \tag{149}$$

where eq. (133) has been used.

We see that

$$\lim_{t\to 0} \overline{\Delta S}^{\vec{a}_0} = -\tfrac{1}{2}\,nk \tag{150}$$

in agreement with eq. (30).

Furthermore we note that $\overline{\Delta S}^{\vec{a}_0}$ consists of two parts, of which the first part $\frac{1}{2}\,k\,\vec{\vec{g}} : \vec{\vec{v}}^{-1}$ only is independent of the initial conditions. We shall see in a moment that this first part increases monotonously until it reaches the value $-\frac{1}{2}\,nk$. The second part $\frac{1}{2}\,\vec{\vec{g}} : \overline{\vec{a}}\,\overline{\vec{a}}$ decreases monotonously until its value is zero. Thus for sufficiently large (macroscopic) initial values $\vec{a}_0$, the first part will be negligible for a long period of time:

$$\overline{\Delta S}^{\vec{a}_0} \simeq -\tfrac{1}{2}\,\vec{\vec{g}} : \bar{\vec{a}}^{\vec{a}_0}\,\bar{\vec{a}}^{\vec{a}_0}. \tag{151}$$

One can easily evaluate the length of time for which (151) holds in good approximation under macroscopic initial conditions. Consider the case of a single variable $\alpha$. $\overline{\Delta S}^{\vec{a}_0}$ is initially of order $N_\alpha k$ where $N_\alpha$ is the number of particles in the subsystem to which $\alpha$ refers. In the usual macroscopic experiments this number will be at least $10^{12}$ to $10^{16}$. It then follows from inspection of (149) that the first part may be neglected within an interval of time of the order of several (say 10) relaxation times $M^{-1}$. The time derivative of the average $\overline{\Delta S}^{\vec{a}}$ is obtained from (149)

$$\frac{\partial \overline{\Delta S}^{\vec{a}_0}}{\partial t} = \tfrac{1}{2}\,k\,\vec{\vec{g}} : \frac{\partial \vec{\vec{v}}^{-1}}{\partial t} + \vec{\vec{L}} : \bar{\vec{X}}^{\vec{a}_0}\,\bar{\vec{X}}^{\vec{a}_0} =$$

$$= -\vec{\vec{L}} : (\exp[-\vec{\vec{g}}\cdot\vec{\vec{L}}t]\cdot\vec{\vec{g}}\cdot\exp[-\tilde{\vec{\vec{L}}}\cdot\vec{\vec{g}}t]) + \vec{\vec{L}} : \bar{\vec{X}}^{\vec{a}_0}\,\bar{\vec{X}}^{\vec{a}_0}, \tag{152}$$

where the explicit form (133) for $\vec{\vec{v}}^{-1}$ has been used.

Thus $\partial\overline{\Delta S}^{\vec{a}_0}/\partial t$ also consists of two parts. The first part is negative definite, the second part positive definite. This follows from the positive definite character of the matrices $\vec{\vec{L}}$ and $\vec{\vec{g}}$. Both parts vanish as $t \to \infty$. The second part, however, alone depends on the initial values $\vec{a}_0$. Again for sufficiently large (macroscopic) initial values $\vec{a}_0$ the first part will be negligible for a long period of time. (In fact for a proper choice of $\vec{a}_0$ this will be so for all times.) One then has

$$\frac{\partial \overline{\Delta S}^{\vec{a}_0}}{\partial t} \simeq \vec{\vec{L}} : \bar{\vec{X}}^{\vec{a}_0}\,\bar{\vec{X}}^{\vec{a}_0}, \tag{153}$$

and

$$\frac{\partial \overline{\Delta S}^{\vec{a}_0}}{\partial t} \geq 0 \,. \tag{154}$$

For small $\vec{a}_0$, the inequality (154) will not necessarily be satisfied. In particular one may even have for all times

$$\frac{\partial \overline{\Delta S}^{\vec{a}_0}}{\partial t} \leq 0 \,. \tag{155}$$

This will for instance be the case if $\vec{a}_0 = 0$.

Clearly only (154) has the expected form of the second law of thermodynamics. The possibility (155) also exists as a consequence of the fact that entropy has been defined for a single state according to (147). Indeed if the initial state is the most probable state $\vec{a}_0 = 0$, i.e. $\Delta S = 0$ initially, there will be a finite probability to find states with lower entropy in the course of time. This can be seen from the form of the distribution function (131). Therefore for such initial conditions the average entropy must obviously decrease until the equilibrium value $-\frac{1}{2}nk$ has been reached. However, for macroscopic initial conditions the inequality (154), which does have the form of the second law of thermodynamics, is valid. It has already been stated in section 3 that from the point of view of thermodynamics no distinction is made between values of the entropy which differ by an amount of the order $nk$. Therefore within a purely macroscopic description the inequality (155) has no meaning. In other words, the behaviour of entropy is in complete agreement with the laws of macroscopic theory.

The previous discussion was based on Boltzmann's entropy postulate. Entropy was there a random variable, so that only an average of this quantity represents the relevant macroscopic quantity. In statistical mechanics it is also possible to give a different definition of entropy. Entropy is then defined from the outset in terms of a distribution over possible states. According to the so-called Gibbs entropy postulate:

$$S^{\mathrm{G}} = -k \int P(\vec{a}_0|\vec{a};t) \ln \left\{\frac{P(\vec{a}_0|\vec{a};t)}{f(\vec{a})\Omega}\right\} \mathrm{d}\vec{a} \,, \tag{156}$$

where $f(\vec{a})$ is the equilibrium distribution. $\Omega$ is a constant determined by the value of the entropy at equilibrium (for $t \to \infty$):

$$S^{\mathrm{G}} = k \ln \Omega \,. \tag{157}$$

In fact $\Omega$ is the structure function of the system (cf. (16) and (99)).

From eqs. (156) and (157) we find for the deviation $\Delta S^{\mathrm{G}}$ of the entropy from its equilibrium value:

$$\Delta S^{\mathrm{G}} = -k \int P(\vec{a}_0|\vec{a};t) \ln \left\{\frac{P(\vec{a}_0|\vec{a};t)}{f(\vec{a})}\right\} \mathrm{d}\vec{a} \,. \tag{158}$$

Introducing into this expression the explicit forms (21) and (131) of the distribution functions $f(\vec{a})$ and $P(\vec{a}_0|\vec{a};t)$, this becomes:

$$\Delta S^{G} = -\tfrac{1}{2}\,k\,\ln\left(\frac{|\vec{\vec{v}}|}{|\vec{\vec{g}}|}\right) - \tfrac{1}{2}\,\vec{\vec{g}} : \{\exp[-\vec{\vec{M}}t]\,.\,(\vec{a}_0\,\vec{a}_0 - k\,\vec{\vec{g}}^{-1})\,.\,\exp[-\tilde{\vec{\vec{M}}}t]\}\,. \tag{159}$$

The first term on the right hand side of this equation can be written explicitly with eq. (133) as:

$$-\tfrac{1}{2}\,k\,\ln\left(\frac{|\vec{\vec{v}}|}{|\vec{\vec{g}}|}\right) = \tfrac{1}{2}\,k\,\ln\,|\vec{\vec{U}} - \vec{\vec{g}}\,.\,\exp[-\vec{\vec{M}}t]\,.\,\vec{\vec{g}}^{-1}\,.\,\exp[-\tilde{\vec{\vec{M}}}t]|\,. \tag{160}$$

We thus have finally

$$\Delta S^{G} = \tfrac{1}{2}\,k\,\ln\,|\vec{\vec{U}} - \vec{\vec{g}}\,.\,\exp[-\vec{\vec{M}}t]\,.\,\vec{\vec{g}}^{-1}\,.\,\exp[-\tilde{\vec{\vec{M}}}t]| + \\ + \tfrac{1}{2}\,k\,\vec{\vec{g}} : (\exp[-\vec{\vec{M}}t]\,.\,\vec{\vec{g}}^{-1}\,.\,\exp[-\tilde{\vec{\vec{M}}}t]) - \tfrac{1}{2}\,\vec{\vec{g}} : \overline{\vec{a}}^{\vec{a}_0}\,\overline{\vec{a}}^{\vec{a}_0}\,. \tag{161}$$

This expression consists of three parts of which the first two are independent of the initial values $\vec{a}_0$. For simplicity's sake we shall discuss the relative order of magnitude of the various parts for the case of a single variable $a$. We assume again $\frac{1}{2}\,g\vec{a}^{2}$ to be initially of order $10^{12}\,k$ to $10^{16}\,k$ as discussed previously. Then the logarithmic term which is initially infinite will be of order $10^{4}\,k$ to $10^{10}\,k$ within times of order $10^{-10^{4}}\,M^{-1}$ to $10^{-10^{10}}\,M^{-1}$ that is within times which are completely irrelevant compared to the macroscopic relaxation time $M^{-1}$. The whole theory of the Gaussian Markoff processes loses its meaning for such extremely small time intervals. After such times the logarithmic term may therefore already be neglected with respect to $\frac{1}{2}\,g\vec{a}^{2}$. The second term of (161) can be neglected for all times. The situation is similar for several variables. Thus for macroscopic initial conditions

$$\Delta S^{G} \simeq -\tfrac{1}{2}\,\vec{\vec{g}} : \overline{\vec{a}}^{\vec{a}_0}\,\overline{\vec{a}}^{\vec{a}_0}\,. \tag{162}$$

From the point of view of macroscopic theory, one is not interested in the behaviour of $\Delta S^{G}$ within the short times mentioned, so that (162) represents the relevant result. The result (158) is identical with (152), found on the basis of Boltzmann's entropy postulate.

The time derivative of $\Delta S^{G}$ can be obtained by observing that

$$\Delta S^{G} = -\,k\,H(t)\,, \tag{163}$$

where $H(t)$ is the function defined in (134). It follows from (143) that

$$\frac{\partial \Delta S^{G}}{\partial t} = k^{2}\int P\,\vec{\vec{L}} : \frac{\partial \ln (P/f)}{\partial \vec{a}}\,\frac{\partial \ln (P/f)}{\partial \vec{a}}\,\mathrm{d}\vec{a} \geq 0\,. \tag{164}$$

This inequality now represents the second law of thermodynamics.

With the present definition of entropy it is valid for arbitrary initial conditions. Explicit evaluation of the integral in (164) leads to

$$\frac{\partial \Delta S^G}{\partial t} = k \vec{\vec{L}} : (\vec{\vec{v}} + \vec{\vec{g}} \cdot \vec{\vec{v}}^{-1} \cdot \vec{\vec{g}} - 2 \vec{\vec{g}}) + \vec{\vec{L}} : \overline{\vec{X}}^{\vec{a}_0} \overline{\vec{X}}^{\vec{a}_0} . \qquad (165)$$

Here also only the last part $\vec{\vec{L}} : \overline{\vec{X}}\,\overline{\vec{X}}$ of the right hand side depends on the initial conditions. For macroscopic initial conditions, one can again establish that the order terms may be beglected after time intervals which are irrelevant on a macroscopic scale, i.e. which are negligible compared to the macroscopic relaxation times. The expression (165) then becomes

$$\frac{\partial \Delta S^G}{\partial t} \simeq \vec{\vec{L}} : \overline{\vec{X}}^{\vec{a}_0} \overline{\vec{X}}^{\vec{a}_0} . \qquad (166)$$

This result is again identical with (153) found on the basis of Boltzmann's entropy postulate.

Let us finally turn our attention to the point of view taken in macroscopic phenomenological theory. Here no distinction is made between the behaviour of a single system on a single occasion and the average behaviour of a system. Therefore if the entropy of a non-equilibrium state is given by (cf. ew. (28)):

$$\Delta S = - \tfrac{1}{2} \vec{\vec{g}} : \vec{a}\,\vec{a} , \qquad (167)$$

the second law of thermodynamics is expressed by:

$$\frac{d\Delta S}{dt} = - \vec{\vec{g}} : \vec{a} \left(\frac{d\vec{a}}{dt}\right) = \vec{X} \cdot \left(\frac{d\vec{a}}{dt}\right) = \vec{\vec{L}} : \vec{X}\,\vec{X} \geq 0 , \qquad (168)$$

where it has been assumed, in agreement with the point of view just mentioned, that the $\vec{a}$-variables obey themselves the phenomenological laws. It is obvious that (167) and (168) are precisely of the form of the macroscopic relevant results derived in this section.

We may note that, according to the phenomenological theory of irreversible thermodynamics, the positive definite character of the coefficient matrix $\vec{\vec{L}}$ follows from the expression for the second law of thermodynamics (168). On the other hand in the considerations of this section it is the second law which is a consequence of the positive definite character of $\vec{\vec{L}}$.

To conclude this discussion on some aspects of the statistical basis of non-equilibrium thermodynamics, I may add that such a discussion can also be given with the help of the kinetic theory of dilute gases. It then turns out that in the so-called first Enskog approximation, corresponding to linear transport phenomena, one can again justify the postulates of thermodynamics of irreversible processes.

*Bibliography*

The material of these lectures is also dealt with in the monograph
S. R. de Groot and P. Mazur, Non-equilibrium thermodynamics, North-Holland Publishing Company, Amsterdam (to appear in 1962).
For sections 8 and 9, see also:
P. Mazur, Rendiconti della Scuola Istituto di Fisica "Enrico Fermi", Varenna (1959).
For the original papers on the reciprocal relations, see:
L. Onsager, Phys. Rev. 37 (1931) 405; 38 (1931) 2265.
For a discussion of the principle of microscopic reversibility, see, for instance:
E. P. Wigner, Journ. Chem. Phys. 22 (1954) 1912.
For a discussion of Markoff processes in ocnnection with the description of irreversible phenomena, see:
M. S. Green, Journ. Chem. Phys. 20 (1952) 1281.
N. Hashitsume, Progr. Theor. Phys. 8 (1952) 461.
N. G. van Kampen, Fortschritte d. Phys. 4 (1956) 405.
For properties of the time correlation function (in harmonic oscillator systems) see, e.g.:
P. Mazur and E. M. Montroll, J. Math. Phys. 1 (1960) 70.
The fluctuation dissipation theorem was derived by:
H. B. Callen and R. F. Greene, Phys. Rev. 86 (1952) 702; 88 (1952) 1387.
H. B. Callen, M. L. Barasch and J. L. Jackson, Phys. Rev. 88 (1952) 1382.
See also:
R. Kubo, Lectures in Theoretical Physics, Boulder Summer School, 1 (1958).
For a derivation of the "master equation" see:
L. Van Hove, Physica 21 (1955) 517.
Gaussian Markoff processes were extensively considered by:
L. Onsager and S. Machlup, Phys. Rev. 91 (1953) 1505.
For a discussion of the validity of thermodynamic relations outside equilibrium by means of the kinetic theory of gases, see:
I. Prigogine, Physica 15 (1959) 272.
H. C. Reik, Zeits. f. Phys. 148 (1957) 156, 333.

# SOME REMARKS ON THE INTEGRAL EQUATIONS OF STATISTICAL MECHANICS

ELLIOTT W. MONTROLL
Instituut Lorentz, University of Leiden, Holland and
and
International Business Machines Corporation,
Thomas J. Watson Research Center, P.O. Box 218,
Yorktown Heights, New York

*Introduction*

In the past few years considerable progress has been made in the theory of quantum statistics of interacting particles as well as in the theory of non-equilibrium processes. Much of this progress has resulted from the systematic use of perturbation theory and diagram techniques. Although these techniques were first developed by Feynman [1)] for quantum electrodynamics, they have proven to be very appropriate for the many-body problem and statistical mechanics.

One of the unfortunate aspects of perturbation theory and diagramatic techniques in the many-particle problem is that a certain amount of unpleasant combinatorial analysis is always involved in their application. The whole subject of combinatorics and diagramatics occupies a rather special place in mathematical physics in that it divides people into two classes - those who pratice it and those who don't. To the first group it becomes almost trivial, but they can never quite successfully explain it to the members of the second class who find it repulsive and probably not worth learning. A certain amount of unpleasant labor is necessary to pass from the latter to the former class and most members of the latter feel (possibly correctly) that "since all things that can be accomplished by diagramatics and combinatorics can probably be accomplished by more straightforward and civilized methods", one should devote that labor to something else.

This seminar has two aims. The first is to derive and exhibit the integral equations which form the basis for much of the recent work in statistical mechanics of many interacting particles (especially in the perturbation theory mentioned above). The second is, as a concession to some of the lecturer's friends who think that diagramatics and combinatorics should be outlawed, to show through an example how these techniques can sometimes be avoided. The specific example chosen is the derivation of an exact differential equation for the diagonal elements of the density matrix. These elements give the variation with time of the occupation number of the various quantum states of an assembly of interest. Certain aspects of this example have been carefully investigated by Van Hove [2]. The material presented here is somewhat similar to that given by Zwanzig [3] and the lecturer [4] in their Boulder Summer School lectures of last year.

Of course, avoiding diagrams is an old game, having been practiced especially by Kirkwood [5], Born and Green [6], Yvon [7], Bogolubov [8] and others in classical statistical mechanics and by Martin and Schwinger [9] in quantum statistics. The relevant researches of the above-mentioned authors are concerned either with hierarchies of equations for multiparticle distribution functions or Green's functions. In order to solve a particular problem, one must usually make some hypothesis concerning the relation of a function involving a given number of particles to one involving one more (for example, the Kirkwood super-position principle). The scheme discussed here is somewhat different, and perhaps in spirit is more similar to recent work of Caianiello [10] in quantum field theory than to the papers of the authors quoted above.

Since much of the discussion here is rather general, we will come to no physical conclusions. Some remarks will be made as to how equations such as the Pauli or "master" equation might follow. Prof. P. Mazur and the author are attempting to apply the methods of section 4 to some special problems; and if all goes well, these will be reported on elsewhere.

## 1. *Fundamental equations of dynamics and statistical mechanics*

We begin by exhibiting the fundamental equations of dynamics and statistical mechanics. The time variation of any dynamical operator $B$ associated with an assembly of particles is given by Heisenberg equation

$$\frac{\partial B}{\partial t} = -\frac{1}{i\hbar}[H, B] . \tag{1}$$

$H$ being the Hamiltonian operator of the assembly. The density matrix $\rho$ satisfies a similar equation

$$\frac{\partial \rho}{\partial t} = \frac{1}{i\hbar}[H, \rho] . \tag{2}$$

Notice the absence of the minus sign on the right-hand side of (2). In the stationary canonical ensemble at temperature $T = 1/k\beta$

$$\rho = \exp(-\beta H) \tag{3}$$

so that

$$\partial \rho / \partial \beta = -H\rho . \tag{4}$$

The stationary density matrix $\rho$ is related to the partition function

$$Z = \text{trace } (\exp - \beta H) \tag{5}$$

from which all thermodynamic quantities can be found in the well known manner. In any assembly stationary or otherwise the expectation value of the operator $B$ is given by

$$\langle B\rangle = \text{trace } B\rho \tag{6}$$

while the $j$-th diagonal element of $\rho$, $\rho_{jj}(t)$ is the occupation number (or probability) of the $j$-th state at time $t$.

Classical dynamics and statistics can be written in a similar concise form in terms of the Liouville operation $L$ which is related to the classical Hamilton function $H$ through

$$i\, L = \sum_j \left(\frac{\partial H}{\partial q_j}\frac{\partial}{\partial p_j} - \frac{\partial H}{\partial p_j}\frac{\partial}{\partial q_j}\right) \tag{7}$$

the $p_j$'s and $q_j$'s being respectively the set of momenta and position coordinates of our assembly. If we define $(A, B)$ to be the Poisson bracket of the functions $A$ and $B$,

$$(A, B) = \sum_j \left(\frac{\partial A}{\partial q_j}\frac{\partial B}{\partial p_j} - \frac{\partial A}{\partial p_j}\frac{\partial B}{\partial q_j}\right) . \tag{8}$$

The effect of operating $i\,L$ on a function $g(p,q)$ is

$$i\, L\, g = (H, g) . \tag{9}$$

The set of Hamilton's equations

$$\dot{g}_j = \partial H/\partial p_j \quad \text{and} \quad \dot{p}_j = - \partial H/\partial q_j , \quad j = 1, 2, \ldots, n \tag{10}$$

can be summarized in the differential equation

$$\dot{f} = - i\, L\, f \tag{11}$$

where $f$ is a vector with components

$$f = \{p_1, q_1, p_2, q_2, \ldots, p_n, q_n\} . \tag{12}$$

Clearly, if we operate $- i\, L$ on this vector, we find

$$- i\, L\, f = \{-\frac{\partial H}{\partial q_1}, \frac{\partial H}{\partial p_1}, -\frac{\partial H}{\partial q_2}, \frac{\partial H}{\partial p_2}, \ldots\} .$$

If we then compare the individual components of the left and right hand side of (11), we obtain Hamilton's equations.

The classical analogue of (1) is easily obtained when $B$ is an arbitrary function of $p$ and $q$, which does not depend explicitly on the time (the condition also for the validity of (1)). Note that Hamilton's equations imply

$$\frac{\partial}{\partial t} B(p,q) = \sum \left(\frac{\partial B}{\partial q_j} \dot{q}_j + \frac{\partial B}{\partial p} \dot{p}_j\right)$$
$$= \sum \left(\frac{\partial B}{\partial q_j}\frac{\partial H}{\partial p_j} - \frac{\partial B}{\partial p_j}\frac{\partial H}{\partial q_j}\right)$$
$$= -(H,B) = -i\,L\,B(p,q)$$

so that

$$\dot{B} = -i\,L\,B\ . \tag{13}$$

Liouville's equation for the distribution function of representative points of assemblies in phase space

$$\dot{\rho} = (H,\rho) = i\,L\,\rho \tag{14}$$

is the classical analogue of (2). The analogue of (6) is

$$\langle B\rangle = \int B(p,q)\,\rho(p,q)\,d^3p\,d^3q\ . \tag{15}$$

The Liouville operator was first introduced by Koopman and has been used extensively by Kirkwood and Prigogine [11)] and their collaborators in their work on non-equilibrium statistical mechanics.

When $H$ and $L$ do not depend explicitly on the time the formal solution of Heisenberg's equation is

$$B(t) = e^{itH/\hbar}\,B(0)\,e^{-itH/\hbar} \tag{16}$$

while that of the density matrix equation (2) is

$$\rho(t) = e^{-itH/\hbar}\,\rho(0)\,e^{itH/\hbar}\ . \tag{17}$$

The formal solution of Hamilton's equations is

$$f(t) = e^{-iLt}\,f(0) \tag{18}$$

while that of the Liouville equation is

$$\rho(t) = e^{iLt}\,\rho(0)\ . \tag{19}$$

It is seldom that these formal solutions of our fundamental differential equations are directly applied to a specific problem although they are frequently employed in intermediate steps.

Most of the recent developments in statistical mechanics are concerned with situations in which $H$ has two components

$$H = H_0 + \lambda\,H_I(t)\ . \tag{20}$$

All properties of the unperturbed component $H_0$ are assumed to be known and the problem is to determine the influence of the "perturbation" $\lambda\,H_I(t)$ on these properties. We now derive integral equations for our various quantities of interest when $H$ has the form (20). These are equivalent to the differential equations 1, 2, 4, 11, 13 and 14.

## 2. *Integral equation form for fundamental equations*

Let us suppose that

$$H_I(t) \to 0 \quad \text{as} \quad t \to 0 . \tag{1}$$

This corresponds to a statement that no perturbation existed in the distant past. A particular form of $H_I(t)$ which is of considerable interest is one in which the interaction is switched on suddenly at $t = 0$. Then

$$H_I(t) = \theta(t)\, U \tag{2a}$$

where $\theta(t)$ is the step function

$$\theta(t) = \begin{cases} 1 & \text{if} \quad t > 0 \\ 0 & \text{if} \quad t < 0 \end{cases} \tag{2b}$$

and $U$ is a time independent interaction.

Under the general condition (1) we can easily show that the integral equation

$$\rho(t) = \rho_o(t) - \frac{i\lambda}{\hbar}\int_{-\infty}^{t} e^{-i(t-t')H_o/\hbar}\,[H_I(t'), \rho(t')]\, e^{i(t-t')H\ /\hbar}\, dt' \tag{3}$$

is equivalent to the differential equation (1.2) provided that

$$\frac{\partial \rho_o}{\partial t} = \frac{1}{i\hbar}[H_o, \rho_o] . \tag{3a}$$

For differentiation of (3) with respect to time yields

$$\dot\rho(t) = \dot\rho_o(t) - \frac{i\lambda}{\hbar}[H_I(t), \rho(t)] + \frac{i}{\hbar}[H_o, \frac{i\lambda}{\hbar}\int_{-\infty}^{t} e \cdots] \tag{3b}$$

where the integral in the second commutator is just the integral given in (3) so that substitution of (3) and (3a) into (3b) yields

$$\dot\rho(t) = \frac{1}{i\hbar}[H_o, \rho_o] - \frac{i\lambda}{\hbar}[H_I(t), \rho(t)] + \frac{i}{\hbar}[H_o, \rho_o(t) - \rho(t)] = \frac{1}{i\hbar}[H, \rho]$$

which is exactly (1.2). When $H_I(t)$ has the special form (2), (3) becomes

$$\rho(t) = \rho_o(t) - \frac{i\lambda}{\hbar}\int_0^t e^{-i(t-t')H_o/\hbar}[U, \rho(t')]\, e^{i(t-t')H_o/\hbar} . \tag{4}$$

In a similar manner it is easily shown that (1.1) is equivalent (under condition (1)) to

$$B(t) = B_o(t) + \frac{i\lambda}{\hbar}\int_{-\infty}^{t} e^{i(t-t')H_o/\hbar}[H_I(t'), B(t')]\, e^{-i(t-t')H_o/\hbar} . \tag{5}$$

The classical analogue of this equation is (while in (5) $B$ is an operator, in (6) it is a function of the dynamical $p$'s and $q$'s)

$$B(t) = B_0(t) - i\lambda \int_{-\infty}^{t} e^{-i(t-\tau)L_0} L_1(\tau)\, B(\tau)\, d\tau \tag{6}$$

where the Liouville operator $L$ is postulated to be of the form

$$L(t) = L_0 + \lambda L_I(t) \tag{7a}$$

while
$$L_I(t) \to 0 \quad \text{as} \quad t \to -\infty \tag{7b}$$

and $B_0$ is defined to be a solution of

$$\dot{B}_0 = -i L_0 B_0 \tag{7c}$$

with $L_0$ not depending explicitly on the time. The equivalence of (6) and (1.13) is immediately apparent from differentiating (6). Then

$$\dot{B}(t) = \dot{B}_0(t) - i L_1(t)\, B(t) + i \{i \lambda L_0 \int_{-\infty}^{t} e^{-i(t-\tau)\lambda_0} L_1(\tau)\, B(\tau)\, d\tau\} . \tag{8}$$

When the integral on the right hand side of this equation is replaced by $[B_0(t) - B(t)]$ (see eq. (6)), then (7c) implies

$$\begin{aligned} \dot{B}(t) &= -i L_0 B_0 - i L_1(t)\, B(t) - i L_0 (B_0 - B) \\ &= -i L B \end{aligned}$$

as is required.

The integral equation equivalent to Hamilton's equations is

$$f(t) = f_0(t) - i\lambda \int_{-\infty}^{t} e^{-i(t-\tau)L_0} L_1(\tau)\, f(\tau)\, d\tau \tag{9}$$

while the equivalent of Liouville's equation is

$$\rho(t) = \rho_0(t) + i\lambda \int_{-\infty}^{t} e^{i(t-\tau)L_0} L_1(\tau)\, \rho(\tau)\, d\tau . \tag{10}$$

When the Hamiltonian has the form (2) which corresponds to the sudden switching on of an interaction at time $t = 0$, the Liouville operator has the same form and (6) and (10) become respectively

$$B(t) = B_0(t) - i\lambda \int_{0}^{t} e^{-i(t-\tau)L_0} L_1\, B(\tau)\, d\tau \tag{11}$$

and
$$\rho(t) = \rho_0(t) + i\lambda \int_{0}^{t} e^{i(t-\tau)L_0} L_1(\tau)\, \rho(\tau)\, d\tau . \tag{12}$$

When $B$ is replaced by $f$ in (11), one has an integral equation which can be used as a basis for classical mechanics. It can probably be applied advantageously to celestial mechanics.

The integral equation equivalent to the differential equation for the stationary density matrix in the case

$$H = H_0 + \lambda H_I$$

with $H_o$ and $H_I$ both not depending explicitly on time or temperature is

$$\rho(\beta) = \rho_o(\beta) - \lambda \int_0^\beta e^{-(\beta-\beta')H_o} H_I \, \rho(\beta') \, d\beta' \tag{13}$$

provided that at $\beta = 0$ (infinite temperature) $\rho(\beta) \equiv \rho_o(\beta)$. If, for example, $H_o$ represents the kinetic energy of an assembly and $H_I$ the potential energy of interaction of its constituent particles, it is well known that at infinite temperature $H_I$ does not contribute to the equation of state (which is in that limit the perfect gas law, $pv = kT$). Then

$$\lim_{\beta\to 0} \rho(\beta) \to \exp - \beta H_o = \rho_o(\beta)$$

so that in that limit $d\rho_o/d\beta = - H_o \rho_o$.

Note that all time-dependent equations which correspond to the sudden switching on of an interaction at time $t = 0$ (or if one wishes, the switching off of an interaction) are of the "Faltung type". They can be re-expressed as ordinary equations in a Laplace transform representation. The Faltung theorem of Laplace transforms is important in the use of that representation. Let $\{f_j(t)\}$ be a set of functions of the time and let

$$F_j(s) = \int_0^\infty e^{-st} f_j(t) \, dt = \mathcal{L} f_j \,, \qquad j = 1, 2, \ldots \tag{14}$$

be the set of Laplace transforms of the $f_j$'s. The Faltung theorem states that if

$$f_3(t) = \int_0^t f_2(t - \tau) \, f_1(\tau) \, d\tau \tag{15}$$

then its Laplace transform is

$$F_3(s) = F_2(s) \, F_1(s) \,.$$

If an operator $A$, which does not depend explicitly on the time, is inserted between $f_1$ and $f_2$ in (15), the Laplace transform of the new expression is

$$F_3(s) = F_2(s) \, A \, F_1(s) \,.$$

If we apply the above formulae to the "switch on interaction form" of Hamilton's equation (eq. (11) with $B$ replaced by $f$) we find

$$F(s) = F_o(s) - i \, [s + i \, L_o]^{-1} L_I \, F_I(s) \tag{16}$$

(here $F(s) = \mathcal{L} f(t)$) so that

$$\begin{aligned} F(s) &= [I + i(s + i \, L_o)^{-1} \lambda \, L_I]^{-1} F_o(s) \\ &= (s + i \, L_o) \, [(s + i \, L_o) + i \, \lambda \, L_I]^{-1} \, F_o(s) \end{aligned} \tag{17}$$

$I$ being the identity operator.

The formal solution of the equation for the stationary density matrix equation (13) is (with $P(s) = \mathcal{L}\, \rho(t)$)

$$P(s) = P_o(s) - \lambda (s + H_o)\, H_I$$

$$= P_o(s) - \lambda\, P_o(s)\, H_I\, P(s)$$

or

$$P(s) = [1 + \lambda\, P_o(s)\, H_I]^{-1}\, P_o(s)\ . \tag{18}$$

These use of Laplace transforms in the discussion of (4) is presented in section 4.

## 3. *Perturbation theory*

The various integral equations given above have been the subject of many investigations in recent years in the study of the many body problem and in non-equilibrium statistical mechanics. Generally, one solves these equations by iteration to obtain solutions which are power series in the coupling constant $\lambda$. This gives a very concise form of perturbation theory and leads immediately to various diagramatic analyses. Although it is not the purpose of this lecture to discuss perturbation theory and diagrams, we make a few remarks about this subject.

First, we outline briefly the Kubo [12)] theory of transport coefficients, which is a first-order perturbation theory. Suppose $H_o$ represents the Hamiltonian of a closed assembly including all interaction between constituent particles of the assembly and let us suppose that the assembly is driven by an external driving force so that the interaction with these driving forces can be represented by

$$\lambda\, H_I(t) = -\, E(t)\, A \tag{1}$$

where $E(t)$ is a time-dependent scalar quantity and $A$ is an operator that does not depend explicitly on the time. The response of a dynamical variable $B$ to the external driving force is

$$\langle B \rangle_{Av} = \text{trace}\ B\rho \tag{2}$$

where $\rho$ is the solution of the density matrix equation (2.3) with (1) employed. The solution which is correct to first order in $E(t)$ is obtained by iterating (2.3) once and retaining only terms linear in $E$. Then

$$\rho(t) = \rho_o(t) + \frac{i}{\hbar}\int_{-\infty}^{t} e^{-i(t-t')H_o/\hbar}\ E(t')\,[A, \rho_o]\, e^{i(t-t')H_o/\hbar}\, dt'\ . \tag{3}$$

If $B$ is some kind of current which vanishes in the absence of an external driving force, trace $B\rho_o = 0$ so that if we let $(t - t') = \tau$

$$\langle B \rangle_{Av} = \frac{i}{\hbar}\,\text{trace}\int_{0}^{\infty} B\, e^{-i\tau H_o/\hbar}\ E(t-\tau)\,[A, \rho_o]\, e^{i\tau H_o/\hbar}\, d\tau\ . \tag{4}$$

In particular, if

$$E(t) = e^{(i\omega+\alpha)t} \tag{5}$$

where $\alpha$ is a very small positive number (so that as $t \to -\infty$, $E(t) \to 0$ as is required in the derivation of (2.3)) and $\omega$ is any positive number we can find as $\alpha \to 0$ the response to a periodic driving force of circular frequency $\omega$. Then

$$<B>_{Av} = \frac{i}{\hbar} E(t) \lim_{\alpha\to 0} \text{trace} \int_0^\infty B\, e^{-i\tau H_o/\hbar}\, e^{-(i\omega+\alpha)\tau}\, [A, \rho_o]\, e^{i\tau H_o/\hbar}\, d\tau\ . \tag{6}$$

This is Kubo's formula for the linear response to an external driving force. A special case is ohms law. Let $E(t)$ be an external electric field acting on an assembly of charged particles and let $B$ be the electric current which results from the external field

$$B = J = \sum e_j\, v_j = \sum e_j\, p_j/m_j\ .$$

In this case the interaction (1) has the form $-E(t)A$ with

$$A = \sum e_j\, r_j\ .$$

Then (6) is exactly ohms law

$$<J>_{Av}/E(t) = \sigma(\omega)$$

where the electrical conductivity $\sigma(\omega)$ is now

$$\sigma(\omega) = \frac{i}{\hbar} \lim_{\alpha\to 0} \text{trace} \int_0^\infty J\, e^{-i\tau H_o/\hbar}\, e^{-(i\omega+\alpha)\tau}\, [A, \rho_o]\, e^{i\tau H_o/\hbar}\, d\tau\ . \tag{7}$$

If our assembly was part of a canonical ensemble before the external field was switched on

$$\rho_o = Z^{-1} \exp - \beta H_o\ .$$

Then, using Kubo's identity

$$[A, e^{-\beta H_o}] = - e^{-\beta H_o} \int_0^\infty e^{\lambda H_o}\, [A, H_o]\, e^{-\lambda H_o}\, d\lambda$$

$$= \frac{\hbar}{i} e^{-\beta H_o} \int_0^\beta e^{\lambda H_o}\, \dot{A}\, e^{-\lambda H_o}\, d\lambda$$

and the fact that

$$\dot{A} = \sum e_j\, v_j = J$$

we obtain

$$\sigma(\omega) = \lim_{\alpha\to 0} \text{trace} \frac{1}{Z} \int_0^\infty J\, e^{-(i\omega+\alpha)\tau}\, d\tau$$

$$\times \int_0^\beta d\lambda\, e^{-H_o[\beta-\lambda+(i\tau/\hbar)]}\, J\, e^{H_o[(i\tau/\hbar)-\lambda]}\ . \tag{8}$$

If we apply the formula

$$J(\tau) = e^{i\tau H_o/\hbar} J e^{-i\tau H_o/\hbar} \tag{9}$$

we find for the conductivity tensor

$$\sigma_{\mu\nu}(\omega) = \lim_{\alpha\to 0} \text{trace} \frac{1}{Z} \int_0^\infty e^{-(i\omega+\alpha)\tau} d\tau \int_0^\beta d\lambda\, e^{-\beta H_o} J_\mu(-i\lambda\hbar) J_\nu(\tau) \tag{10}$$

which is the integral of the correlation function of the $\nu$-th component of the current at time $\tau$ and the $\mu$-th at the imaginary time $(-i\lambda\hbar)$. Actually, this can be rewritten as a correlation function of the current at two times (cf. Kubo's Boulder Summer School lecture notes of 1958 [13]). Similar formulae exist for other transport coefficients. A discussion of the cluster integral expansion of transport coefficients and the connection of (10) with the Boltzmann equation in the classical low density limit is discussed by Montroll and Ward [14] (see also reference 4).

Another application of perturbation theory is in the calculation of the partition function. Consider the iteration of the integral equation for the density matrix (2.13). Then

$$\rho(\beta) = \rho_o(\beta) - \lambda \int_0^\beta \rho_o(\beta-\beta') H_I \rho_o(\beta') d\beta' +$$

$$+ \lambda^2 \int_0^\beta \int_0^{\beta'} d\beta' d\beta'' \rho_o(\beta-\beta') H_I \rho_o(\beta'-\beta'') H_I \rho_o(\beta'') + \ldots \tag{11}$$

with $\rho_o(\beta) = \exp - \beta H_o$. The partition function is

$$Z = \text{trace } \rho(\beta) . \tag{12}$$

Two types of representations are commonly used for $\rho$ in the calculation of $Z$. The first is a position or momentum representation with the matrix elements of:

$$(r_1', r_2', \ldots, r_N' | \rho | r_1, r_2, \ldots, r_N)$$

being associated with a transition $(r_1, r_2, \ldots, r_N) \to (r_1', r_2', \ldots, r_N')$ or in momentum space

$$(p_1', p_2', \ldots, p_N' | \rho | p_1, p_2, \ldots, p_N)$$

the wave functions used in computing the matrix elements being free particle wave functions. The second type of representation is in terms of occupation numbers ($n_j$ representing the occupation number of the $j$-th state) with

$$(n_1', n_2', \ldots | \rho | n_1, n_2, \ldots)$$

corresponding to a transition $(n_1, n_2, \ldots) \to (n_1', n_2', \ldots)$. In the first case, $H_o$ and $H_I$ are expressed in terms of position coordinates; in the second, in terms of creation and annihilation operators. In both cases one ends with a diagramatic interpretation of various events which can occur through interactions. Considerable literature has developed on this subject. Reviews and bibliographies exist in the Les Houches lecture notes of 1958 and 1959 and the Proceedings of the Utrecht Many Body Conference of 1960.

The Laplace transform representation has also been used as a basis of perturbation theory. For example (18) can be expanded in powers of $\lambda$ to give

$$P(S) = P_o(S) - \lambda\, P_o(S)\, H_I\, P_o(S) + \lambda^2\, P_o(S)\, H_I\, P_o(S)\, H_I\, P_o(S) \dots .$$

## 4. *Diagonal elements of the density matrix*

The occupation number of various states at a given time is proportional to the diagonal element of the density matrix which corresponds to that state. On the basis of various statistical hypotheses (in particular, a random phase approximation) which are made at all times during a process, Pauli and others have derived an equation for the diagonal elements alone of $\rho$:

$$\frac{d\rho_{jj}}{dt} = \sum_k (P_{jk}\,\rho_{kk} - P_{kj}\,\rho_{jj}) . \tag{1}$$

In recent approaches to the theory of non-equilibrium processes, one attempts to derive equations such as (1) by making a statistical hypothesis at time $t = 0$ and using dynamics to deduce the development of the assembly as time goes on. Certain postulates must also be made concerning the interaction Hamiltonian (for example, Van Hove's diagonal singularity condition) in order to derive (1).

The aim of this section is to derive an exact equation for the diagonal elements of the density matrix. A technique will be discussed for the elimination of the off diagonal elements of $\rho$. Similar ideas have been discussed by Zwanzig [13]. No diagrams or perturbation theory are necessary. It will be shown, however, how certain terms are related to sums over various diagrams which appear in the perturbation theory.

We consider an assembly whose Hamiltonian can be divided into two parts

$$H = H_o + \lambda\, U\, \theta(t) \tag{2}$$

with

$$\theta(t) = \begin{cases} 0 & t < 0 \\ 1 & t > 0 \end{cases}$$

and we examine the density matrix in a representation in which $H_o$ is diagonal. Furthermore, we postulate that $U$ has no diagonal elements in this representation (its diagonal elements being absorbed in $H_o$). Then for $t = 0$, we postulate $\rho$ to be a diagonal matrix with $\rho(0) = \rho^{(o)}$ which commutes with $H_o$ so that

$$[\rho^{(o)}, H_o] = 0 \quad \text{and} \quad \rho^{(o)}_{jk} = \rho^{(o)}_{jj}\ \delta_{jk}\ . \tag{3}$$

Then, for $t \geq 0$, eq. (1.2) implies

$$\frac{\partial \rho}{\partial t} = -\frac{i}{\hbar}[H, \rho] = -\frac{i}{\hbar}[H_o + \lambda U, \rho]\ . \tag{4}$$

If we substitute the integral equation (2.4) into (4) and note that the commutators $[\rho^{(o)}, A]$ and $[H_o, A]$ vanish for any $A$ when $\rho^{(o)}$ and $H_o$ are diagonal (as postulated above), we obtain the exact differential equation for the diagonal elements of $\rho$:

$$\frac{\partial \rho_{jj}}{\partial t} = -\frac{\lambda^2}{\hbar^2}\int_0^t [U, e^{-i(t-t_1)H_o/\hbar}[U, \rho(t_1)]\, e^{i(t-t_1)H_o/\hbar}]\ dt_1. \tag{5}$$

Furthermore, if we let

$$\rho = \rho^{(1)} + \rho^{(2)}$$

where $\rho^{(1)}$ is postulated to be the diagonal part of $\rho$ and $\rho^{(2)}$ the off diagonal part, we find

$$\begin{aligned}\frac{\partial \rho_{jj}}{\partial t} = &-\frac{\lambda^2}{\hbar^2}\int_0^t dt_1\,[U, e^{-i(t-t_1)H_o/\hbar}[U, \rho^{(1)}(t_1)]\, e^{i(t-t_1)H_o/\hbar}]_{jj} \\ &-\frac{\lambda^2}{\hbar^2}\int_0^t dt_1\,[U, e^{-i(t-t_1)H_o/\hbar}[U, \rho^{(2)}(t_1)]\, e^{i(t-t_1)H_o/\hbar}]_{jj}\ .\end{aligned} \tag{6}$$

Let the $j$-th energy level of $H_o$ be $E_j$ and $\omega_{jk}$ be

$$\omega_{jk} = (E_j - E_k)/\hbar\ . \tag{7}$$

Then the first integral in (6) is

$$\begin{aligned}&-\frac{\lambda^2}{\hbar^2}\int_0^t dt_1 \sum_k{}' \{U_{jk}\, U_{kj}[\rho_{jj}(t_1) - \rho_{kk}(t_1)] \exp i(t - t_1)\,\omega_{jk} + \text{c.c.}\} \\ &= -\frac{2\,\lambda^2}{\hbar}\int_0^t d\tau \sum_k{}' \,|U_{jk}|^2\,[\rho_{jj}(t-\tau) - \rho_{kk}(t-\tau)] \cos \tau\omega_{jk}\end{aligned} \tag{8}$$

where the prime indicates that terms with $k = j$ are to be omitted. After the second term of (6) is expanded in terms of matrix elements of the density matrix, (6) becomes

$$\begin{aligned}\frac{d\rho_{jj}}{dt} = &-\frac{2\,\lambda^2}{\hbar}\int_0^t d\tau \sum_k{}' \,|U_{jk}|^2\,[\rho_{jj}(t-\tau) - \rho_{kk}(t-\tau)] \cos \tau\omega_{jk} \\ &-\sum_{l,m}{}'' \frac{\lambda^2}{\hbar^2}\int_0^t dt' \,\{U_{jl}\, e^{i(t-t_1)\omega_{jl}}\,[U_{lm}\,\rho_{mj}(t_1) - \rho_{lm}(t_1)\, U_{mj}] \\ &- e^{-i(t-t_1)\omega_{jl}}\,[U_{jm}\,\rho_{ml}(t_1) - \rho_{jm}(t_1)\, U_{ml}]\, U_{lj}\}\ .\end{aligned} \tag{9}$$

The double prime indicates that terms with $l = j$, $m = j$, and $m = l$ are to be omitted.

Now in order to obtain a differential equation for diagonal elements of the density matrix, one must find a scheme for solving off diagonal elements in terms of diagonal ones and thus eliminating off diagonal elements from (9). This can be done by referring back to the original integral equation (2.4) and examining a typical off diagonal element. Then if $j \neq k$

$$\rho_{jk}(t) = -\frac{i\lambda}{\hbar}\int_0^t e^{i(t-t_1)\omega_{kj}}\, U_{jk}[\rho_{kk}(t_1) - \rho_{jj}(t_1)]\,dt_1$$

$$-\frac{i\lambda}{\hbar}\int_0^t e^{i(t-t_1)\omega_{kj}} \sum_l |U_{jl}\,\rho_{lk}(t_1) - \rho_{jl}(t_1)\,U_{lk}]\,dt_1 . \qquad (10)$$

We define the following Laplace transforms ($\mathcal{L}$ being the Laplace integral operator

$$f_k(p) = \mathcal{L}\,\rho_{kk}(t)\ ; \qquad F_{lk}(p) = \mathcal{L}\,\rho_{lk}(t) \qquad l \neq k . \qquad (11)$$

Then upon application of the Faltung theorem,

$$\mathcal{L}\int_0^t g(t_1)\,u(t-t_1)\,dt_1 = \mathcal{L}\,g\,\mathcal{L}\,u ,$$

we find by taking Laplace transforms of (10), that the $F_{lk}$'s satisfy a set of $N(N-1)$ linear algebraic equations in the case in which our assembly has $N$ energy levels

$$(p + i\omega_{jk})\,F_{jk}(p) + \frac{i\lambda}{\hbar}\sum_l{}' [U_{jl}\,F_{lk}(p) - F_{jl}(p)\,U_{lk}] = \frac{i\lambda}{\hbar}[f_j(p) - f_k(p)]\,U_{jk}$$

This set of equations can be solved for the $F_{jk}$'s in terms of the $f_j$'s. Let $D(p)$ be the determinant of the coefficients of the $F_{jk}$'s and let the minor at the element $(\alpha\beta|\mu\nu)$ be

$$D(\alpha\beta|\mu\nu:p)$$

and, finally, let

$$\mathcal{D}(\alpha\beta|\mu\nu:p) = D(\alpha\beta|\mu\nu:p)\ /\ D(p) . \qquad (12)$$

Then

$$F_{jk}(p) = \frac{i\lambda}{\hbar}\sum_{\alpha\beta}{}' U_{\alpha\beta}[f_\alpha(p) - f_\beta(p)]\ \mathcal{D}(\alpha\beta|jk:p) \qquad (13)$$

where terms with $\alpha = \beta$ are to be excluded.

The Laplace transform of the second term on the right hand side of (9) is

$$-\frac{\lambda^2}{\hbar^2}\sum_{l,m}{}'' \Big\{\frac{U_{jl}}{p + i\,\omega_{lj}}[U_{lm}\,F_{mj}(p) - F_{lm}(p)\,U_{mj}]$$

$$-\frac{U_{lj}}{p + i\,\omega_{jl}}[U_{jm}\,F_{ml}(p) - F_{jm}(p)\,U_{ml}]\Big\} . \qquad (14)$$

If we substitute (13) into (14) and take the Laplace inverse, we will have eliminated the off diagonal elements in the density matrix; and when the Laplace inverse of (14) is substituted into (9), we will have an integro-differential equation which contains only diagonal elements of the density matrix. The first term of (14) becomes

$$-\frac{i\lambda^3}{\hbar^3}\sum_{l,m}''\sum_{\alpha,\beta}'\{U_{jl}\,U_{lm}\,U_{\alpha\beta}[f_\alpha(p)-f_\beta(p)]\frac{\mathcal{D}(\alpha\beta|mj:p)}{(p+i\omega_{lj})}\}\,.$$

Its Laplace inverse, which is just what is required in (9), is obtained by applying the Faltung theorem in reverse and using the well-known formula for the Laplace inverse. One finds it to be

$$-i\frac{\lambda^3}{\hbar^3}\sum_{lm}''\sum_{\alpha\beta}'\{U_{jl}\,U_{lm}\,U_{\alpha\beta}\int_0^t[\rho_{\alpha\alpha}(t-\tau)-\rho_{\beta\beta}(t-\tau)]\frac{d\tau}{2\pi i}$$
$$\times\int_{c-i\infty}^{c+i\infty}e^{p\tau}\frac{\mathcal{D}(\alpha\beta|mj:p)}{(p+i\,\omega_{lj})}dp\}$$

where $c$ is chosen so that the line of integration in the $p$ plane lies to the right of all singularities in the last integrand. This analysis can be applied to each term in (12) and the results substituted into (9). Then (9) becomes

$$\frac{d\rho_{jj}}{dt}=-\frac{2\lambda^2}{\hbar^2}\int_0^t d\tau\sum_k'|U_{jk}|^2[\rho_{jj}(t-\tau)-\rho_{kk}(t-\tau)]\cos\tau\omega_{jk}$$
$$-\frac{\lambda^3}{2\pi\hbar^3}\sum_{l,m}''\sum_{\alpha,\beta}'\int_0^t[\rho_{\alpha\alpha}(t-\tau)-\rho_{\beta\beta}(t-\tau)]\,d\tau$$
$$\times\int_{c-i\infty}^{c+i\infty}dp\,e^{p\tau}\,U_{\alpha\beta}\{\frac{U_{jl}}{p+i\,\omega_{lj}}[U_{lm}\,\mathcal{D}(\alpha\beta|mj:p)-U_{mj}\,\mathcal{D}(\alpha\beta|lm:p)]$$
$$+\frac{U_{lj}}{p+i\,\omega_{jl}}[U_{lm}\,\mathcal{D}(\alpha\beta|jm:p)-U_{jm}\,\mathcal{D}(\alpha\beta|ml:p)]\}\,. \tag{15}$$

This equation is exact and depends only on the fact that the off diagonal elements of $\rho$ are postulated to vanish when $t = 0$.

At this stage we might compare our results with those which would would have been obtained by perturbation theory. In a direct application of perturbation theory one would have solved (2.4) by iteration to obtain a formula for $\rho$. Then one would differentiate this to determine $d\rho/dt$. By calculating $d\rho/dt$ directly, one essentially eliminates the necessity of finding all terms in the expansion of an exponential and he needs only to find an expression for the quantity in the exponent. As will be shown in the next section, each term in the perturbation expansion for $d\rho_{jj}/dt$ is a product of closed cycles of transitions. They are merely cycles which appear in the expansion of the determinant of eq. (15).

## 5. *The Pauli equation*

We close our discussion with some remarks concerning the connection between eq. (4.15) and the Pauli equation. No derivation of the Pauli equation is given, but some problems which occur in the derivation will be discussed. The Pauli equation has the form

$$\frac{d\rho_{jj}}{dt} = 2\pi \frac{\lambda^2}{\hbar} \sum_k{}' |U_{jk}|^2 \, \delta(E_j - E_k)\,(\rho_{kk} - \rho_{jj}) \,. \tag{1}$$

One would like to know under what conditions it is the weak coupling limit ($\lambda \to 0$) of (4.5) or indeed of (1.2) when $H = H_o + \lambda H_I$. The work of Van Hove is concerned with the derivation of (1) from (1.2) in the limit

$$\lambda \to 0\,, \qquad t \to \infty$$

$$\lambda^2 t = O(1)\,. \tag{2}$$

His derivation employs a postulate, the so-called diagonal singularity condition, on the interaction matrix elements.

Expression (4.15) for $d\rho_{jj}/dt$ is composed of the sum two terms, one which seems to be proportional to $\lambda^2$ and the other to $\lambda^3$. The lecturer has shown that under conditions (2), the term proportional to $\lambda^2$ in (15) (reference 4, p. 310) becomes just the right-hand side of (1) in the limit in which the energy spectrum becomes dense ($N \to \infty$, $V \to \infty$, particle density finite). At first glance, one might expect that in the limit $\lambda \to 0$, the term proportional to $\lambda^3$ might become negligible compared to that proportional to $\lambda^2$. We now show that because of the time dependence of these terms, considerable care must be taken before one can arrive at such a conclusion.

Let us first construct the perturbation expansion of the right-hand side of (4.15) in powers of $\lambda$. This expansion is obtained directly as follows. As before, let

$$\rho = \rho^{(1)} + \rho^{(2)} \tag{4}$$

where $\rho^{(1)}$ is the diagonal and $\rho^{(2)}$ the off diagonal part of $\rho$. Then the off diagonal part is related to the diagonal part through (4.10) (here we always write $tH_o$ in units of $\hbar$)

$$\rho^{(2)}(t) = -i\lambda \int_0^t e^{-i(t-t_1)H_o}\,[U, \rho^{(1)}(t_1)]\, e^{i(t-t_1)H_o}\, dt_1$$

$$-i\lambda \int_0^t e^{-i(t-t_1)H_o}\,[U, \rho^{(2)}(t_1)]\, e^{i(t-t_1)H_o}\, dt_1 \,. \tag{5}$$

By iteration, $\rho^{(2)}(t)$ is expressed in terms of $\rho^{(1)}(t)$ through

$$\rho^{(2)}(t) = -i\lambda \int_0^t e^{-i(t-t_1)H_o}[U,\rho^{(1)}(t_1)]\, e^{i(t-t_1)H_o}\, dt_1$$

$$+ (i\lambda)^2 \int_0^t dt_1 \int_0^t dt_2\, e^{-i(t-t_1)H_o}[U, e^{-i(t_1-t_2)H_o}$$

$$\times [U,\rho^{(1)}(t_2)]\, e^{i(t_1-t_2)H_o}]\, e^{i(t-t_1)H_o}$$

$$- (i\lambda)^3 \int_0^t dt_1 \int_0^{t_1} dt_2 \int_0^{t_2} dt_3\, e^{-i(t-t_1)H_o}[U, e^{-i(t_1-t_2)H_o}$$

$$\times [U, e^{-i(t_2-t_1)H_o}[U,\rho^{(1)}(t_3)]\, e^{-i(t_2-t_3)H_o}]$$

$$\times e^{i(t_1-t_2)H_o}]\, e^{i(t-t_1)H_o}] + \ldots \tag{6}$$

the $n$-th order term contains an $n$-th order commutator. Substituting (6) into (4.6) we have an alternative form of (4.15)

$$\frac{d\rho_{jj}}{dt} = -(i\lambda)^2 \int_0^t dt_1 [U, e^{-i(t-t_1)H_o}[U,\rho^{(1)}(t_1)]\, e^{i(t-t_1)H_o}]_{jj}$$

$$+ (i\lambda)^3 \int_0^t dt_1 \int_0^{t_1} dt_2 [U, e^{-i(t-t_1)H_o}[U, e^{-i(t_1-t_2)H_o}$$

$$\times [U,\rho^{(1)}(t_2)]\, e^{i(t_1-t_2)H_o}]\, e^{i(t-t_1)H_o}]_{jj} + \ldots . \tag{7}$$

We might now examine some of these terms explicitly. The term proportional to $\lambda^2$ has already been discussed in section 4. The third order term is

$$-(i\lambda)^3 \int_0^t dt_1 \int_0^{t_1} dt_2 \sum_{k,l}'' U_{jk}\, U_{kl}\, U_{lj}\, \{e^{i[\omega_{jk}t+\omega_{kl}t_1+\omega_{lj}t_2]}\rho_{jj}(t_2)$$

$$- e^{i[\omega_{jk}t+\omega_{kl}t_1+\omega_{lj}t_2]}\rho_{ll}(t) - e^{i[\omega_{jk}t+\omega_{kl}t_2+\omega_{lj}t_1]}\rho_{ll}(t_2)$$

$$+ e^{i[\omega_{jk}t+\omega_{kl}t_2+\omega_{lj}t_1]}\rho_{kk}(t_2)\} + \text{c.c.} \tag{8}$$

Note that the $U$'s and the $\omega$'s run through a cycle $j \to k \to l \to j$ with the '' (and fact that diagonal parts of $U$ are postulated to vanish) indicating that $j \neq k \neq l$.

If $E_j$ is the $j$-th energy level, the matrix elements $U_{jk}$ can be expressed as functions of $E_j$ and $E_k$ instead of $j$ and $k$; i.e.

$$U_{jk} = U(E_j, E_k)$$

so that $U_{jk}\, U_{kl}\, U_{lj}$ is a function of $E_j$, $E_k$, and $E_l$, say

$$F\,(E_j, E_k, E_l)\ ,$$

which might also be written as

$$f(E_k - E_j,\ E_l - E_j;\ E_j) = f_j\,(\omega_1, \omega_2)$$

where

$$\omega_1 = \omega_{kj} = (E_k - E_j)/\hbar$$

$$\omega_2 = \omega_{lj} = (E_l - E_j)/\hbar\ .$$

In the limit of an assembly of a large number of degrees of freedom ($N \to \infty$, $V \to \infty$ with density $\rho = N/V$ fixed), the energy levels form a dense set. Hence, the summation with respect to $k$ and $l$ can be converted to an integral with respect to $\omega_1$ and $\omega_2$. Then the first term in (8) is proportional to

$$\lambda^3 \int_0^t dt_1 \int_0^{t_1} dt_2 \iint d\omega_1\, d\omega_2 f_j\,(\omega_1, \omega_2)\, e^{i[\omega_2(t_2 - t_1) + \omega_1(t_1 - t)]}\ . \tag{9}$$

Now let us express both $t_1$ and $t_2$ in units of $t$ with $t_j = u_j t$. Then

$$0 \le u_2 \le u_1 \le 1$$

so that (9) becomes

$$\lambda^3\, t^2 \int_0^1 du_1 \int_0^{u_1} du_2 \iint d\omega_1\, d\omega_2 f_j(\omega_1, \omega_2)\, \exp i[t\ \omega_2(u_2 - u_1) + t\ \omega_1(u_1 - 1)]\ . \tag{10}$$

Furthermore, if we let $\vartheta_1 = t\,\omega_1$, and $\vartheta_2 = t\,\omega_2$, we have

$$\lambda^2(\lambda\, t^{\frac{1}{2}}) \int_0^1 du_1 \int_0^{u_1} du_2\ \{t^{-\frac{1}{2}} \iint d\vartheta_1\, d\vartheta_2 f_j(\vartheta_1/t\,, \vartheta_2/t)$$

$$\times \exp i\ [\vartheta_2(u_2 - u_1) + \vartheta_1(u_1 - 1)]\}\ . \tag{11}$$

The term $(\lambda t^{\frac{1}{2}})$ is, in view of (2), of order 1. Hence, if as $t \to \infty$ the $\vartheta$ integral becomes independent of $t$ or increases with $t$ more slowly than $ct^{\frac{1}{2}}$, (11) is small compared to the term of $O(\lambda^2)$ in (7) and can be neglected. In the discussion which follows, we assume that as $t \to \infty$ the $\vartheta$ integral becomes independent of $t$. This would be the case if $f_j$ is a product of delta functions. Under this condition all other terms proportional to $\lambda^3$ would also be negligible compared to the $\lambda^2$ terms.

We now show that even in this favorable situation of the $\vartheta$ integral being time independent as $t \to \infty$, troubles still arise in higher order terms. Since the diagonal matrix elements $U_{jj}$ were postulated to be zero

$$\lim_{t \to \infty} f_j(\vartheta_1/t\,, \vartheta_2/t)$$

is not to be interpreted as $f_j(0, 0)$ (which would vanish), but rather as a limit which, in general, does not vanish. A typical fourth order term is

$$\lambda^4 \int_0^t dt_1 \int_0^{t_1} dt_2 \int_0^{t_2} dt_3 \sum_{klm}{}' U_{jk} U_{kl} U_{lm} U_{mj} \times \exp i \left[t\,\omega_{jk} + t_1\,\omega_{kl} + t_2\,\omega_{lm} + t_3\,\omega_{mj}\right] \tag{12}$$

where the summation does not include terms with $j = k$, $k = l$, $l = m$ or $m = j$. Note that it is quite possible for $l$ to achieve the value $j$ when the 4-cycle $(j \to m \to l \to k \to j)$ factors into two 2-cycles $(j \to m \to \to j)$ and $(j \to k \to j)$ in the first case, three independent $\omega$'s exist, while in the second there are only two. Hence (12) is the sum of two terms

$$\lambda^4 \int_0^t dt_1 \int_0^{t_1} dt_2 \int_0^{t_2} dt_3 \sum_{\substack{klm\\(l\neq j)}}{}' U_{jk} U_{kl} U_{lm} U_{mj} \times \exp i \left[(t - t_2)\,\omega_1 + (t_1 - t_2)\,\omega_2 + (t_3 - t)\,\omega_3\right]$$
$$+ \lambda^4 \int \cdots \int dt_1\, dt_2\, dt_3 \sum_{km}{}' |U_{jk}|^2\, |U_{jm}|^2 \exp i\left[(t - t_1)\,\omega_1 + (t_2 - t_3)\,\omega_3\right] \tag{13}$$

where $\omega_1 = \omega_{jk}\,, \quad \omega_2 = \omega_k\,, \quad \omega_3 = \omega_{jm}\,.$

As in the case of third order terms, we characterize the $U_{jk}$'s by $E_j$ and $E_k$ rather than $j$ and $k$. In the limit $N \to \infty$ in which case the $\omega$'s become continuous variables, the summation over $k$, $l$, $m$ becomes an integration over $\omega_1$, $\omega_2$ and $\omega_3$ so that (13) has the form (after the substitution $t_j = tu_j$)

$$\lambda^4 t^3 \int_0^1 du_1 \int_0^{u_1} du_2 \int_0^{u_2} du_3 \int \cdots \int d\omega_1\, d\omega_2\, d\omega_3\, f(\omega_1, \omega_2, \omega_3) \times \exp i\left[(1 - u_2)\, t\,\omega_1 + (u_1 - u_2)\, t\,\omega_2 + (u_3 - u_2)\, t\,\omega_3\right]$$
$$+ \lambda^4 t^3 \int_0^1 du_1 \int_0^{u_1} du_2 \int_0^{u_2} du_3 \iint d\omega_1\, d\omega_3\, f_1(\omega_1)\, f_1(\omega_2) \times \exp i\left[(1 - u_1)\, t\,\omega_1 + (u_2 - u_3)\, t\,\omega_3\right]. \tag{14}$$

Again, if we make the transformation $t\omega_j = \vartheta_j$, this expression becomes

$$\lambda^4 \int_0^1 du_1 \int_0^{u_1} du_2 \int_0^{u_2} du_3 \iiint d\vartheta_1\, d\vartheta_2\, d\vartheta_3\, f(\vartheta_1/t, \vartheta_2/t, \vartheta_3/t) \times \exp i\left[(1 - u_1)\,\vartheta_1 + (u_2 - u_3)\,\vartheta_3\right]$$
$$+ \lambda^2 (t\,\lambda^2) \int_0^1 du_1 \int_0^{u_1} du_2 \int_0^{u_2} du_3 \iint d\vartheta_1\, d\vartheta_3\, f_1(\vartheta_1/t)\, f_1(\vartheta_2/t) \times \exp i\left[(1 - u_1)\,\vartheta_1 + (u_2 - u_3)\,\vartheta_3\right]. \tag{15}$$

If we again make the hypothesis that the $\vartheta$ integrals become time independent, the first term in (15) is of order $\lambda^4$ which is negligible compared with $\lambda^2$ as $\lambda \to 0$. On the other hand, since $\lambda^2 t$ is of order unity in our limits of interest, the second term in (15) is of order $\lambda^2$

and *cannot* be neglected without further analysis when compared to the first term to the right of the equal sign in (7).

A similar analysis can be made for higher order terms in our iteration scheme. A term proportional to $\lambda^{2n}$ has $2n$ $U$'s. Suppose that it factors into $m$ cycles. There are then $2n-m$ independent $\omega$'s, since one loses an independent $\omega$ with the completion of each cycle. Since the number of $t$ integrations is $2n-1$, the integral associated with our term is of the form

$$\lambda^{2n} \int \ldots \int dt_1 \ldots dt_{2n-1} \int \ldots \int d\omega_1 \ldots d\omega_{2n-m} \{\ldots\} \tag{16}$$

where the bracket represents our product of $U$'s, exponentials and a diagonal element of $\rho$. If, as above, we let $t_\alpha = tu_\alpha$ and $t\omega_\beta = \vartheta_\beta$, this becomes

$$\lambda^2[\lambda^{2(n-1)}\, t^{m-1}] \int \ldots \int du_1 \ldots du_{2n-1} \int \ldots \int d\vartheta_1 \ldots d\vartheta_{2n-m} \{\quad\} \tag{17}$$

If the $\vartheta$ integration is again postulated to become time independent as $t \to \infty$ and if $n > m$, then the expression (17) is vanishingly small compared to $\lambda^2$ under condition (2). On the other hand, if $n = m$, the factor in the square bracket becomes $(t\,\lambda^2)^{n-1}$ which is of order unity so that the complete term is of $O(\lambda^2)$ in our required limit. This case can only exist when the $U$'s appear as products of $n$ 2-cycles. Thus unless other reasons can be discovered for discarding the contribution of the sum of all terms which are composed only of products of 2-cycles, they can all be considered to contribute to terms of $O(\lambda^2)$ in (7). The validity of the Pauli equation might be achieved by all the terms composed only of 2-cycles canceling each other, by the complete sum of all terms being of higher order than $\lambda^2$ or by the $\vartheta$ integrals vanishing in the limit of large $t$ (or perhaps by some combination of all these effects existing). In any case one must include more information about the laws of interaction or the matrix elements $U_{jk}$ before one can proceed further. Using the diagonal singularity hypothesis, Van Hove has derived the Pauli equation. Hence he has implicitly carried out the detailed required analysis concerning the contribution of various cycles to terms of $O(t^2)$. We hope to discuss these cycles elsewhere as they appear in certain examples.

In conclusion, the lecturer wishes to thank Professors P. Mazur and L. Van Hove for several interesting discussions concerning the material here presented.

*References*


1. R. P. Feynman, Phys. Rev. 76 (1949) 749.
2. L. Van Hove, La théorie des gaz neutres et ionizés (Hermann 1960), p. 151.
3. R. W. Zwanzig, Lect. in Theor. Phys. (Boulder) 3 (1960) 106.
4. E. W. Montroll, Lect. in Theor. Phys. (Boulder) 3 (1960) 221.
5. T. G. Kirkwood, J. Chem. Phys. 3 (1935) 300; 15 (1947) 72.
6. M. Born and H. S. Green, A general kinetic theory of liquids, Cambridge Univ. Press (London), 1949.
7. J. Yvon, Actualités scientifiques et industrielles (Hermann, Paris), 1935.
8. N. N. Bogolubov, Problems of a dynamical theory in statistical mechanics.
9. P. Martin and J. Schwinger, Phys. Rev. 115 (1959) 1342.
10. E. R. Caianiello, Nuovo Cim. Series 13 (1959) 637.
11. I. Prigogine, A survey of the work of Prigogene and his collaborators will appear in afforthcoming book. See also R. Balescu, Lect. in Theor. Phys. (Boulder) 3 (1960) 382.
12. R. Kubo, J. Phys. Soc. of Japan 12 (1957) 570.
13. R. Kubo, Lect. in Theor. Phys. (Boulder) 1 (1958) 120.
14. E. W. Montroll and J. C. Ward, Physica 25 (1959) 423.
15. The many body problem (Les Houches Lectures 1958) and Theory of neutral and ionized gases (Les Houches Lectures 1959) (Wiley, New York).
16. Physica (Dec. 1960).